Lecture Notes in Networks and Systems

Volume 452

The series "Lecture Notes in Networks and Systems" publishes the latest developments in Networks and Systems—quickly, informally and with high quality. Original research reported in proceedings and post-proceedings represents the core of LNNS.

Volumes published in LNNS embrace all aspects and subfields of, as well as new challenges in, Networks and Systems.

The series contains proceedings and edited volumes in systems and networks, spanning the areas of Cyber-Physical Systems, Autonomous Systems, Sensor Networks, Control Systems, Energy Systems, Automotive Systems, Biological Systems, Vehicular Networking and Connected Vehicles, Aerospace Systems, Automation, Manufacturing, Smart Grids, Nonlinear Systems, Power Systems, Robotics, Social Systems, Economic Systems and other. Of particular value to both the contributors and the readership are the short publication timeframe and the world-wide distribution and exposure which enable both a wide and rapid dissemination of research output.

The series covers the theory, applications, and perspectives on the state of the art and future developments relevant to systems and networks, decision making, control, complex processes and related areas, as embedded in the fields of interdisciplinary and applied sciences, engineering, computer science, physics, economics, social, and life sciences, as well as the paradigms and methodologies behind them.

Indexed by SCOPUS, INSPEC, WTI Frankfurt eG, zbMATH, SCImago.

All books published in the series are submitted for consideration in Web of Science.

For proposals from Asia please contact Aninda Bose (aninda.bose@springer.com).

More information about this series at https://link.springer.com/bookseries/15179

Andrzej Dzielinski · Dominik Sierociuk ·
Piotr Ostalczyk
Editors

Proceedings of the International Conference on Fractional Differentiation and its Applications (ICFDA'21)

Springer

Editors
Andrzej Dzielinski
Institute of Control
and Industrial Electronics
Warsaw University Technology
Warsaw, Poland

Piotr Ostalczyk
Institute of Control
and Industrial Electronics
Warsaw University of Technology
Warsaw, Poland

Dominik Sierociuk
Faculty of Electrical Engineering,
Institute of Control
and Industrial Electronics
Warsaw University of Technology
Warsaw, Poland

ISSN 2367-3370 ISSN 2367-3389 (electronic)
Lecture Notes in Networks and Systems
ISBN 978-3-031-04382-6 ISBN 978-3-031-04383-3 (eBook)
https://doi.org/10.1007/978-3-031-04383-3

This Springer imprint is published by the registered company Springer Nature Switzerland AG
The registered company address is: Gewerbestrasse 11, 6330 Cham, Switzerland

Preface

This volume is touching upon various aspects of a very interesting and growing in popularity category of models of dynamical systems. These are the so-called fractional-order systems. Such models not only are relevant for many fields of science and technology, but may also find numerous applications in other disciplines applying the mathematical modelling tools. Thus, the book is intended for a very wide audience of professionals who want to expand their knowledge of systems modelling and its applications.

The book includes the selections of papers presented at the International Conference on Fractional Calculus and its Applications organized by the Warsaw University of Technology, held online on 6–8 September 2021.

The International Conference on Fractional Calculus and its Applications (ICFDA) has almost twenty years of history. It started in Bordeaux (France) in 2004, followed by Porto (Portugal) 2006, Istanbul (Turkey) 2008, Badajoz (Spain) 2010, Nanjing (China) 2012, Catania (Italy) 2014, Novi Sad (Serbia) 2016, Amman (Jordan) 2018. Next ICFDA was planned in 2020 in Warsaw (Poland), but COVID-19 pandemic shifted it to 6–8 September 2021. Hence, the organizers were forced to change the form of the conference to the online one.

In the volume, twenty-eight high-quality research papers presented during the ICFDA 2021 eleven regular sessions with an additional online discussion session are presented.

The presented papers are scientifically inspiring, leading to new fruitful ideas. They cover a very broad range of many disciplines. Nowadays, and especially in such a subject as fractional calculus it is very difficult to assign papers to specific scientific areas. So, many of the papers included have an interdisciplinary character.

February 2022

Andrzej Dzielinski
Dominik Sierociuk
Piotr Ostalczyk

Contents

Some Proposals for a Renewal in the Field of Fractional Behaviour Analysis and Modelling

Jocelyn Sabatier(✉)

Bordeaux University, 15, Rue Naudet - CS 10 207, 33175 Gradignan Cedex, France
jocelyn.sabatier@u-bordeaux.fr

Abstract. This paper first highlights the drawbacks associated to classical fractional operators (integration and differentiation) and to classical fractional order models (fractional pseudo state space descriptions or fractional differential equations). To solve these drawbacks, it is then proposed

– to replace the singular kernel used in the definition of the fractional order integration operator by new non-singular kernels, but which still make it possible to generate power law behaviours;
– to use new models such as Volterra equations of the first kind, nonlinear models, distributed time delay models or partial differential equations with spatially variable coefficients.

It is shown how these new kernels and models solve the drawbacks mentioned.

Keywords: Fractional integration · Fractional differentiation · Fractional models · Non singular kernels · Volterra equations · Nonlinear models · Distributed time delay models · Diffusion equations

1 Introduction

After a period of very intense progression and unity that lasted more than 40 years [1], the community of researchers in the field of fractional operators and fractional models now seems to be split in two: conservatives and reformers. This situation gives rise in particular to rather harsh publication duels (described at the end of Sect. 2). The conservatives seek to progress along the existing path and criticise (sometimes wrongly [2]) the weaknesses of the reformers' new ideas, while the reformers have grasped the limits of fractional operators and try to offer alternatives [3–7].

This article falls into the second group of approaches that propose reformulations of fractional operators in the definition of dynamical models, using new kernels in Volterra type operators as done in [3, 4, 8–12]. Criticisms of this way of thinking are quite recent, even though reformulation proposals have been made for some time [13–21].

Of course we cannot ignore the fact that, for a continuous time random walk process with a power-law waiting time, probability density function and jump lengths with finite variance, the corresponding probability density function to find the particle at position x

A. Dzielinski et al. (Eds.): ICFDA 2021, LNNS 452, pp. 1–25, 2022.
https://doi.org/10.1007/978-3-031-04383-3_1

at time t in the Fourier–Laplace domain is the same as the solution of the time fractional diffusion equation with the Riemann–Liouville fractional derivative [22, 23]. We cannot ignore similar approaches followed with other continuous time random walk processes with different waiting time and jump length probability density functions.

These approaches might suggest that a physical interpretation of fractional operators and models is possible, but it is not. It should be pointed out that these results were obtained under spaces of infinite dimension and asymptotic time analysis (time tends towards infinity). This link of the fractional Riemann–Liouville operator (and others) with an infinite space domain can also be found in [24, 25] and this is what induces the distribution of time constants on $]-\infty, 0]$ inherent in fractional Riemann–Liouville and other operators.

This constitutes one of the drawbacks of fractional operators [7] (infinitely slow and infinitely fast time constants, the former inducing an infinite memory) that can be solved by changing the kernels used in the Riemann–Liouville operator (and similar operators).

While the reformulations of the classical fractional differentiation and integration operators proposed in the literature mainly sought to eliminate the singularity of the kernels in their definition, the goal of this paper is to go further and to introduce new kernels and models that solve several other drawbacks associated to classical fractional operators and models.

In this paper, the drawbacks associated to fractional operators (integration and differentiation) and to classical fractional order models (fractional pseudo state space descriptions or fractional differential equations) are highlighted. To solve these drawbacks, it is then proposed

- to replace the singular kernel used in the definition of fractional order integration by new non-singular kernels, but which still enable power law behaviours to be generated in a limited frequency or time domain;
- to use new models such as Volterra equations of the first kind, nonlinear models, distributed time delay models or partial differential equations with spatially variable coefficients.
- It is shown how these new kernels and models solve the drawbacks mentioned.

2 Fractional Order Models and Associated Drawbacks

For more than 40 years, research on derivative or fractional integration operators (non-integer would be more appropriate as the integration and derivative orders considered are not limited to the ratio of two numbers) has continued to increase [1]. There are many reasons for this growth, but we can mention two major ones.

- First of all, many fields have revealed processes or systems whose evolutionary kinetics or dynamic behaviour are power laws, that is to say evolutions or relaxations in $t^{-\nu}$ in the time domain or $f^{\nu-1}$ in the frequency domain, $\nu \in \mathbb{R}$ (see [26] and [7] for more rigorous definitions). By way of example – a non-exhaustive list as the results are numerous – these behaviours exist in electrochemistry [27, 28], thermal science [29], biology [30], mechanics [31], acoustics [32], and electrical engineering [33]. Given

the pervasiveness of these behaviours, effective modelling tools are needed. The idea of using fractional integration and differentiation operators as modelling tools then gradually emerged, because these operators too have power law type behaviours.
- A second reason for this growth in work devoted to fractional operators is the opportunity offered by this field of research to generalize or "fractionalize" simply and quickly existing results for classical dynamical systems.

This was no doubt the starting point for many errors or misunderstandings that are beginning to be revealed in the literature. Indeed, while from a mathematical point of view these operators are now well defined, this is not the case for their physical interpretations or for their use in the definition of fractional models [34].

In order to have models capable of capturing the behaviour of systems with power law behaviours, a first reflex of the community was to generalize well-known tools such as differential equations or state space representations (which are the models mainly used in the literature) to the fractional case.

A fractional differential equation involves fractional derivatives of the model input $u(t) \in \mathbb{R}^m$ and of the model output $y(t) \in \mathbb{R}^p$ and is described by the relation

$$\sum_{k=0}^{N_a} a_k \boldsymbol{D}_{t_0}^{\nu_{a_k}} (y(t)) = \sum_{k=0}^{N_b} b_k \boldsymbol{D}_{t_0}^{\nu_{b_k}} (u(t)) \tag{1}$$

in which the coefficients $a_k \in \mathbb{R}$ and $b_k \in \mathbb{R}$. $\boldsymbol{D}_{t_0}^{\nu_{a_k}}$ and $\boldsymbol{D}_{t_0}^{\nu_{b_k}}$ denote fractional differentiation operators of orders $\nu_{a_k} \in \mathbb{R}$ and $\nu_{b_k} \in \mathbb{R}$ respectively. If the orders ν_{a_k} and ν_{b_k} verify relations $\nu_{a_k} = k/q$ and $\nu_{b_k} = k/q$, $q \in \mathbb{N}^*$, that is if all differentiation orders are multiples of $1/q$, then system (1) is commensurate.

Using the order commensurability condition and for null initial conditions, differential Eq. (1) admits a pseudo state space description of the form:

$$\begin{cases} \boldsymbol{D}_{t_0}^{\nu} x(t) = Ax(t) + Bu(t) \\ y(t) = Cx(t) + Du(t) \end{cases} \tag{2}$$

where $x(t) \in \mathbb{R}^n$ is the pseudo state vector, $\nu = 1/q$ is the fractional order of the system and $A \in \mathbb{R}^{nxn}$, $B \in \mathbb{R}^{nxm}$, $C \in \mathbb{R}^{pxn}$ and $D \in \mathbb{R}^{pxm}$ are constant matrices.

The impulse responses of models (1) and (2), calculated with the residue theorem and using a Bromwich-Wagner contour, can be written in the form [35]:

$$h(t) = h_p(t) + h_d(t) \tag{3}$$

with

$$h_p(t) = \sum_{i=1}^{n} a_i e^{tp_i} \qquad h_d(t) = \int_0^{\infty} \mu(x) e^{-tx} dx \tag{4}$$

The function $h_p(t)$ comes from the poles of models (1) and (2) (residues of the Cauchy method). As indicated in [36], the function $\mu(x)$ in $h_d(t)$ is defined by:

$$\mu(x) = \frac{1}{2i\pi} \left[H\left((-x)^-\right) - H\left((-x)^+\right) \right] \tag{5}$$

where $H(s)$ is the transfer function linking the input and the output of models (1) and (2). The Laplace transform of function $h_d(t)$ is given by:

$$h_d(s) = \int_0^{\infty} \frac{\mu(z)}{s+z} e^{-tz} dz \tag{6}$$

Such a relation highlights that models (1) and (2) have poles distributed from 0 to $-\infty$ leading to the first drawback.

Drawback 1 – *The memory of fractional models of types* (1) *and* (2) *is infinite. These models present infinitely fast and infinitely slow time constants (even if they are attenuated by the function* $\mu()$*, they exist), which induces their infinite memory and excludes the possibility of linking the internal variables of these models to physical variables.*

The infinite memory associated with fractional models can also be highlighted by another interpretation. If an input $u(t)$ is applied to the impulse response submodel $h_d(t)$, the resulting output $y_d(t)$ is given by the relation [25, 37]:

$$\begin{cases} \frac{\partial w(t,z)}{\partial t} = -zw(t,z) + u(t) \\ y_d(s) = \int_0^{\infty} \mu(z) w(t,z) dz \end{cases} \quad \text{with} \quad z \in \mathbb{R}^+, \tag{7}$$

known in the literature as diffusive representation [37]. The inverse spatial Fourier transform applied to (7), leads to:

$$\begin{cases} \frac{\partial \phi(t,\zeta)}{\partial t} = \frac{\partial^2 \phi(t,\zeta)}{\partial \zeta^2} + u(t)\delta(\zeta) \\ y_d(s) = \int_{-\infty}^{\infty} m(\zeta)\phi(t,\zeta) d\zeta \end{cases} \quad \text{with} \quad \zeta \in \mathbb{R} \tag{8}$$

and $\phi(t,\zeta) = \mathcal{F}^{-1}\{w(t, 4\pi^2 z^2)\}$, $m(\zeta) = \mathcal{F}^{-1}\{4\pi^2 \zeta \mu(4\pi^2 \zeta^2)\}$.

Relation (8) therefore allows us to state that a fractional system can be associated with an infinite dimensional system described by a diffusion equation over an infinite domain ($\zeta \in \mathbb{R}$) [25]. It is this infinite domain that generates the infinite memory mentioned above. This can also be illustrated in the thermal field [38, 39]. A fractional integrator is the solution of the heat equation (relating a heat flow to the measured temperature) only if:

- the temperature measurement is made at the point where the flux is applied;
- a medium of infinite dimension is considered.

Other spatial configurations (not infinite domain) can lead to power law type behaviours but cannot be modelled by models (1) or (2) (exponential and hyperbolic functions of the Laplace variable then appear in the transfer functions).

Moreover, in representation (2), a quick analysis of the units leads to doubts about the physical nature of the coefficients of the matrices A and B, which leads to the following drawback.

Drawback 2 – *The units of the parameters associated with representation* (2) *(parameters of matrices A and B) have no physical meaning (*$sec^{-\nu}$ *for the parameters of matrix A for example).*

Finding physical interpretations of fractional operators and fractional models has been a constant concern of researchers in the field. But this led to the next drawback.

Drawback 3 – *In terms of the physical interpretation of fractional integration or derivative operators, we can cite the continuous-time random walk processes with a power law as a function of the probability density of waiting times and the jump lengths with finished variance. The corresponding probability density function of finding the particle at position x at time t is the same as the solution of the time fractional diffusion equation with the Riemann-Liouville fractional derivative* [22, 23]. *But it should be observed that these results were obtained under spaces of infinite dimension and by asymptotic time analysis (time tends towards infinity). Once again we find the link between fractional operators and infinite spatial domains mentioned above, which therefore leads to unrealistic interpretations. The other physical interpretations proposed in the literature are not obtained from the observation of a given phenomenon but result from purely mathematical discussions* [40–47] *and are not very convincing. In the case of non-commensurate orders for an equation such as* (1), *certain interpretations can even invalidate the model obtained* [48].

Representation (2) is known in the literature as a "fractional state space representation". However, this is an improper designation that results from a generalization of concepts dedicated to integer systems without questioning the notion of state. This analysis is demonstrated in [49], and leads to the following drawback.

Drawback 4 – *Representation* (2) *is not a state space description because the variable $x(t)$ does not have the properties of a state. This is why the terms "pseudo state" and "pseudo state description" were introduced* [49] *and are now widely used in the literature.*

While the notion of fractional integration is fairly well defined, this is not the case for the fractional derivative. In representations (1) and (2), the fractional derivative operator $\frac{d^\nu}{dt^\nu}$ is not uniquely defined, which leads to the following drawback.

Drawback 5 – *There are over 30 definitions of the operator* $\frac{d^\nu}{dt^\nu}$ [50].

This multiplicity of definitions has led some authors to develop results by choosing the most convenient definition to obtain them. This is why the Caputo definition has become so popular because it offers the possibility of taking into account the initial conditions without taking into account the whole past of the system. Although from a mathematical point of view the definitions of Caputo, Riemann-Liouville or others pose no problem, their use for the definition of fractional models such as models (1) and (2) is questionable. While fractional models are known to have a long and even infinite memory, the use of the Caputo derivative causes this memory to disappear at a given time (initial time). This paradoxical situation allows us to state the following drawback.

Drawback 6 – *The initial conditions are not taken into account correctly in representations* (1) *and* (2) *if the Caputo or Riemann-Liouville definitions are used* [25, 51, 56].

As a consequence, many erroneous results obtained on the basis of models using these definitions have been reported in the literature. However, questions about taking initial conditions into account are not new. This problem has been analysed in the literature on several occasions. The first studies were carried out by Lorenzo and Hartley [52, 53] who proposed to introduce initialization functions to take into account in a practical way the past of the model in a finite interval, but also to allow the Caputo and Riemann-Liouville definitions to have the same behaviour in the presence of initial conditions. The idea that initialization functions should be used instead of initial values was reinforced in [54]. Later in [55], the "prehistories" of functions before the initial time were introduced to solve the initial value problems of fractional viscoelastic equations, since these functions are crucial for a unique solution. In [24, 25], it was shown on a counterexample that the initial conditions cannot be taken into account in a dynamic model either by the Caputo or the Riemann-Liouville definitions. A representation based on the impulse response was therefore introduced to define the initialization and we concluded as in [51], that the fractional models and the initialization of the fractional derivatives are two distinct questions. A similar analysis was performed for the partial differential equations in [56] using an initial time shift method. Note that a similar time shift was recently used in [57] to demonstrate that the mathematical description of groundwater flow using the time fractional partial derivatives of Caputo or Riemann - Liouville is not objective (in the Newtonian or Einsteinian sense). The authors of this paper did not go further, but I believe that the introduction of an initialization function would restore this objectivity.

Failure to take into account the infinite memory of models (1) and (2), or incorrectly taking into account the initial conditions in the analysis of fractional models, lead to incorrect results. This can be illustrated with the observability property through the results published in [58, 59] (among others) which are simply extensions of well-known results for integer models. But these apparently strong results are wrong. In fact, the exact observability of the model (2) cannot be reached because all the past of the model must be known to predict its future [35]. Similar criticisms can be made about the controllability or flatness properties of fractional models that can be found in the literature and which are in reality much poorer and more restrictive, and often unrelated to the systems they model. This leads to the following drawback.

Drawback 7 *– The poor properties of fractional models (sometimes unrelated to the systems they model) are another consequence of the infinite memory of models* (1) *and* (2)*, sometimes in contradiction with certain results proposed in the literature.*

As for integer order systems, pseudo state variables can be chosen to obtain a controllability or observability canonical form (without allowing conclusions to be drawn in terms of observability). The block diagram of Fig. 1, associated to the pseudo state space representation (2), can thus be obtained.

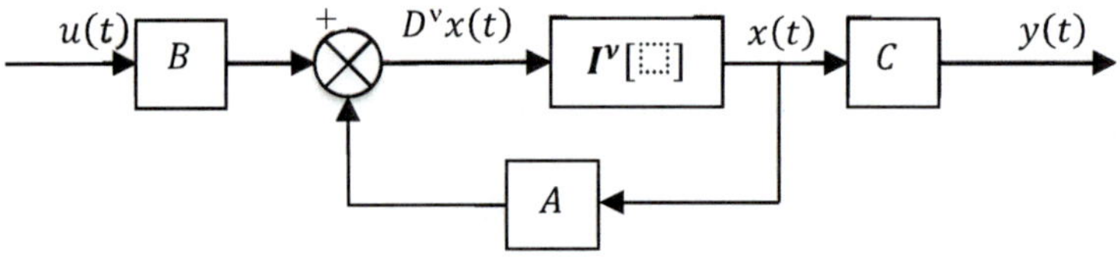

Fig. 1. Block diagram associated to the pseudo state space representation (2)

In Fig. 1, $\boldsymbol{I}^{\nu}$ denotes a fractional order integrator of order ν. With order $\nu \in \mathbb{R}_{+}^{*}$ the fractional integral of order $\nu \in \mathbb{R}_{+}^{*}$ of a real function $f(t)$ is defined by:

$$\boldsymbol{I}_{t_0}^{\nu}\left[f(t)\right] \triangleq \frac{1}{\Gamma(\nu)}\int_{t_0}^{t}(t-\tau)^{\nu-1}f(\tau)d\tau = \int_{t_0}^{t}\eta(t-\tau)f(\tau)d\tau, \tag{9}$$

in which $\Gamma(\nu)$ is the Gamma function defined by

$$\Gamma(\nu) = \int_{0}^{\infty} e^{-x}x^{\nu-1}dx \tag{10}$$

The model resulting from Fig. 1 and that takes into account the infinite memory of fractional operators can thus be defined by the relation

$$\begin{cases} x(t) = \boldsymbol{I}_{t_0\to-\infty}^{\nu}[Ax(t)+Bu(t)] \\ y(t) = Cx(t)+Du(t) \end{cases}. \tag{11}$$

This representation has the advantage of solving drawbacks 5 and 6. However, other drawbacks remain, and in particular this one.

Drawback 8 – *The fractional integration given by relation* (9) *implies a singular kernel; this leads to complications in the solution/simulation of fractional order differential equations* [3, 4].

The same remark holds for relation (1). A similar (but more complex) representation to that in Fig. 1 can be obtained with model (1) by applying the Laplace transform to relation (1) (without considering initial conditions), leading to:

$$\left(\sum_{k=0}^{N_a} a_k s^{\nu_{a_k}}\right)y(s) = \left(\sum_{k=0}^{N_b} b_k s^{\nu_{b_k}}\right)u(s) \tag{12}$$

or by multiplying both sides by $s^{-\nu_{a_{N_a}}}$

$$\left(\sum_{k=0}^{N_a} a_k s^{\nu_{a_k}-\nu_{a_{N_a}}}\right)y(s) = \left(\sum_{k=0}^{N_b} b_k s^{\nu_{b_k}-\nu_{a_{N_a}}}\right)u(s) \tag{13}$$

with $\nu_{a_k} - \nu_{a_{N_a}} < 0, 0 < k < N_a - 1$, and $\nu_{b_k} - \nu_{a_{N_a}} < 0, 0 < k < N_b$.

Relation (13) can be split into two parts, and by introducing a variable $x(t)$, rewritten as follows:

$$\left(\sum_{k=0}^{N_a} a_k s^{\nu_{a_k}-\nu_{a_{N_a}}}\right)y(s) = x(s) \qquad x(s) = \frac{u(s)}{\left(\sum_{k=0}^{N_a} a_k s^{\nu_{a_k}-\nu_{a_{N_a}}}\right)} \tag{14}$$

The second relation in (14) is also equal to

$$x(s)s^{\nu_{a_0}-\nu_{a_{N_a}}} = -x(s)\sum_{k=1}^{N_a}\frac{a_k}{a_0}s^{\nu_{a_k}-\nu_{a_{N_a}}} + \frac{u(s)}{a_0} \tag{15}$$

showing that variable $x(t)$ results in a fractional integrator chain as in Fig. 1.

3 Some New Kernels

Drawback 8 on the singularity of the kernel appearing in the definition of fractional derivatives and integrals has led some authors to propose definitions based on non-singular kernels [3, 4, 5, 60]. Although it was not the first time that modifications of fractional derivation and integration operators were proposed [17, 18], these publications have triggered reactions that are sometimes strong. For example, they have been accused of:

- leading to the definition of operators that are no longer fractional [61, 62],
- making it impossible to capture the dynamic behaviour of certain systems [63]
- leading to restrictions [64–66].

The latter results are however totally false, since they are based on a definition of the initial conditions of the models analogous to the definition of Caputo, therefore by taking into account in a completely erroneous way the past of the model [2, 67].

In spite of these reactions, perhaps dictated by a conservative spirit in an area which has prospered greatly, I believe on the contrary that this is one of the ways to eliminate the drawbacks cited in the previous section, as shown in the sequel.

This section thus proposes several non-singular kernels that solve some of the previous drawbacks and that can be used for power law type long memory behaviour modelling. For each of the kernels $\eta_k^\nu(t)$ defined in the sequel and for a given function $f(t)$, an integral-like power law behaviour can be obtained with the relation

$$\int_0^t f(z)\eta_k^\nu(t-z)dz \qquad k \in \{1, 2, 3, 4\}. \tag{16}$$

Note that a derivative-like power law behaviour is obtained using the relations

$$\frac{d}{dt}\int_0^t f(z)\eta_k^\nu(t-z)dz \quad \text{or} \quad \int_0^t \frac{df(z)}{dz}\eta_k^\nu(t-z)dz \qquad k \in \{1, 2, 3, 4\} \tag{17}$$

3.1 Kernels $\eta_1^\nu(t)$

The following class of kernel is now considered:

$$\eta_1^\nu(t) = \sum_{i=1}^{6}\left(\sum_{k=1}^{N} F_{i,k}(t)\right) \quad \text{with} \quad 0 < \nu_{i,k} < 1 \tag{18}$$

where

$$F_{1,k}(t) = C_{1,k}\frac{t^{\nu_{1,k}-1}}{\Gamma(\nu_{1,k})} \tag{19}$$

$$F_{2,k}(t) = C_{2,k}t^{\nu_{2,k}-1}E_{\nu_{2,k},\nu_{2,k}}\left(-\omega_{2,k}^{\nu_{2,k}}t^{\nu_{2,k}}\right) \tag{20}$$

$$F_{3,k}(t) = C_{3,k}\left(1 - E_{\nu_{3,k},1}\left(-\omega_{3,k}^{\nu_{3,k}}t^{\nu_{3,k}}\right)\right) \tag{21}$$

$$F_{4,k}(t) = C_{4,k} E_{\nu_{4,k},1}\left(-\omega_{4,k}^{\nu_{4,k}} t^{\nu_{4,k}}\right) \tag{22}$$

$$F_{5,k}(t) = C_{5,k} t^{\nu_{5,k}} E_{1,1+\nu_{5,k}}\left(-\omega_{5,k} t\right) \tag{23}$$

$$F_{6,k}(t) = C_{6,k} t^{\nu_{6,k}} E_{1,1-\nu_{6,k}}\left(-\omega_{6,k}^{\nu_{6,k}} t^{\nu_{6,k}}\right) \tag{24}$$

and where $E_{\alpha,\beta}(z)$ is the Mittag-Leffler function defined by [68]

$$E_{\alpha,\beta}(z) = \sum_{k=0}^{\infty} \frac{z^k}{\Gamma(\alpha k + \beta)}. \tag{25}$$

Using Laplace transform tables available in the literature [69], the Laplace transform of kernel $\eta_1^\nu(t)$ is given by

$$\eta_1^\nu(t) = \sum_{i=1}^{6} \left(\sum_{k=1}^{N} F_{i,k}(s)\right) \tag{26}$$

with

$$F_{1,k}(s) = \frac{C_{1,k}}{s^{\nu_{1,k}}} \tag{27}$$

$$F_{2,k}(s) = C_{2,k} \frac{1}{s^{\nu_{2,k}} + \omega_{2,k}^{\nu_{2,k}}} \tag{28}$$

$$F_{3,k}(s) = C_{3,k} \frac{\omega_{3,k}^{\nu_{3,k}}}{s\left(s^{\nu_{3,k}} + \omega_{3,k}^{\nu_{3,k}}\right)} \tag{29}$$

$$F_{4,k}(s) = C_{4,k} \frac{s^{\nu_{4,k}}}{s\left(s^{\nu_{4,k}} + \omega_{4,k}^{\nu_{4,k}}\right)} \tag{30}$$

$$F_{5,k}(s) = C_{5,k} \frac{1}{s^{\nu_{5,k}}\left(s + \omega_{5,k} t\right)} \tag{31}$$

$$F_{6,k}(t) = C_{6,k} \frac{s^{\nu_{6,k}}}{\left(s + \omega_{5,k} t\right)} \tag{32}$$

As an example, the kernel

$$\eta_1^\nu(t) = K\left(H_e(t) + \frac{t^{-\nu}}{\omega_{min}^\nu \Gamma(\nu)}\right) \quad \text{with} \quad K = \frac{1}{\sqrt{1 + \frac{2cos(\nu\pi/2)}{\omega_{min}^\nu} + \frac{cos(\nu\pi/2)^2}{\omega_{min}^{2\nu}}}} \tag{33}$$

is considered, where $H_e(t)$ denotes the Heaviside function and whose transfer function is

$$\eta_1^\nu(s) = K\left(\frac{1}{s} + \frac{1}{\omega_{min}^\nu s^{1-\nu}}\right) = K \frac{\left(\frac{s}{\omega_{min}}\right)^\nu + 1}{s}. \tag{34}$$

This kernel exhibits a power law behaviour of order $-1+\nu$ on the frequency domain $[\omega_{min}, \infty]$ as shown by Fig. 2 for various values of ν. Such a kernel remains singular at $t=0$:

$$\lim_{t \to 0} \eta_1^\nu(t) = \lim_{s \to \infty} s\eta_1^\nu(s) = +\infty. \tag{35}$$

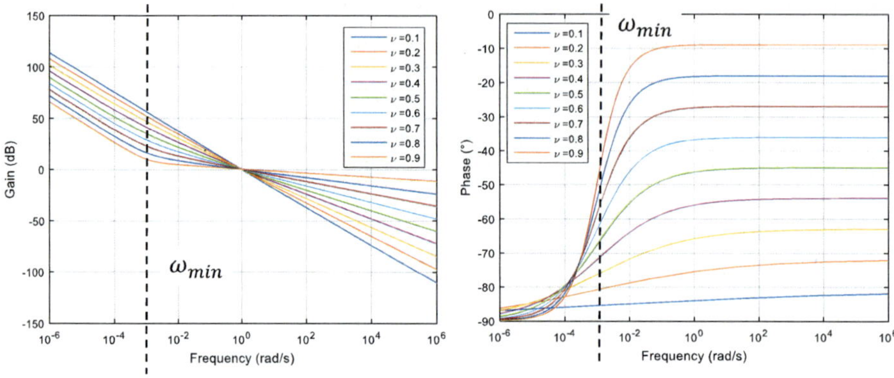

Fig. 2. Frequency response of the non-singular kernel $\eta_1^\nu(t)$ for ν varying from 0.1 to 0.9 with $\omega_{min} = 0.001$ rd/s

Now consider the kernel

$$\eta_1^\nu(t) = K\left(1 + E_{\nu,1}\left(-\omega_{max}^\nu t^\nu\right)\right) + \frac{K\omega_{max}^\nu}{\omega_{min}^\nu} E_{\nu,1}\left(-\omega_{max}^\nu t^\nu\right) \tag{36}$$

with $K = \sqrt{1 + \frac{2cos(\nu\pi/2)}{\omega_{max}^\nu} + \frac{cos(\nu\pi/2)}{\omega_{max}^{2\nu}}^2} / \sqrt{1 + \frac{2cos(\nu\pi/2)}{\omega_{min}^\nu} + \frac{cos(\nu\pi/2)}{\omega_{min}^{2\nu}}^2}$ and whose Laplace transform is

$$\eta_1^\nu(s) = K\frac{\omega_{max}^\nu}{s\left(s^\nu + \omega_{max}^\nu\right)} + \frac{K\omega_{max}^\nu}{\omega_{min}^\nu} \frac{s^\nu}{s\left(s^\nu + \omega_{max}^\nu\right)} = K\frac{\left(\frac{s}{\omega_{min}}\right)^\nu + 1}{s\left(\left(\frac{s}{\omega_{max}}\right)^\nu + 1\right)}. \tag{37}$$

This kernel exhibits a power law behaviour of order $-\nu$ on the frequency domain $[\omega_{min}, \omega_{max}]$ as shown by Fig. 3 for various values of ν. Such a kernel is no longer singular at $t=0$:

$$\lim_{t \to 0} \eta_1^\nu(t) = \lim_{s \to \infty} s\eta_1^\nu(s) = \frac{K\omega_{max}^\nu}{\omega_{min}^\nu}. \tag{38}$$

but does not solve most of the drawbacks listed in Sect. 2, as now discussed.

Computation of the impulse response of the transfer function (26) using Cauchy's theorem gives

$$\eta_1^\nu(t) = f(t) + \int_0^\infty \mu(x) e^{-tx} dx \tag{39}$$

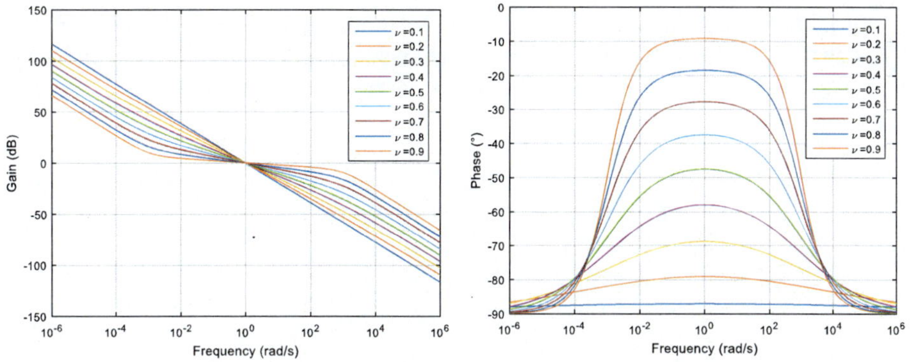

Fig. 3. Frequency response of the non-singular kernel $\eta_1^\nu(t)$ for ν varying from 0.1 to 0.9 with $\omega_{min} = 0.001$ rd/s and $\omega_{max} = 1000$ rd/s

where functions $f(t)$ and $\mu(x)$ are defined in [36, 37]. Laplace transform applied to relation (39) leads to

$$\eta_1^\nu(s) = f(s) + \int_0^\infty \frac{\mu(x)}{s+x} dx. \tag{40}$$

Relation (40) shows that the transfer function $\eta_1^\nu(s)$ exhibits a distribution of poles within the interval $]-\infty, 0]$ (without taking into account the $f(s)$ term). Kernel $\eta_1^\nu(t)$ thus exhibits infinitely slow and infinitely fast time constants. This class of kernel thus undergoes the set of drawbacks described in Sect. 2. The transfer functions (27) to (32) in the kernel are all made up of Laplace variables s with fractional powers (s^ν). This is precisely what induces the distribution of poles within the interval $]-\infty, 0]$ and the resulting drawback. Other types of kernels must therefore be sought in particular based on transfer functions not involving fractional powers of the Laplace variable (such as $(\tau s + 1)^\nu$).

3.2 Kernel $\eta_2^\nu(t)$

The following kernel is now considered:

$$\eta_2^\nu(t) = -C_2(\omega_{min}t)^{-\nu} e^{-\omega_{min}t} E_{1,1-\nu}(\omega_{min}t) \quad \text{with} \quad C_2 = \frac{1}{\left(\frac{1}{\omega_{min}^2} + 1\right)^{\frac{\nu}{2}}} \tag{41}$$

This kernel is the so-called tempered (or truncated) Mittag-Leffler memory kernel, which has been investigated in different contexts and especially in generalised diffusion equations [17, 20], in generalised diffusion-wave equations [21], and in generalised Langevin equations [18, 19].

As $\mathcal{L}\{t^\alpha E_{1,1+\alpha}(at)\} = s^{-\alpha}/(s-a)$ [69], and using the shift rule of the Laplace transform, the Laplace transform of kernel $\eta_2^\nu(t)$ is given by

$$\eta_2^\nu(s) = C_2 \frac{\left(\frac{s}{\omega_{min}} + 1\right)^\nu}{s}. \tag{42}$$

Such a transfer function no longer involves fractional powers of the Laplace variable. This transfer function exhibits a power law behaviour of order $-1+\nu$ on the frequency domain $[\omega_{min}, \infty]$ as shown by Fig. 4 for various values of ν.

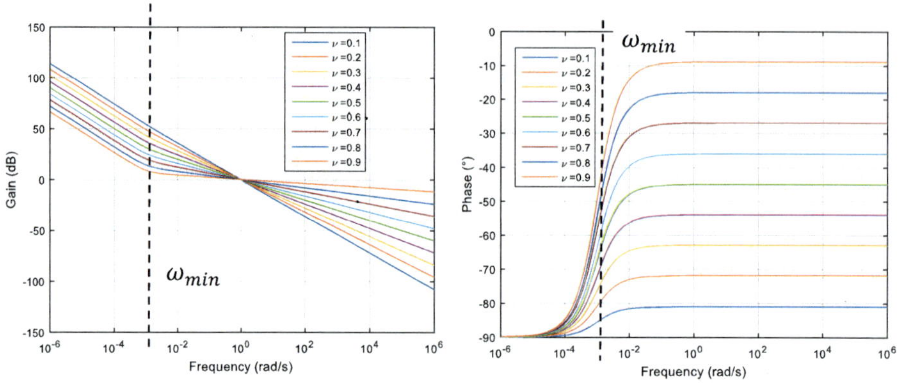

Fig. 4. Frequency response of the non-singular kernel $\eta_2^\nu(t)$ for ν varying from 0.1 to 0.9 with $\omega_{min} = 0.001$ rd/s.

Such a kernel remains singular at $t = 0$. Computation of the impulse response of the transfer function (41) using Cauchy's theorem [8] gives

$$\eta_2^\nu(t) = C_2 H_e(t) + \frac{C_2 sin(\nu\pi)}{\pi}\int_{\omega_{min}}^{\infty}\frac{\left(\frac{x}{\omega_{min}}-1\right)^\nu}{x}e^{-tx}dx. \quad (43)$$

The Laplace transform applied to relation (43) leads to

$$\eta_2^\nu(s) = \frac{C_2}{s} + \frac{C_2 sin(\nu\pi)}{\pi}\int_{\omega_{min}}^{\infty}\frac{\left(\frac{x}{\omega_{min}}-1\right)^\nu}{x(s+x)}dx. \quad (44)$$

Relation (44) shows that the transfer function $\eta_2^\nu(s)$ exhibits a distribution of poles within the interval $]-\infty, -\omega_{min}]$ (without taking into account the $1/s$ term). Kernel $\eta_2^\nu(t)$ thus does not exhibit infinitely slow time constants, which should simplify the initialisation of the equation in which this kernel will be used. However, this kernel still has infinitely fast time constants, which can be problematic for a physical interpretation.

3.3 Kernel $\eta_3^\nu(t)$

As an attempt to reduce the pole distribution domain, the following kernel is considered:

$$\eta_3^\nu(t) = C_3\left((\omega_{min}t)^{\nu-1}e^{-\omega_{min}t}E_{1,\nu}(\omega_{min}t) - \frac{\omega_{min}^{\nu-1}}{\Gamma(\nu)}t^{\nu-1}e^{-\omega_{max}t} + \frac{\omega_{min}^{\nu-1}}{\omega_{max}^{\nu-1}}e^{-\omega_{max}t}\right). \quad (45)$$

$$\text{with}\quad C_3 = \left| \left(\frac{\left(\frac{j}{\omega_{min}}+1\right)^{1-\nu}}{j^\nu} - \frac{\omega_{min}^{\nu-1}}{\omega_{max}^{\nu}} \frac{1}{\left(\frac{j}{\omega_{max}}+1\right)^{\nu}} + \frac{\omega_{min}^{\nu-1}}{\omega_{max}^{\nu}} \frac{1}{\frac{j}{\omega_{max}}+1} \right) \right|^{-1}. \quad (46)$$

The Laplace transform of this kernel is defined by

$$\eta_3^\nu(s) = C_3 \left(\frac{\left(\frac{s}{\omega_{min}}+1\right)^{1-\nu}}{s} - \frac{\omega_{min}^{\nu-1}}{\omega_{max}^{\nu}} \frac{1}{\left(\frac{s}{\omega_{max}}+1\right)^{\nu}} + \frac{\omega_{min}^{\nu-1}}{\omega_{max}^{\nu}} \frac{1}{\frac{s}{\omega_{max}}+1} \right), \quad (47)$$

This kernel exhibits a power law behaviour of order $-\nu$ on the frequency domain $[\omega_{min}, \omega_{max}]$ as shown by Fig. 5 for various values of ν. As time t tends towards 0

$$\lim_{t\to 0} \eta_3^\nu(t) = \lim_{s\to\infty} s\eta_3^\nu(s) = C_3 \frac{\omega_{max}^{1-\nu}}{\omega_{min}^{1-\nu}}. \quad (48)$$

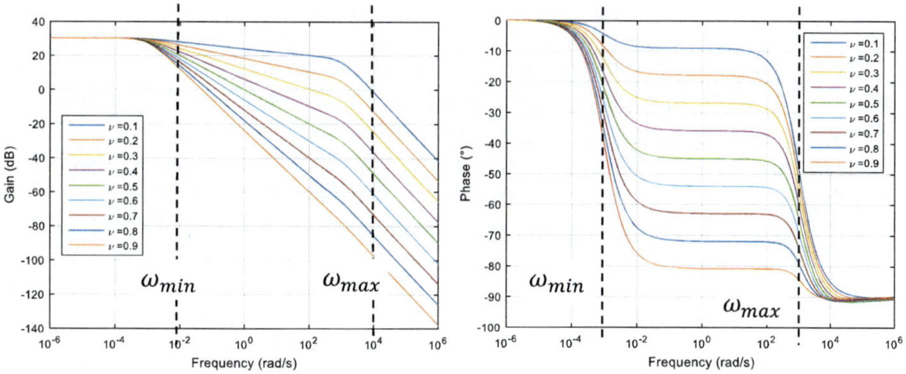

Fig. 5. Frequency response of the non-singular kernel $\eta_3^\nu(t)$ for ν varying from 0.1 to 0.9 with $\omega_{min} = 0.001$ rd/s and $\omega_{max} = 1000$ rd/s

This highlights the non-singularity of kernel (45) that thus overcomes drawback 8.

The impulse response of the previous function can be computed using Cauchy's theorem (see [8]). It is given by:

$$\eta_3^\nu(t) = C_3 \left(\frac{\sin((\nu-1)\pi)}{\pi(\omega_{min})^{1-\nu}} \int_{\omega_{min}}^{\infty} \frac{(x-\omega_{min})^\nu e^{-xt} dx}{x} - \frac{\sin(\nu\pi)}{\pi} \left(\frac{\omega_{min}}{\omega_{max}}\right)^{\nu-1} \int_{\omega_{max}}^{\infty} \frac{e^{-xt} dx}{(x-\omega_{max})^\nu} \right.$$
$$\left. + \left(\frac{\omega_{min}}{\omega_{max}}\right)^{\nu-1} e^{-\omega_{max} t} \right). \quad (49)$$

This relation shows that drawback 1 is partially overcome as the kernel no longer has infinitely slow time constants, and thus exhibits a limited memory, but it still has infinitely fast time constants.

3.4 Kernel $\eta_4^\nu(t)$

The following kernel is now considered:

$$\eta_4^\nu(t) = C_4 \frac{(\omega_{max})^\nu}{(\omega_{min})^\nu} \phi_2(\nu - 1, 1 - \nu, 1; \omega_{min} t, \omega_{max} t) \quad \text{with} \quad C_4 = \frac{\left(\frac{1}{\omega_{min}} + 1\right)^{\frac{1-\nu}{2}}}{\left(\frac{1}{\omega_{max}} + 1\right)^{\frac{1-\nu}{2}}} \tag{50}$$

where $\phi_2(\beta, \beta', \gamma; x, y)$ is a two-variable hypergeometric function defined by [70]

$$\phi_2(\beta, \beta', \gamma; x, y) = \sum\nolimits_{m,n} \frac{(\beta)_m (\beta')_n}{(\gamma)_{m+n} m! n!} x^m y^n. \tag{51}$$

The Laplace transform of this kernel is defined by [70] (p. 238):

$$\eta_4^\nu(s) = C_4 \frac{\left(\frac{s}{\omega_{min}} + 1\right)^{1-\nu}}{s\left(\frac{s}{\omega_{max}} + 1\right)^{1-\nu}} \tag{52}$$

The Bode diagrams of transfer functions $\eta_4^\nu(s)$ are shown in Fig. 6 with $\omega_{min} = 0.001$ rd/s, $\omega_{max} = 1000$ rd/s, and for various values of ν, showing that this kernel exhibits a power law behaviour of order ν in the frequency band $[\omega_{min}, \omega_{max}]$.

A property (and advantage) of this kernel is a distribution of its poles on a limited frequency band. To demonstrate such a property, the impulse response of the transfer function (50) is computed using Cauchy's theorem as demonstrated in [8, 71]. For $t > 0$, it is given by

$$\eta_4^\nu(t) = C_4 \left(H_e(t) + \left(\frac{\omega_{max}}{\omega_{min}}\right)^{1-\nu} \frac{\sin((1-\nu)\pi)}{\pi} \int_{\omega_{min}}^{\omega_{max}} \frac{(x - \omega_{min})^{1-\nu}}{x(\omega_{max} - x)^{1-\nu}} e^{-xt} dx \right) \tag{53}$$

where $H_e(t)$ is the Heaviside function. Laplace transform of relation (53) gives:

$$\eta_4^\nu(s) = C_4 \left(\frac{1}{s} + \left(\frac{\omega_{max}}{\omega_{min}}\right)^{1-\nu} \frac{\sin((1-\nu)\pi)}{\pi} \int_{\omega_{min}}^{\omega_{max}} \frac{(x - \omega_{min})^{1-\nu}}{(s + x)x(\omega_{max} - x)^{1-\nu}} dx \right). \tag{54}$$

Relation (54) demonstrates that the poles (variable x in relation (54)) of the kernel impulse $\eta_4^\nu(t)$ are distributed in a limited frequency band defined by $[\omega_{min}, \omega_{max}]$.

Now, still with kernel $\eta_4^\nu(t)$, the definition of the fractional integration in relation (9) is modified in order to limit its memory. The following modified operator is thus introduced:

$$J_{T_f}^\nu [f(t)] = \int_0^t f(\tau) d\tau + \int_{t-T_f}^t \eta_4(t - \tau) f(\tau) d\tau \tag{55}$$

$$\text{wtih} \quad \chi_4(t) = C_4 \left(\frac{\omega_{max}}{\omega_{min}}\right)^{1-\nu} \frac{\sin((1-\nu)\pi)}{\pi} \sum\nolimits_{\omega_{min}}^{\omega_{max}} \frac{(x - \omega_{min})^{1-\nu}}{x(\omega_{max} - x)^{1-\nu}} e^{-xt} dx. \tag{56}$$

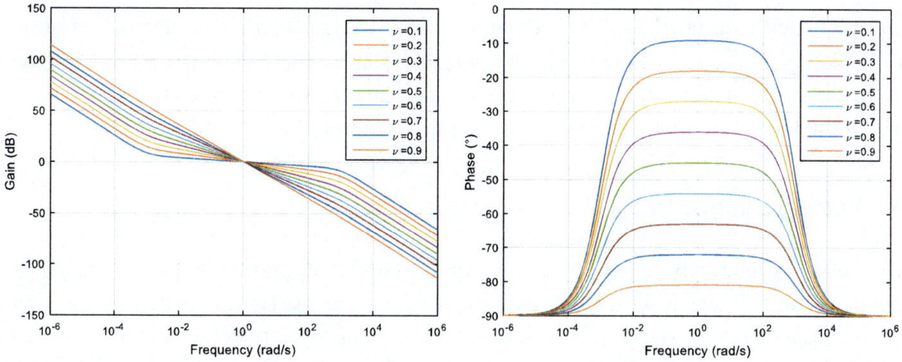

Fig. 6. Gain diagrams of transfer functions $\eta_4^\nu(s)$ with $\omega_{min} = 0.001$ rd/s, $\omega_{max} = 1000$ rd/s, for ν varying from 0.1 to 0.9

Using the change of variable $\xi = t - \tau$ and, thus, $d\xi = -d\tau$, relation (55) becomes

$$\boldsymbol{J}_{T_f}^{\nu}\big[f(t)\big] = \int_0^t f(\tau)d\tau + \int_0^{T_f} \chi_4(\xi)f(t-\xi)d\xi. \tag{57}$$

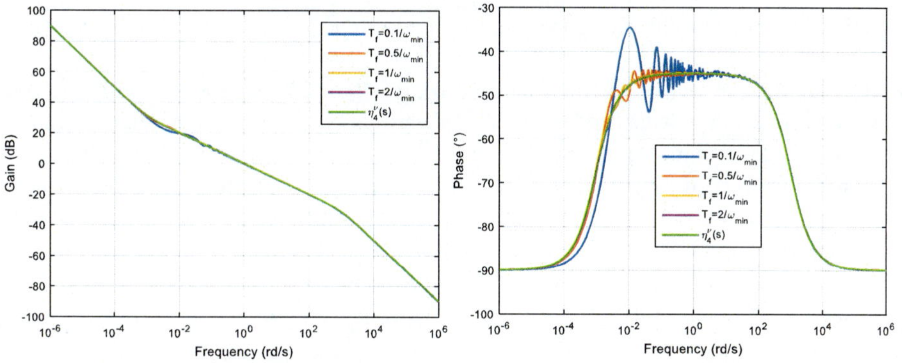

Fig. 7. A comparison of gain diagrams of $\eta_{T_f}^\nu(s)$ and of $\eta_4^\nu(s)$ in the domain of interest for $\nu = 0.5$, $\omega_{min} = 0.001$ rd/s, and $\omega_{max} = 1000$ rd/s and various values of T_f

The Laplace transform of relation (56) without considering initial conditions is

$$\mathcal{L}\Big\{\boldsymbol{J}_{T_f}^{\nu}\big[f(t)\big]\Big\} = \frac{F(s)}{s} + F(s)\int_0^{T_f} \chi_4(\xi)e^{-\xi s}d\xi \tag{58}$$

and thus, the operator corresponding to this modified fractional integrator is:

$$\eta_{T_f}^{\nu}(s) = \frac{1}{s} + \int_0^{T_f} \chi_4(\xi)e^{-\xi s}d\xi. \tag{59}$$

Figure 7 compares the gain diagram of $\eta_{T_f}^\nu(s)$ (computed numerically) with the gain diagram of $\eta_4^\nu(s)$ given by relation (54) in the domain of interest. It shows that the frequency response of $\eta_{T_f}^{\nu}(s)$ remains similar to that of $\eta_4^{\nu}(s)$ under the condition $T_f > 1/\omega_{min}$. In such a situation, parameter T_f can be viewed as the memory length of this kernel. Note that the memory length is connected to the corner frequency below which the kernel frequency response has a pure integrator behaviour.

4 Volterra Equations

The kernels previously listed make it possible to define operators that exhibit a power-law behaviour but can also be used to define new models involving a Volterra equation of the first kind. This class of equations, associated to appropriate kernels, is more general than the pseudo state space model very frequently used in the literature and given by relation (2). As shown in [72] model (2) can indeed be rewritten

$$\int_0^t \left(\frac{t^{-\nu}}{\Gamma(1-\nu)}I_n - A\right)x(\tau)d\tau = v(t) \quad \text{with} \quad v(t) = \int_0^t Bu(\tau)d\tau, \; y(t) = Cx(t), \tag{60}$$

where I_n denotes an identity matrix of dimension n. Relation (60) demonstrates that a pseudo state space description is a particular case of a Volterra equation of the first kind, as the kernel $\eta(t) = \left(\frac{t^{-\nu}}{\Gamma(1-\nu)}I_n - A\right)$ in (60) has a fixed structure. Using a Volterra equation, it is thus possible to generalise a pseudo state space description in two ways:

- By adapting the kernel $\eta(t)$ in relation (60), it is possible to produce, with the same kind of equation, power law behaviours of various types (denoted explicit, implicit), but also many other long memory behaviours;
- In relation (60), if $x(t) \in \mathbb{R}^n$, then $\eta(t)$ is a matrix of kernels such that $\eta(t) = \left[\eta_{i,j}(t)\right]$, thus permitting great flexibility in the tuning of relation (60). The case $\eta(t) = diag[\eta_i(t)]$ comes closer to the non-commensurate fractional pseudo state space representation case, but it should be remembered that physical interpretations invalidate this kind of model [48].

Description (60) has another important advantage. Model memory can be limited by introducing a parameter T_f in the integral bounds such that

$$\int_{t-T_f}^{t} \eta(t-\tau)x(\tau)d\tau = v(t). \tag{61}$$

Using the change of variable $\xi = t - \tau$, relation (61) becomes

$$\int_0^{T_f} \eta(\xi)x(t-\xi)d\xi = v(t). \tag{62}$$

Relation (62) explicitly shows that knowledge of the model state $x(t)$ is required on $\left[0, T_f\right]$ to compute its future. Modification of the lower bound of relation (60) to produce relation (62) is thus of interest in the initialisation problem. Initialisation of relation (62) only requires knowledge of the past of variable $x(t)$ on the interval $\left[t_0 - T_f, t_0\right]$ if t_0

denotes the initial time, while knowledge of the past on $]-\infty, t_0]$ is required for the fractional models (1) or (2).

For all these reasons, in a modelling approach it is better to work with model (62) than with models (1) or (2), since it is more general as previously demonstrated. This can be done by searching the kernel $\eta(t)$ directly without any assumption on its structure.

By choosing an appropriate kernel $\eta(t)$ in relation (62), it is possible to produce power law behaviours of various kinds. To highlight this, the following model involving a Volterra equation of the first kind is considered:

$$\int_0^t \eta(t-\tau)x(\tau)d\tau = v(t) \text{ with } v(t) = \int_0^t u(\tau)d\tau, \quad y(t) = x(t). \tag{63}$$

It is assumed that $u(t) \in \mathbb{R}$ denotes the input of the model and that $y(t) \in \mathbb{R}$ is its output. The Laplace transform of relation (63) is, without considering initial conditions,

$$\eta(s)x(s) = v(s) \text{ with } v(s) = \frac{1}{s}u(s), \quad y(s) = x(s), \tag{64}$$

and from an input–output point of view, the following transfer function is thus obtained:

$$\frac{y(s)}{u(s)} = \frac{1}{s\eta(s)}. \tag{65}$$

Several kernels producing power law behaviours to transfer function (65) were proposed in [8].

5 Other Alternative Models to Capture Power Law Type Behaviours

Volterra equations are one of the ways to define models for power-law behaviours without the drawbacks cited in Sect. 2. Other solutions exist and are now reviewed.

5.1 Distributed Time Delay Systems

The following class of time delay system was considered in [73].

$$\frac{d}{dt}x(t) = A_0x(t) + A_1\int_0^{T_f} \eta(\tau)x(t-\tau)d\tau + Bu(t). \tag{66}$$

The Laplace transform of relation (66) is given by

$$sx(s) = A_0x(s) + A_1\int_0^{T_f} \eta(\tau)x(s)e^{-\tau s}d\tau + Bu(s) \tag{67}$$

and thus

$$sx(s) = A_0x(s) + A_1x(s)\int_0^{T_f} \eta(\tau)e^{-\tau s}d\tau + Bu(s). \tag{68}$$

The transfer function linking the system input $u(t)$ to the system output $x(t)$ is thus defined by:

$$\frac{x(s)}{u(s)} = \frac{B}{s + A_0 + A_1 \int_0^{T_f} \eta(\tau)e^{-\tau s}d\tau}. \tag{69}$$

As an example, among the infinity of possible functions, some of which are given in [73], kernel $\eta(t)$ in (66) is assumed to be defined by:

$$\eta(t) = C_0\left(\frac{\omega_{min}^{\nu}}{\Gamma(\nu)}t^{\nu-1}e^{-\omega_{min}t} - \frac{\omega_{min}^{\nu}}{\Gamma(\nu)}t^{\nu-1}e^{-\omega_{max}t} + \omega_{min}^{\nu}\omega_{max}^{1-\nu}e^{-\omega_{max}t}\right) \tag{70}$$

$$\text{with} \quad C_0 = \left|\left(\frac{1}{\left(\frac{j}{\omega_{min}}+1\right)^{\nu}} - \left(\frac{\omega_{min}}{\omega_{max}}\right)^{\nu}\frac{1}{\left(\frac{j}{\omega_{max}}+1\right)^{\nu}} + \left(\frac{\omega_{min}}{\omega_{max}}\right)^{\nu}\frac{1}{\frac{j}{\omega_{max}}+1}\right)^{-1}\right|. \tag{71}$$

As t tends towards 0, the following relation holds (using Taylor expansions of the exponential function in relation (70)):

$$\eta(t) \underset{t\to 0}{\approx} C_0\left(\frac{\omega_{min}^{\nu}}{\Gamma(\nu)}t^{\nu-1}(-\omega_{min}t + \omega_{max}t) + \omega_{min}^{\nu}\omega_{max}^{1-\nu}\right) \tag{72}$$

and thus

$$\lim_{t\to 0}\eta(t) = C_0\omega_{min}^{\nu}\omega_{max}^{1-\nu}. \tag{73}$$

This highlights the non-singularity of kernel (70) as time t tends towards 0. Integral

$$I(s) = \int_0^{T_f}\eta(\tau)e^{-\tau s}d\tau \tag{74}$$

is defined by:

$$\begin{aligned}I(s) = K\Bigg(&((s+\omega_{min})T_f)^{-\frac{\nu}{2}}e^{-\frac{(s+\omega_{min})T_f}{2}}T_f^{\nu}(s+\omega_{max})W_M\left(\frac{\nu}{2},\frac{\nu}{2}+\frac{1}{2},\right.\\ &\left.(s+\omega_{min})T_f\right) - ((s+\omega_{max})T_f)^{-\frac{\nu}{2}}e^{-\frac{(s+\omega_{max})T_f}{2}}T_f^{\nu}(s+\omega_{max})W_M\left(\frac{\nu}{2},\frac{\nu}{2}+\frac{1}{2},\right.\\ &\left.(s+\omega_{max})T_f\right) - (1+\nu)\left(\left(\omega_{max}{}^{1-\nu}\Gamma(\nu)\nu + T_f^{\nu}(s+\omega_{max})\right)e^{-(s+\omega_{max})T_f} + T_f^{\nu}\right.\\ &\left.(s+\omega_{max})e^{-(s+\omega_{min})T_f} - \omega_{max}^{1-\nu}\Gamma(\nu)\right)\Bigg)\end{aligned} \tag{75}$$

with

$$K = \frac{C_0\left(s^2\omega_{min}^{\nu} + 2s\omega_{min}^{1+\nu} + \omega_{min}^{2+\nu}\right)}{(s+\omega_{max})\Gamma(\nu)(s+\omega_{min})^2\nu(1+\nu)} \tag{76}$$

where $W_M(a, b, z)$ is the Whittaker M function defined by

$$W_M(a, b, z) = z^{b+\frac{1}{2}} e^{-\frac{z}{2}} \sum_{n=0}^{\infty} \frac{\left(b - a + \frac{1}{2}\right)_n}{n!(2b+1)_n} z^n$$
$$= z^{b+1/2} e_1^{-z/2} F_1\left(\frac{1}{2} + b - a, 1 + 2b, z\right) \quad (77)$$

Figure 8 shows the gain and phase diagrams of integral $I(s)$ for various values of ν and with $\omega_{min} = 10^{-3}$ rd/s, $\omega_{max} = 10^4$ rd/s and $T_f = 10/\omega_{min}$.

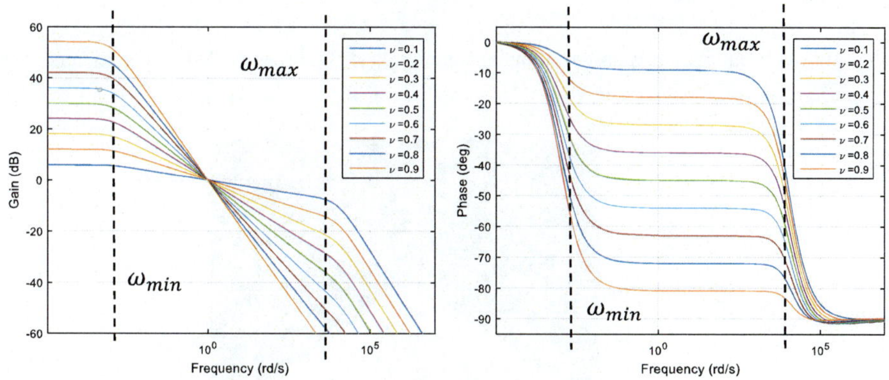

Fig. 8. Frequency response of integral $I(s)$ for ν varying from 0.1 to 0.9 and with $\omega_{min} = 10^{-3}$ rd/s, $\omega_{max} = 10^4$ rd/s and $T_f = 10/\omega_{min}$.

Parameters A_0, A_1 and B in model (66), can be used to control the frequency band on which the power law behaviour takes place. As an example, Fig. 9 shows the Bode diagram of the transfer function $x(s)/u(s)$ given by relation (69) under the following conditions:

- $\nu = 0.7$,$\omega_{min} = 10^{-3}$ rd/s, $\omega_{max} = 10^4$ rd/s, $T_f = 10/\omega_{min}$, $A_0 = -36969$, $A_1 = -10^8$, $B = 10^8$
- $\nu = 0.3$, $\omega_{min} = 10^{-3}$ rd/s, $\omega_{max} = 10^4$ rd/s, $T_f = 10/\omega_{min}$, $A_0 = -3686$, $A_1 = -10^5$, $B = 10^5$

These Bode diagrams exhibit a power law behaviour of order $1 - \nu$ in a limited frequency band.

5.2 Nonlinear Models

The idea of using non-linear models to model systems with power-law behaviours came from analysing the evolution of the top of a pile of sand $h(t)$ under a constant flow $Q(t)$ of sand, and assuming the angle of repose $\alpha(t)$ constant (see Fig. 10).

The kinetics of the evolution of this heap is of the form $Kt^{1/3}$, and can be very well fitted by a fractional transfer function model $Ks^{\frac{1}{3}-1}$. But such a model is no longer

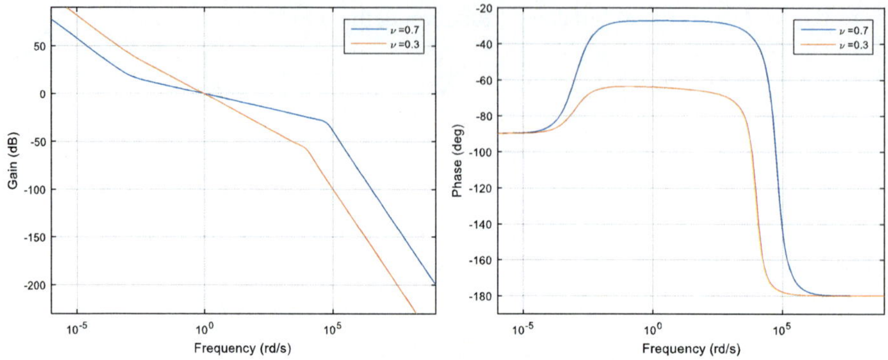

Fig. 9. Frequency response of transfer function $x(s)/u(s)$ for two sets of parameters

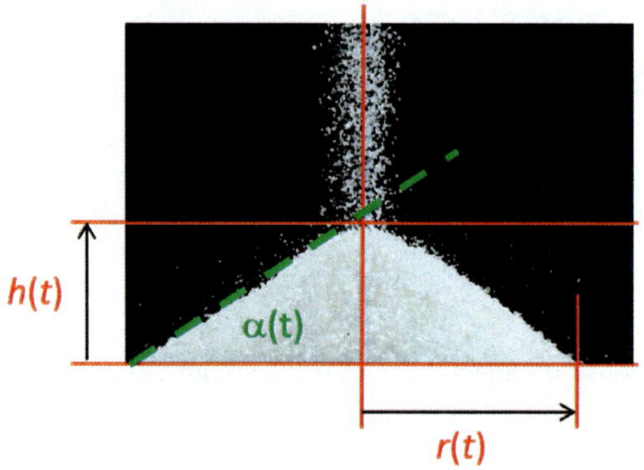

Fig. 10. Notations for the characterisation of sand heap growth

suitable if the flow of falling sand is doubled or if this flow is stopped (the fractional model relaxes while the height of the top remains constant). By analysing this system a little more closely [7], we can analytically demonstrate that a model of this system is of the form

$$\frac{dh(t)}{dt} = \frac{K}{h(t)^2} Q(t), \tag{78}$$

that is to say a non-linear input affine (or distributional) model form:

$$\frac{dx(t)}{dt} = f(x(t))u(t), \tag{79}$$

where $x(t)$ is the state and the output of the system, and where $u(t)$ denotes its input.

Such modelling solves a large part of the problems described in Sect. 2 and has been applied to the modelling of adsorption/desorption phenomena [74, 75] which are used in many fields and in particular for the design of Love wave based sensors [76].

5.3 Partial Differential Equations (Diffusion Equations) with Spatially Variable Coefficients

Starting from the observation that the recursivity (or more exactly geometric distribution) popularized by Oustaloup [47] (and introduced by Manabe [77]), of the parameters of electrical networks allowing the creation of power-type behaviors is a particular case among an infinity of possible distribution, it is shown that diffusion equations whose coefficients are spatially variable, allow to generate power law-type behaviours [72]. The coefficients must be chosen suitably, but an infinite number of combinations are possible. It is also possible to generalize this results to partial differential equations other than diffusion equations [72]. These results seem promising for modelling phenomena in fractal environments, in order to relate the geometry with the spatially variable coefficients of the equation.

6 Conclusion

In this paper, new kernels and new models are proposed to solve the drawbacks of the fractional (integration and differentiation) operators and fractional order models commonly used in the literature [68]. These drawbacks are first recalled. Some are linked to the singularity of the kernel used in their definition and some result from the doubly infinite dimension of the considered fractional operators (and therefore fractional models are also of doubly infinite dimension). As shown in this paper, these fractional operators and models are of infinite dimension as they can be described by a diffusion equation, and defined over ***an infinite domain*** [25]. It is this infinite domain that generates several drawbacks because it gives them their infinite memory property or equivalently the distribution of their poles over an infinite domain. This also explains that all the past must be taken into account for their initialization (or in an equivalent way all the infinite domain).

To solve these drawbacks, two different approaches are proposed in this paper:

- new kernels for the definition of operators and fractional models,
- new models capable of capturing power law behaviours.

The proposed new kernels are based on transfer functions ***not involving fractional powers of the Laplace variable*** (s^{ν}) as it is a solution to suppress the infinite memory evoked above. These new kernels can be used in Volterra type equations of the first kind to define a new class of models that is more general than the usual fractional models. But as shown in the paper, other classes of models such as nonlinear models, distributed time delay models or partial differential equations with spatially variable coefficients can also be used to capture power law type behaviours.

References

1. Tenreiro, M.J., Kiryakova, V., Mainardi, F.: Recent history of fractional calculus. Commun. Nonlinear Sci. Numer. Simul. **16**(3), 1140–1153 (2011)

2. Sabatier, J.: Fractional-order derivatives defined by continuous kernels: are they really too restrictive? Fractal Fract. **4**, 40 (2020)
3. Caputo, M., Fabrizio, M.: A new definition of fractional derivative without singular kernel. Prog. Fract. Differ. Appl. **1**, 73–85 (2015)
4. Atangana, A., Baleanu, D.: New fractional derivatives with nonlocal and non-singular kernel: theory and application to heat transfer model. Therm. Sci. **20**, 763–769 (2016)
5. Gao, F., Yang, X.J.: Fractional Maxwell fluid with fractional derivative without singular kernel. Therm. Sci. **20**(Suppl. 3), S873–S879 (2016)
6. Losada, J., Nieto, J.J.: Properties of a new fractional derivative without singular kernel. Prog. Fract. Differ. Appl. **1**, 87–92 (2015)
7. Sabatier, J., Farges, C., Tartaglione, V.: Some alternative solutions to fractional models for modelling power law type long memory behaviours. Mathematics **8**(2), 196 (2020)
8. Sabatier, J.: Non-singular kernels for modelling power law type long memory behaviours and beyond. Cybern. Syst. **51**(4), 383–401 (2020)
9. Sabatier, J.: Fractional state space description: a particular case of the Volterra equation. Fractal Fract. **4**(23), 2020 (2020)
10. Zhao, D., Luo, M.: Representations of acting processes and memory effects: general fractional derivative and its application to theory of heat conduction with finite wave speeds. Appl. Math. Comput. **346**, 531–544 (2019)
11. Fernandez, A., Özarslan, M.A., Baleanu, D.: On fractional calculus with general analytic kernels. Appl. Math. Comput. **354**, 248–265 (2019)
12. Baleanu, D., Fernandez, A.: On fractional operators and their classifications. Mathematics **7**, 830 (2019)
13. Lutz, E.: Fractional Langevin equation. Phys. Rev. E **64** (2001). https://doi.org/10.1103/PhysRevE.99.052125
14. Pottier, N.: Aging properties of an anomalously diffusing particle. Phys. Stat. Mech. Appl. **317**, 371–382 (2003)
15. Viñales, A.D., Despósito, M.A.: Anomalous diffusion induced by a Mittag-Leffler correlated noise. Phys. Rev. E **75**, 042102 (2007). https://doi.org/10.1103/physreve.75.042102
16. Viñales, A.D., Wang, K.-G., Despósito, M.A.: Anomalous diffusive behavior of a harmonic oscillator driven by a mittag-leffler noise. Phys. Rev. E **80**, 011101 (2009). https://doi.org/10.1103/physreve.80.011101
17. Sandev, T., Chechkin, A., Kantz, H., Metzler, R.: Diffusion and Fokker-Planck-Smoluchowski equations with generalized memory kernel. Fract. Calc. Appl. Anal. **18**(4), 1006–1038 (2015). https://doi.org/10.1515/fca-2015-0059
18. Liemert, A., Sandev, T., Kantz, H.: Generalized Langevin equation with tempered memory kernel. Physica A **466**, 356–369 (2017)
19. Sandev, T.: Generalized langevin equation and the prabhakar derivative. Mathematics **5**, 66 (2017)
20. Sandev, T., Deng, W., Xu, P.: Models for characterizing the transition among anomalous diffusions with different diffusion exponents. Phys. A Stat. Mech. Appl. **51**, 405002 (2018)
21. Sandev, T., Tomovski, Z., Dubbeldam, J.L.A., Chechkin, A.: A. Phys. A Stat. Mech. Appl. **52**, 015101 (2018)
22. Metzler, R., Klafter, J.: The random walk's guide to anomalous diffusion: a fractional dynamics approach. Phys. Rep. **339**(1), 1–77 (2000)
23. Barkai, E., Metzler, R., Klafter, J.: From continuous time random walks to the fractional Fokker-Planck equation. Phys. Rev. E **61**(1), 132–138 (2000)
24. Sabatier, J., Merveillaut, M., Malti, R., Oustaloup, A.: On a representation of fractional order systems: interests for the initial condition problem. In: 3rd ed.; IFAC Workshop: Ankara, Turkey (2008)

25. Sabatier, J., Merveillaut, M., Malti, R., Oustaloup, A.: How to impose physically coherent initial conditions to a fractional system? Commun. Nonlinear Sci. Numer. Simul. **15**, 1318–1326 (2010)
26. Oppenheim, A.V., Alan, S. Willsky, A.S., Hamid, S.: Signals and Systems, Pearson New International Edition (1996)
27. Rodrigues, S., Munichandraiah, N., Shukla, A.K.: A review of state-of-charge indication of batteries by means of A.C. impedance measurements. J. Power Sources **87**, 12–20 (2000)
28. Sabatier, J., Aoun, M., Oustaloup, A., Grégoire, G., Ragot, F.: Fractional system identification for lead acid battery sate charge estimation. Signal Process. **86**, 2645–2657 (2006)
29. Battaglia, J.L., Cois, O., Puigsegur, L., Oustaloup, A.: Solving an inverse heat conduction problem using a noninteger identified model. Int. J. Heat Mass Transf. **44**, 2671–2680 (2001)
30. Magin, R.L.: Fractional Calculus in Bioengineering. Begell House Publishers Inc., Danbury (2006)
31. Mainardi, F.: Fractional Calculus and Waves in Linear Viscoelasticity. Imperial College Press, London (2010)
32. Matignon, D., D'Andréa Novel, B., Depalle, P., Oustaloup, A.: Viscothermal Losses in Wind Instruments: A Non Integer Model; International Symposium on the Mathematical Theory of Networks and Systems (MTNS), Regensburg, Germany (1993)
33. Enacheanu, O.: Modélisation fractale des réseaux électriques, Ph.D. thesis, Université Joseph Fourier—Grenoble I (2008). http://www.theses.fr/2008GRE10159
34. Tarasov, V.E.: Rules for fractional-dynamic generalizations: difficulties of constructing fractional dynamic models. Mathematics **7**, 554 (2019)
35. Sabatier, J., Farges, C., Merveillau, M., Fenetau, L.: On observability and pseudo state estimation of fractional order systems. Eur. J. Control **18**(3), 1–12 (2012)
36. Matignon, D.: Stability properties for generalized fractional differential systems. ESAIM Proc. **5**, 145–158 (1998)
37. Montseny, G.: Diffusive representation of pseudo-differential time-operators. ESAIM Proc. **5**, 159–175 (1998)
38. Malti, R., Sabatier, J., Akçay, H.: Thermal modeling and identification of an aluminum rod using fractional calculus, IFAC Proc. Vol. (IFAC-PapersOnline) **15**(Part 1), 958–963 (2009)
39. Sabatier, J., et al.: Fractional models for thermal modeling and temperature estimation of a transistor junction. Adv. Differ. Equ. **2011**(1), 1–12 (2011). https://doi.org/10.1155/2011/687363
40. Nigmatullin, R.R.: A fractional integral and its physical interpretation. Theoret. Math. Phys. **90**(3), 242–251 (1992)
41. Rutman, R.S.: On physical interpretations of fractional integration and differentiation. Theor. Math. Phys. **105**(3), 393–404 (1995)
42. Ben, A.F.: Geometric interpretation of the fractional derivative. J. Fract. Calc. **11**, 21–52 (1997)
43. Gorenflo, R.: Afterthoughts on interpretation of fractional derivatives and integrals. In: Rusev, P., Dimovski, I., Kiryakova, V. (eds.) Transform Methods and Special Functions, Varna 1996, Institute of Mathematics and Informatics, Bulgarian Academy of Sciences, Sofia (1998)
44. Mainardi, F.: Considerations on fractional calculus: interpretations and applications. In: Rusev, P., Dimovski, I., Kiryakova, V. (eds.) Transform Methods and Special Functions, Varna 1996, Institute of Mathematics and Informatics, Bulgarian Academy of Sciences, Sofia (1998)
45. Podlubny, I.: Geometric and physical interpretation of fractional integration and fractional differentiation. J. Fract. Calc. Appl. Anal. **5**(4), 357–366 (2002)
46. Tenreiro Machado, J.A.: A probabilistic interpretation of the fractional-order differentiation. J. Fract. Calc. Appl. Anal. **6**(1), 73–80 (2003)

47. Oustaloup, A.: Diversity and Non-Integer Differentiation for System Dynamics. Wiley, Hoboken (2014)
48. Dokoumetzidis, A., Magin, R., Macheras, P.: A commentary on fractionalization of multi-compartmental models. Pharmacokinet Pharmacodyn **37**, 203–207 (2010)
49. Sabatier, J., Farges, C., Trigeassou, J.C.: Fractional systems state space description: some wrong ideas and proposed solutions. J. Vib. Control **20**(7), 1076–1084 (2014)
50. De Oliveira, E.C., Tenreiro Machado, J.A.: A review of definitions for fractional derivatives and integral. Math. Probl. Eng. **2014** (2014). Article ID 238459
51. Ortigueira, M.D., Coito, F.J.: System initial conditions vs derivative initial conditions. Comput. Math. Appl. **59**(5), 1782–1789 (2010)
52. Lorenzo, C.F., Hartley, T.T.: Initialized fractional calculus. Int. J. Appl. Math. **3**(3), 249–265 (2000)
53. Lorenzo, C.F., Hartley, T.T.: Initialization in fractional order systems. In: European Conference on Control, ECC, Porto, Portugal, pp. 1471–1476 (2001)
54. Ortigueira, M.D.: On the initial conditions in continuous-time fractional linear systems. Signal Process. **83**, 2301–2309 (2003)
55. Fukunaga, M., Shimizu, N.: Role of prehistories in the initial value problems of fractional viscoelastic equations. Nonlinear Dyn. Syst. **38**, 207–220 (2004)
56. Sabatier, J., Farges, C.: Comments on the description and initialization of fractional partial differential equations using Riemann-Liouville's and Caputo's definitions. J. Comput. Appl. Math. **339**, 30–39 (2018)
57. Balint, A.M., Balint, S.: Mathematical description of the groundwater flow and that of the impurity spread, which use temporal caputo or Riemann-Liouville fractional partial derivatives, is non-objective. Fractal Fract. **4**, 36 (2020)
58. Bettayeb, M., Djennoune, S.: A note on the controllabity and the observability of fractional dynamical systems. In: IFAC-FDA workshop, Porto, Portugal (2006)
59. Matignon, D., D'Andrea-Novel, B.: Some results on controllability and observability of finite-dimensional fractional differential systems. In: Computational Engineering in Systems Applications, Lille, France, vol. 2, pp. 952–956 (1996)
60. Yang, X.J., Gao, F., Machado, J.A.T., Baleanu, D.: A new fractional derivative involving the normalized sinc function without singular kernel. arXiv:1701.05590 (2017)
61. Tarasov, V.E.: No nonlocality. No fractional derivative. Commun. Nonlinear Sci. Numer. Simul. **62**, 157–163 (2018)
62. Ortigueira, M.D., Tenreiro Machado, J.A.T.: A critical analysis of the Caputo–Fabrizio operator. Commun. Nonlinear Sci. Numer. Simul. **59**, 608–611 (2019)
63. Ortigueira, M.D., Martynyuk, V., Fedula, M., Machado, J.: The failure of certain fractional calculus operators in two physical models. Fract. Calc. Appl. Anal. **22**(2), 255–270 (2019)
64. Stynes, M.: Fractional-order derivatives defined by continuous kernels are too restrictive. Appl. Math. Lett. **85**, 22–26 (2018). https://doi.org/10.1016/j.aml.2018.05.013
65. Hanyga, A.: A comment on a controversial issue: a generalized fractional derivative cannot have a regular kernel. Fract. Calc. Appl. Anal. **23**(1), 211–223 (2020)
66. Diethelm, K., Garrappa, R., Giusti, A., Stynes, M.: Why fractional derivatives with nonsingular kernels should not be used. Fract. Calc. Appl. Anal. **23**(3), 610–634 (2020)
67. Sabatier, J., Farges, C.: Initial value problems should not be associated to fractional model descriptions whatever the derivative definition used. AIMS Math. **6**(10), 11318–11329 (2021)
68. Samko, S.G., Kilbas, A.A., Marichev, O.I.: Fractional Integrals and Derivatives: Theory and Applications. Gordon and Breach Science Publishers, London (1993)
69. Monje, C.A., Chen, Y., Vinagre, B.M., Xue, D., Feliu-Batlle, V.: Fractional-Order Systems and Controls: Fundamentals and Applications. Springer, Heidelberg (2010). https://doi.org/10.1007/978-1-84996-335-0

70. Erdelyi, A.: Tables of Integral Transforms, vol. 1. McGraw-Hill Book Company, New York (1954)
71. Sabatier, J., Rodriguez Cadavid, S., Farges, C.: Advantages of limited frequency band fractional integration operator. In: 6th International Conference on Control, Decision and Information Technologies (Codit 2019), 23–26 April 2019, Paris, France (2019)
72. Sabatier, J.: Beyond the particular case of circuits with geometrically distributed components for approximation of fractional order models: application to a new class of model for power law type long memory behaviour modelling. J. Adv. Res. **25**, 243–325 (2020)
73. Sabatier, J.: Power law type long memory behaviors modeled with distributed time delay systems. Fractal Fract **4**(1), 1 (2020)
74. Tartaglione, V., Farges, C., Sabatier, J.: Dynamical modelling of random sequential adsorption. In: Proceedings of European Control Conference (ECC 2020), Saint Petersburg, Russia (2020)
75. Tartaglione, V., Sabatier, J., Farges, C.: Adsorption on fractal surfaces: a non linear modeling approach of a fractional behaviour. Fractal Fract. **5**, 65 (2021)
76. Hallil, H., Menini, P., Aubert, H.: Novel microwave gas sensor using dielectric resonator with SnO2 sensitive layer. Procedia Chem. **1**(1), 935–938 (2009)
77. Manabe, S.: The non-integer Integral and its application to control systems. ETJ Jpn. **6**, 83–87 (1961)

Initial Value Problem Should Not Be Associated to a Fractional Model Description Whatever the Derivative Definition Used

Jocelyn Sabatier[1(✉)] and Christophe Farges[2]

[1] Bordeaux University, 15, Rue Naudet - CS 10 207, 33175 Gradignan, France
jocelyn.sabatier@u-bordeaux.fr
[2] IMS Laboratory UMR 5218 CNRS, Bordeaux University, 351 Cours de la Liberation, 33405 Talence, France
christophe.farges@u-bordeaux.fr

Abstract. The paper shows that the Caputo definition of fractional differentiation is problematic if it is used in the definition of a fractional model and if initial conditions are taken into account. The demonstration is done using simple examples (or counterexamples). The analysis is extended to the Riemann-Liouville and Grünwald-Letnikov definitions. These results thus question the validity of results produced in the field of fractional model analysis in which initial conditions are involved.

Keywords: Fractional derivative · Caputo definition · Riemann-Liouville definition · Grünwald-Letnikov definition · Initial conditions · Memory

1 Introduction

A Subsection Sample In much of the literature, fractional models are defined using the Caputo definition, in which fractional models are models described by fractional differential equations or pseudo state space descriptions. The Caputo definition is widely acclaimed because it makes it possible to define initial conditions that relate to the integer derivatives of the derived functions in the models considered. However, this paper shows that this definition does not take initial conditions properly into account if used to define a fractional model.

The problem was analysed for the first time by Lorenzo and Hartley [1, 2]. To take the past of the model into account in a convenient way in a finite interval, they introduced an initialization function. The idea of replacing the commonly used initial values by an initial function was further developed in [3]. In [4], the need to consider the "prehistories" before the initial instant of the derivate functions was shown, making it possible to address the initialisation of fractional visco-elastic equations to reach a unique solution. In [5, 6], a counter example was used to demonstrate that initial conditions cannot be correctly taken into account in a dynamical model whether by Caputo or Riemann–Liouville definitions. This led to the conclusion in [7] that fractional derivative and

A. Dzielinski et al. (Eds.): ICFDA 2021, LNNS 452, pp. 26–37, 2022.
https://doi.org/10.1007/978-3-031-04383-3_2

fractional model initializations are two distinct problems. Still using an initial time shifting method, counter examples were proposed in [8] to show similar initialisation problems with the Caputo definition for partial differential equations. A time shifting technique was also recently used in [9] to analyse a groundwater flow model with time Caputo or Riemann–Liouville fractional partial derivatives. The non-objectivity of these models was demonstrated in this paper. The authors in [9] did not address the problem of initialization, but this objectivity can be restored by also introducing an initialization function (instead of initial conditions).

In this paper, two examples are used to show that the Caputo definition does not enable initial conditions to be correctly handled ***when this definition is used to define a fractional model***. In the first example, the response of a simple model, assumed to be at rest, is calculated analytically on a given time interval. Then inside this interval, a second response is computed by considering initial conditions resulting from the first simulation, and ignoring the model past before the considered initial time. This is the initialisation currently found in the literature and this example shows that it is unable to ensure the correct model trajectories. In the second example, two different histories are generated that produce the same initial conditions for the model. This example shows that in spite of equal initial conditions, the model response is different, thus showing that all the model past must be taken into account to define its future. A similar analysis is also carried out with the Riemann-Liouville and the Grünwald-Letnikov's definitions, suggesting that other definitions should also be problematic.

2 Problem Analysis with Caputo's Definition

The fractional integral of order ν, $0 < \nu < 1$, of a function $y(t)$ is defined by [10]:

$$\boldsymbol{I}_{t_0}^{\nu} y(t) = \frac{1}{\Gamma(\nu)} \int_{t_0}^{t} \frac{y(\tau)}{(t-\tau)^{1-\nu}} d\tau. \tag{1}$$

$\Gamma(.)$ being Euler's gamma function. From this definition, the Caputo derivative definition of order ν, $0 < \nu < 1$, of a function $y(t)$ is defined by [11]:

$$_C\boldsymbol{D}_{t_0}^{\nu} y(t) = \boldsymbol{I}_{t_0}^{1-\nu}\left(\frac{d}{dt} y(t)\right) = \frac{1}{\Gamma(1-\nu)} \int_{t_0}^{t} \frac{1}{(t-\tau)^{\nu}} \frac{dy(\tau)}{d\tau} d\tau. \tag{2}$$

Laplace transform applied to relation (2) reveals how initial conditions are associated to this definition:

$$\mathcal{L}\left\{_C\boldsymbol{D}_{t_0}^{\nu} y(t)\right\} = \frac{1}{s^{1-\nu}}(sY(s) - y(t_0)) = s^{\nu} Y(s) - s^{\nu-1} y(t_0). \tag{3}$$

To demonstrate that Caputo's definition is not able to take initial conditions correctly into account when used to define a fractional model (a fractional differential equation or a pseudo state space description), the following model is considered

$$\boldsymbol{D}^{\nu} y(t) = -a y(t) + u(t) \qquad 0 < \nu < 1 \quad a > 0. \tag{4}$$

In relation (4), $\boldsymbol{D}^{\nu}$ denotes the Caputo definition in this section but denotes the Riemann-Liouville or Grünwald-Letnikov definitions in the next section. Then, the following algorithm is used to study model (4).

Algorithm 1

1 – Simulation on the time interval $[0, t_1]$ *of the fractional model (for instance model (4)) with null initial conditions (for* $t \in]-\infty, 0]$*). Let* S_1 *denote this simulation.*

2 – Record the model output $y(t)$ *and the integer derivatives of* $y(t)$ *(denoted* $y'(t)$*,* $y''(t), \ldots.$*) at time* t_0 *such that* $0 < t_0 < t_1$*.*

3 - Simulate the model again on $[t_0, t_1]$*, using* $y(t_0)$*,* $y'(t_0)$*,* $y''(t_0)\ldots$ *as initial conditions. Let* S_2 *denote this simulation.*

4 – Compare S_1 *and* S_2 *on* $[t_0, t_1]$ *and notice if they are different.*

■

Algorithm 1 is now applied to model (4) with $a = 0$. The model is assumed to be at rest before $t = 0$, and the input $u(t)$ is assumed to be a Heaviside function $H(t)$. In such conditions, relation (4) is equivalent to [11]

$$y(t) = y(t_0) + \boldsymbol{I}_{t_0}^{\nu}\{H(t)\} \qquad 0 < \nu < 1. \tag{5}$$

As a consequence, the simulation defined in Algorithm 1 provides the following solutions:

$$S_1 : y(t) = \frac{t^{\nu}}{\Gamma(\nu + 1)} \qquad 0 < t < t_1 \tag{6}$$

$$S_2 : y(t) = \frac{(t - t_0)^{\nu}}{\Gamma(\nu + 1)} + y(t_0) \quad t_0 < t < t_1. \tag{7}$$

Figure 1 proposes a comparison of S_1 and S_2 and reveals a difference, thus demonstrating that the Caputo definition does not correctly take initial conditions into account.

Another way to illustrate this result is to consider two different input signals $u_1(t)$ and $u_2(t)$ that create two different histories with:

$$u_i(t) = A_i H(t + t_i) - A_i H(t) \quad \text{with} \quad t_i > 0, \quad i = \{1, 2\}. \tag{8}$$

The model is assumed to be at rest on $t \in]-\infty, t_i]$. A constraint is also imposed on these signals so that at $t = 0$, the two resulting model outputs coincide:

$$y_1(0) = y_2(0). \tag{9}$$

The output $y_i(t)$ is thus defined by:

$$y_i(t) = \frac{A_i}{a}\Big(1 - E^1_{\nu,1}\big(-a(t + t_i)^{\nu}\big)\Big)H(t + t_i) - \frac{A_i}{a}\Big(1 - E^1_{\nu,1}\big(-at^{\nu}\big)\Big)H(t). \tag{10}$$

where $E^{\gamma}_{\alpha,\beta}(z)$ is the Mittag-Leffler function defined by [12]:

$$E^{\gamma}_{\alpha,\beta}(z) = \sum\nolimits_{k=0}^{\infty} \frac{\Gamma(\gamma + k)}{\Gamma(\alpha k + \beta)\Gamma(\gamma)} \frac{z^k}{k!}. \tag{11}$$

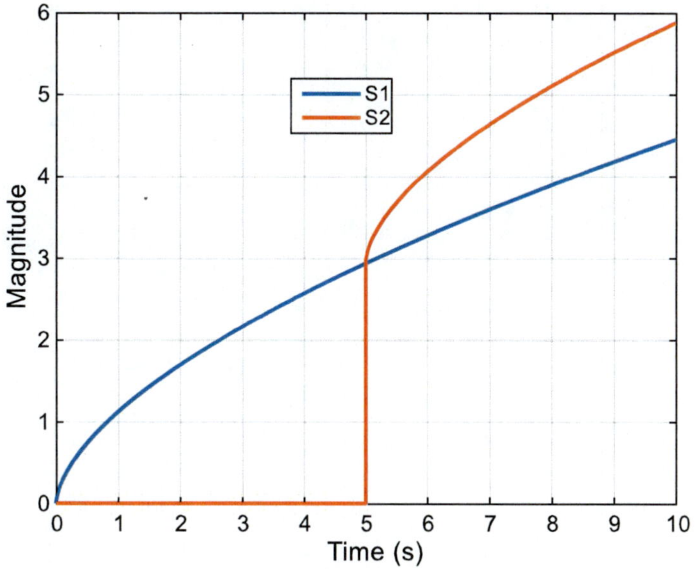

Fig. 1. Comparison of the exact response of model (4) with the responses obtained with Caputo definitions with initial conditions (($t_0 = 5$ s, $a = 0$, $\nu = 0.6$)

Condition (9) thus leads to

$$\frac{A_1}{a}\left[\left(1 - E^1_{\nu,1}\left(-a(t_1)^{\nu}\right)\right) - \left(1 - E^1_{\nu,1}(0)\right)\right]$$
$$= \frac{A_2}{a}\left[\left(1 - E^1_{\nu,1}\left(-a(t_2)^{\nu}\right)\right) - \left(1 - E^1_{\nu,1}(0)\right)\right] \tag{12}$$

thus leading to the condition:

$$A_1 = A_2 \frac{1 - E^1_{\nu,1}(-a(t_2)^{\nu})}{1 - E^1_{\nu,1}(-a(t_1)^{\nu})}. \tag{13}$$

With $\nu = 0.4$, $a = 1$, $t_1 = -8s$, $t_2 = -2s$, $A_2 = 5$ and thus $A_1 \approx 4.17$, Fig. 2 shows the signal inputs $u_1(t)$ and $u_2(t)$ used for the analysis and proposes a comparison of the resulting outputs. This figure shows that the two responses have the same values at $t = 0$, but that the evolutions for $t > 0$ are not the same. The information at $t = 0$ is thus not enough to predict the future of the model. All the past must be taken into account to predict the future of the model, which confirms that initialization as defined by the Caputo definition is not acceptable if used to define a fractional model such as (4).

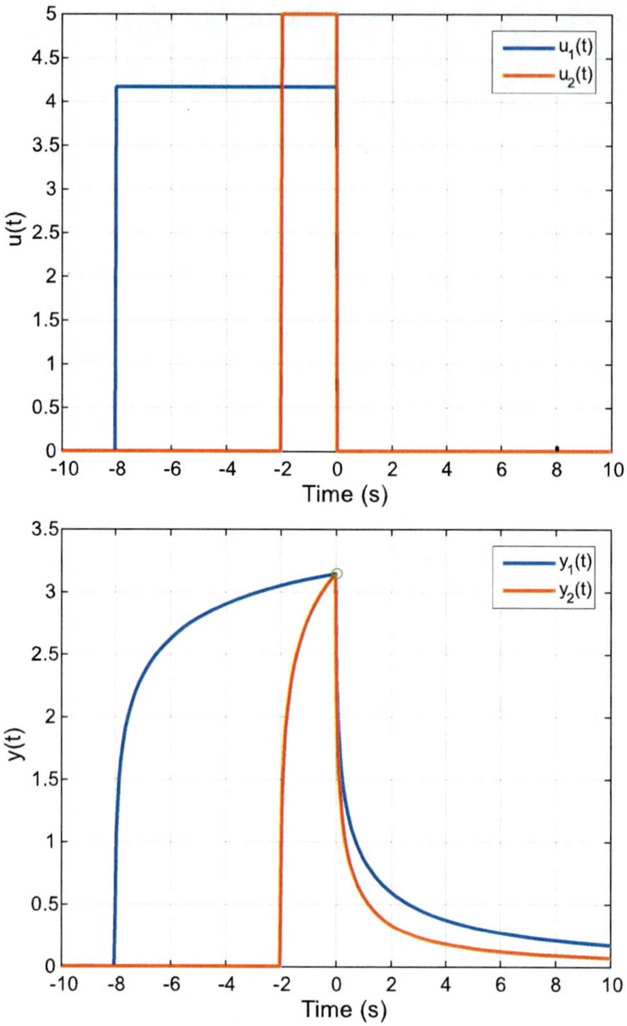

Fig. 2. Comparison of the responses $y_1(t)$ and $y_2(t)$ of model (4) to two inputs that provide the same initial conditions

3 Analysis with Other Definitions

The previous section showed that the Caputo definition should no longer be used to define fractional models such as (4). What about other definitions?

3.1 The Riemann-Liouville Definition

The Riemann-Liouville derivative of order ν, $0 < \nu < 1$, of a function $y(t)$ is defined by [11]:

$$ {}_{RL}\boldsymbol{D}_{t_0}^{\nu} y(t) = \frac{1}{\Gamma(1-\nu)} \frac{d}{dt} \int_{t_0}^{t} \frac{y(\tau)}{(t-\tau)^{\nu}} d\tau. \tag{14} $$

Laplace transform applied to relation (14) reveals how initial conditions are associated to this definition:

$$ \mathcal{L}\left\{ {}_{RL}\boldsymbol{D}_{t_0}^{\nu} y(t) \right\} = \mathcal{L}\left\{ \frac{d}{dt} \left(\frac{1}{\Gamma(1-\nu)} \int_{t_0}^{t} \frac{y(\tau)}{(t-\tau)^{\nu}} d\tau \right) \right\} = s \frac{1}{s^{1-\nu}} Y(s) - \left[\boldsymbol{I}_{t_0}^{1-\nu} y(t) \right]_{t=t_0} \tag{15} $$

As a consequence, in [11, 13], the initialisation of relation (4) is defined by

$$ \frac{d^{\nu}}{dt^{\nu}} y(t) = -a y(t) + u(t) \qquad I_{t_0}^{1-\nu} \{y(t)\} \Big|_{t=t_0} = y_0. \tag{16} $$

and thus the initialisation problem of relation (4) is equivalent to the integral equation

$$ y(t) = \frac{y_0}{\Gamma(\nu)} (t-t_0)^{\nu-1} + I_{t_0}^{\nu} \{-a y(t) + u(t)\}. \tag{17} $$

Algorithm 1 is applied again to model (4) with $a = 0$. The model is assumed to be at rest before $t = 0$, and the input $u(t)$ is assumed to be a Heaviside function $H(t)$. Algorithm 1 provides the following solutions:

$$ S_1 : y(t) = \frac{t^{\nu}}{\Gamma(\nu+1)} \qquad 0 < t < t_1 \tag{18} $$

$$ S_2 : y(t) = \frac{(t-t_0)^{\nu}}{\Gamma(\nu+1)} + \frac{y_0}{\Gamma(\nu)} (t-t_0)^{\nu-1} \qquad t_0 < t < t_1. \tag{19} $$

Relation (19) seems to say that any value of y_0 can be chosen, but whatever the value selected, for S_2 $y(t)$ tends toward infinity as t tends toward t_0 if $y_0 \neq 0$ whereas $y(t) = t_0^{\nu}/\Gamma(\nu+1)$ for S_1. The two simulations thus give different results. This is illustrated by Fig. 3 for various values of y_0.

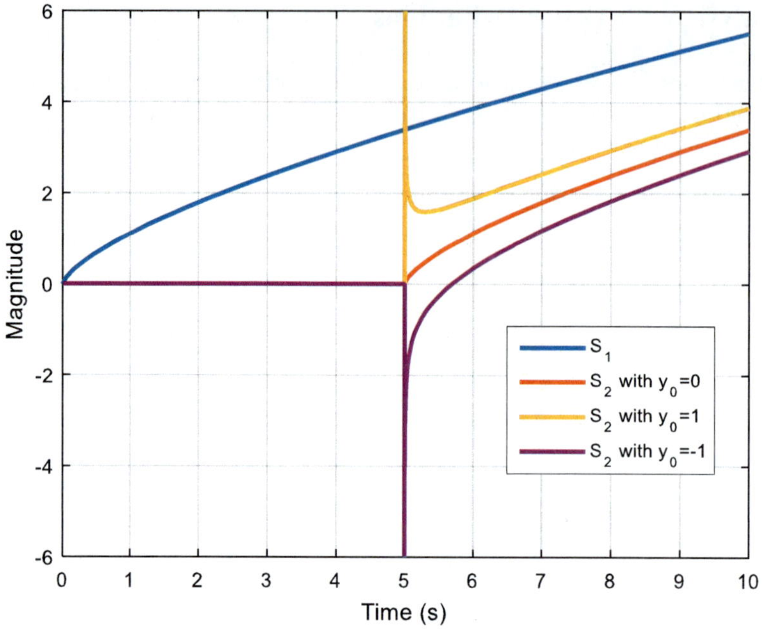

Fig. 3. Comparison of the exact response of model (4) with the responses obtained with the Riemann-Liouville definition ($t_0 = 5$ s, $a = 0$, $\nu = 0.7$)

3.2 The Grünwald-Letnikov Definition

The Grünwald-Letnikov derivative of order ν, $0 < \nu < 1$, of a function $y(t)$ is defined by:

$$ {}_{GL}\boldsymbol{D}^{\nu}_{t_0} y(t) = \lim_{h \to 0} \frac{1}{h^{\nu}} \sum\nolimits_{0 \le m < \infty} (-1)^m \binom{\nu}{m} f(t - mh) \qquad t > t_0 \tag{20} $$

with $\binom{\nu}{m} = \frac{\Gamma(\nu+1)}{m!\Gamma(\nu-m+1)} = \frac{\nu(\nu-1)(\nu-2)\ldots(\nu-m+1)}{m(m-1)(m-2)\ldots(m-m+1)}$.

This definition is often used in the literature as it provides a simple numerical scheme for fractional derivative implementation. In some research [14, 15, 16], these numerical schemes are used to solve the initialisation problem:

$$ \begin{aligned} \boldsymbol{D}^{\nu}_{t_0} y(t) = -ay(t) + u(t) \qquad 0 < \nu < 1 \quad a > 0 \quad \text{for} \quad t_0 < \nu < T, \\ y(t_0) = y_0. \end{aligned} \tag{21} $$

In this case, it is not the Grünwald-Letnikov derivative definition which is questionable, but the idea that a fractional model can be initialized solely with information on the initial moment. From relation (20), it is possible to observe that variable m goes from 0 to infinity, and thus this definition is able to take into account the past of the derivative function, prior to t_0. In (21), the problem is the way the initial conditions are defined.

To illustrate this problem, Algorithm 1 is applied to model (4) with $a = 1$. The model is assumed to be at rest before $t = 0$, and the input $u(t)$ is assumed to be a Heaviside function $H(t)$. In such conditions, the simulation S_1 defined in Algorithm 1 provides the following solution:

$$S_1 : y(t) = \frac{t^{\nu}}{\Gamma(\nu + 1)} \qquad 0 < t < t_1. \tag{22}$$

Simulation S_2 is done using the Grünwald-Letnikov formula (20) and provides

$$S_2 : y(t) = \frac{\frac{1}{h^{\nu}} \sum_{1 \leq m < \infty} (-1)^m \binom{\nu}{m} y(t - mh) + H(t)}{\frac{1}{h^{\nu}} + 1} \qquad t > t_0. \tag{23}$$

This simulation is done under two conditions:

- S_{21}: by taking into account all the past of the model (all the values of $y(t)$ on $t \in [0, t_0]$, provided by S_1)
- S_{22}: by considering only an initial condition at t_0 (value of $y(t)$ at t_0 provided by S_1).

The comparison of the three simulations is done in Fig. 4 and reveals that the Grünwald-Letnikov definition produces an exact solution provided that all the past of the model is taken into account.

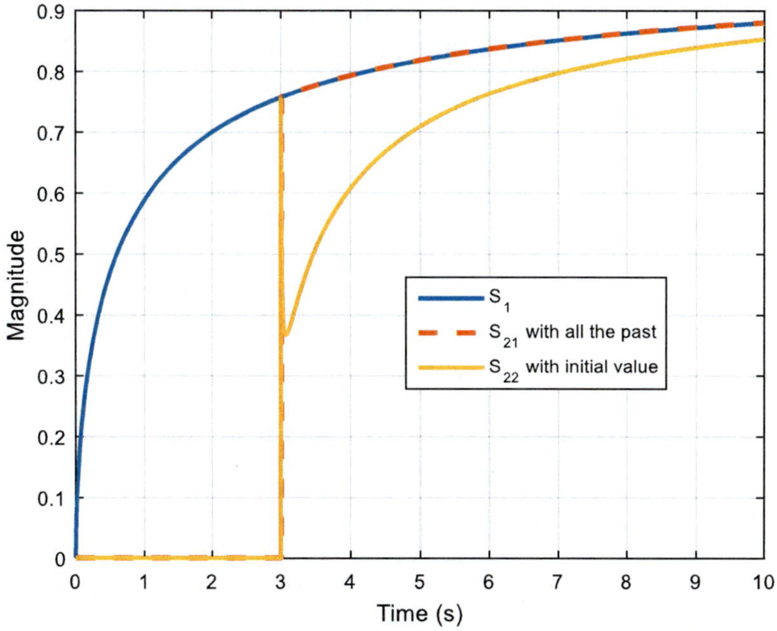

Fig. 4. Comparison of the exact response of model (4) with the responses obtained with the Grünwald-Letnikov definition ($t_0 = 3$ s, $a = 1$, $\nu = 0.6$).

4 Need to Take into Account All of the Model Past

The need to take into account the all past of a fractional model and not just the knowledge of its pseudo state at a single point in the past can be demonstrated quite simply on relation (4) (a particular case of fractional differential equation or of pseudo state space description). Contrary to what relation (4) might suggest, Fig. 5 highlights that the implementation of fractional differential equations does not explicitly involve the fractional differentiation operator but the fractional order integration operator $\boldsymbol{I}^{\nu}$. Thus in practice, it is not necessary to specify which particular definition is used for $\boldsymbol{D}^{\nu}$ in Eq. (4). Moreover, even if the system is assumed to have zero initial conditions at $t = 0$, namely if the system is supposed at rest ($u(t) = y(t) = 0, \forall t < 0$), it is important to note that $y(t)$ cannot be considered as a state for the fractional model and that all the past of $y(t)$ is required to compute the model evolution.

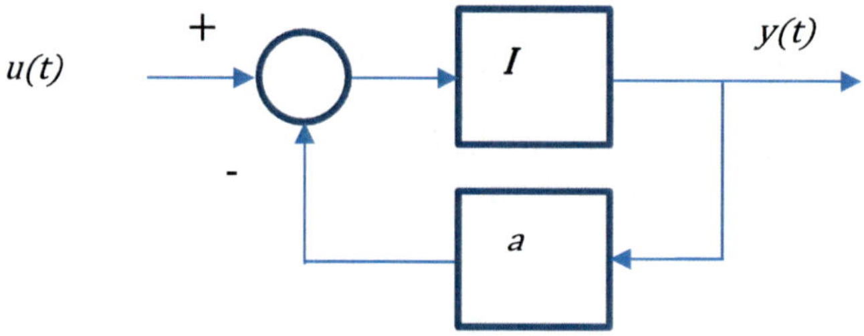

Fig. 5. Block diagram Eq. (4)

To better illustrate such a concept, a simple fractional model is used: a fractional integrator supposed at rest at $t = 0$. The corresponding block diagram is shown in Fig. 6.

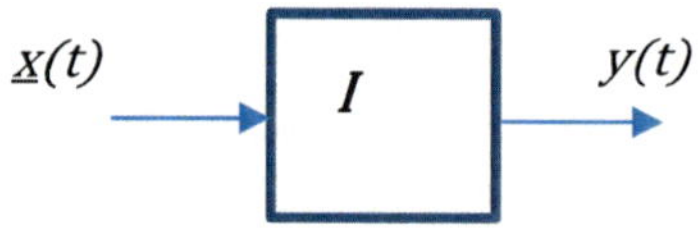

Fig. 6. Block diagram of an order ν fractional integrator

For an integer integrator, $\nu = 1$, relation (4) is really a state space description. At $t_1 > 0$, state $y(t)$ can be computed if the input between 0 and t_1 is known:

$$y(t_1) = \int_0^{t_1} \underline{x}(\tau)d\tau = y_1 = cst. \tag{24}$$

Values of $y(t)$ at later times than t_1 are given by:

$$y(t) = \int_0^{t} \underline{x}(\tau)d\tau = \underbrace{\int_0^{t_1} \underline{x}(\tau)d\tau}_{y_1=cst} + \int_{t_1}^{t} \underline{x}(\tau)d\tau, \qquad t > t_1. \tag{25}$$

Thus, $y(t)$ can be computed if $\underline{x}(t)$ is known within t_1 and t. Integrator output at time t_1 thus summarizes the whole model past. $y(t)$ is really the state of the dynamic model, in agreement with the definition given in [26].

Let us apply the same reasoning to the fractional integrator case of order ν. From the definition of fractional integration, value of $y(t)$ at $t_1 > 0$ can be computed if the input between $t = 0$ and t_1 is known:

$$y(t_1) = \frac{1}{\Gamma(\nu)} \int_0^{t_1} (t_1 - \tau)^{\nu-1} \underline{x}(\tau) d\tau = y_1 = cst. \tag{26}$$

Variable $y(t)$, $\forall t > t_1$, is thus given by:

$$y(t) = \frac{1}{\Gamma(\nu)} \int_0^{t} (t - \tau)^{\nu-1} \underline{x}(\tau) d\tau = \\ \underbrace{\frac{1}{\Gamma(\nu)} \int_0^{t_1} (t - \tau)^{\nu-1} \underline{x}(\tau) d\tau}_{\alpha(t) \neq y_1} + \frac{1}{\Gamma(\nu)} \int_{t_1}^{t} (t - \tau)^{\nu-1} \underline{x}(\tau) d\tau. \tag{27}$$

Two notable differences can be highlighted with respect to the integer case. First, term $\alpha(t)$ in Eq. (27) is not a constant but depends on the considered time t. Moreover, even if $y_1 = y(t_1)$ is known, it is not enough to compute $\alpha(t)$. Output $y(t)$ of the fractional integrator is thus not a state. The same analysis can be held for the general case of a pseudo state description or a fractional differential equation.

Beyond discussions on the concept of state, computation of $\alpha(t)$ in relation (27) whatever time t, requires to know $y(t)$ $\forall t \in [0..t_1]$, thus all the model past. This clearly shows that knowledge of $y(t)$ at a unique point of the past is not enough.

5 Conclusion

Fractional operators and the resulting fractional models are known for their memory property. However, for the following two reasons, many studies proposed in the literature seem to ignore this property when the model initialization problem is considered:

- they use the Caputo definition that involves only integer derivatives of the derivate function at the initial time,
- they use other definitions but initialization is done by taking only an initial value for the initial time into consideration.

This kind of initialization means that the operator or model memory exists everywhere on the time axis, except at the initial time. This is not consistent. Memory is an intrinsic property that exists all the time and that is proved in this paper with very simple examples. If from a mathematical point of view, most of the fractional derivative definitions encountered in the literature [17] are not problematic, this paper shows that the Caputo and Riemann-Liouville definitions are not able to ensure a proper initialization when used in a model definition. The paper also shows that this problem is not encountered with the Grünwald-Letnikov definition, provided that all the past of the model (from $t \to -\infty$) is taken into account. And this is precisely one of the drawbacks of fractional models that induces a physical inconsistency and many analysis problems [17].

What are the possible solutions? One solution can be to add an initialization function to the definition of the model. This is what was proposed by Lorenzo and Hartley [1, 2]. Yet again however, it requires the knowledge of all the model past (from $t \to -\infty$). Another solution consists in introducing new kernels for the definition of fractional integration as in [19]. But the goal would not be to solve only a singularity problem as in [19], but to reach a finite memory length as was done for instance in [20]. Note that while it was claimed in [21] that this class of kernels was too restrictive, it is linked to the problem analyzed in this paper: the inability of the Caputo definition to take into account initial conditions properly if used to define a fractional model [22]. The other solution is to introduce new solutions for fractional behavior modeling, without the drawbacks associated to fractional models [18]:

- distributed time delay models [23];
- non-linear models [24];
- partial differential equation with spatially varying coefficients [25].

References

1. Lorenzo, C.F., Hartley, T.T.: Initialized fractional calculus. Int. J. Appl. Math. **3**, 249–265 (2000)
2. Lorenzo, C.F., Hartley, T.T.: Initialization in fractional order systems. In: Proceedings of the European Conference on Control ECC, Porto, Portugal, pp. 1471–1476 (2001)
3. Ortigueira, M.D.: On the initial conditions in continuous-time fractional linear systems. Signal Process. **83**, 2301–2309 (2003)
4. Fukunaga, M., Shimizu, N.: Role of prehistories in the initial value problems of fractional viscoelastic equations. Nonlinear Dyn. **38**, 207–220 (2004)
5. Sabatier, J., Merveillaut, M., Malti, R., Oustaloup, A.: On a representation of fractional order systems: interests for the initial condition problem. In: IFAC Workshop, 3rd edn., Ankara, Turkey (2008)
6. Sabatier, J., Merveillaut, M., Malti, R., Oustaloup, A.: How to impose physically coherent initial conditions to a fractional system? Commun. Nonlinear Sci. Numer. Simul. **15**, 1318–1326 (2010). https://doi.org/10.1016/j.cnsns.2009.05.070
7. Ortigueira, M.D., Coito, F.: System initial conditions vs derivative initial conditions. Comput. Math. Appl. **59**, 1782–1789 (2010). https://doi.org/10.1016/j.camwa.2009.08.036
8. Sabatier, J., Farges, C.: Comments on the description and initialization of fractional partial differential equations using Riemann–Liouville's and Caputo's definitions J. Comput. Appl. Math. **339**, 30–39 (2018). https://doi.org/10.1016/j.cam.2018.02.030
9. Balint, A.M., Balint, S.: Mathematical description of the groundwater flow and that of the impurity spread, which use temporal Caputo or Riemann–Liouville fractional partial derivatives, is non-objective. Fractal Fract. **4**, 36 (2020). https://doi.org/10.3390/fractalfract4030036
10. Samko, S.G., Kilbas, A.A., Marichev, O.I.: Fractional Integrals and Derivatives. Gordon and Breach Science Publishers, London (1993)

11. Li, C., Qian, D., Chen, Y.Q.: On Riemann-Liouville and Caputo derivatives. Discret. Dyn. Nat. Soc. **2011**, 1–15 (2011). Article ID 562494
12. Abramowitz, M., Stegun I.: Handbook of Mathematical Functions. Dover Publications, New York (1964)
13. Bai, Z., Zhang, S., Sun, S., Yin, C.: Monotone iterative method for fractional differential equations. Electron. J. Differ. Eqn. **2016**(06), 1–8 (2016)
14. Diethelm, K., Ford, N.J.: Analysis of fractional differential equations. J. Math. Anal. Appl. **265**(2), 229–248 (2002)
15. Chena, H., Holland, F., Stynes, M.: An analysis of the Grünwald-Letnikov scheme for initial-value problems with weakly singular solutions. Appl. Numer. Math. **139**, 52–61 (2019)
16. Garrappa, R.: Numerical solution of fractional differential equations: a survey and a software tutorial. Mathematics **6**(16) (2018)
17. Oliveira, E.C.D., Machado, J.A.T.: A review of definitions for fractional derivatives and integral. Math. Phys. Eng. **2014**, 1–6 (2014). Article ID 238459
18. Sabatier, J., Farges, C., Tartaglione, V.: Some alternative solutions to fractional models for modelling long memory behaviors. Mathematics **8**, 196 (2020)
19. Atangana, A., Baleanu, D.: New fractional derivatives with nonlocal and non-singular kernel: theory and application to heat transfer model. Therm. Sci. **20**, 763–769 (2016)
20. Sabatier, J.: Non-singular kernels for modelling power law type long memory behaviours and beyond. Cybern. Syst. **51**, 383–401 (2020)
21. Stynes, M.: Fractional-order derivatives defined by continuous kernels are too restrictive. Appl. Math. Lett. **85**, 22–26 (2018)
22. Sabatier, J.: Fractional-order derivatives defined by continuous kernels: are they really too restrictive? Fractal Fract. **4**, 40 (2020)
23. Sabatier, J.: Power law type long memory behaviors modeled with distributed time delay systems. Fractal Fract. **4**, 1 (2019)
24. Tartaglione, V., Farges, C., Sabatier, J.: Nonlinear dynamical modeling of adsorption and desorption processes with power-law kinetics: application to CO_2 capture. Phys. Rev. E **102**, 052102 (2020)
25. Sabatier, J.: Beyond the particular case of circuits with geometrically distributed components for approximation of fractional order models: application to a new class of model for power law type long memory behaviour modelling. J. Adv. Res. **25**, 243–255 (2020)
26. Hinrichsen, D. Pritchard, A.: Mathematical Systems Theory I: Modelling, State Space Analysis, Stability and Robustness. Texts in Applied Mathematics, vol. 48. Springer, Heidelberg (2005). 10.1007/b137541

The Complex Order Fractional Derivatives and Systems are Non Hermitian

Manuel D. Ortigueira[1,2(✉)]

[1] CTS-UNINOVA, Caparica, Portugal
[2] NOVA School of Science and Technology of NOVA University of Lisbon, Caparica, Portugal
mdo@fct.unl.pt

Abstract. This paper discusses the concept of fractional derivative with complex order from the application point of view. It is shown that a fractional derivative is hermitian, if and only if the order is real. The hermitian part of complex order derivatives behave like allpass filters with almost logarithmic phase.

Keywords: Hermitian derivative · Complex order · Liouville derivative

1 Introduction

Complex order fractional derivatives have been presented in the most interesting books on the subject [1,2]. From a pure mathematical point of view, this is correct, because there are no conceptual differences among real and complex order derivatives. From the applicational point of view, the situation is not so simple, because the complex order derivatives are not hermitian [3–5]. Adams et al. [6] made a time domain analysis of the complex order derivative and, on being aware of the non hermitian behaviour, tried to overcome it by proposing the use of the sum or difference of the derivative and its conjugate. Makris [7,8] proposed the use of the "analytical signal" and more recenly [9] returned to the study of the involved problems and concluded that the fractional complex order derivatives do not lead to physically realizable systems.

In this paper, we study the non hermitian problem of the complex order derivative in the frequency domain, showing that the corresponding frequency response has neither even amplitude nor odd phase. The decomposition hermitian/anti-hermitian of the derivative is performed obtaining two operators coinciding with the ones proposed by Adams et al. [6]. The study of the hermitian operator shows that its amplitude is bounded on $\mathbb{R}$, but it does not decrease as the frequency increases, exhibiting an allpass character.

Funded by Portuguese National Funds through the FCT-Foundation for Science and Technology within the scope of the CTS Research Unit - Center of Technology and Systems/UNINOVA/FCT/NOVA, under the reference UIDB/00066/2020.

A. Dzielinski et al. (Eds.): ICFDA 2021, LNNS 452, pp. 38–44, 2022.
https://doi.org/10.1007/978-3-031-04383-3_3

2 The Hermitian Property of Real Order Fractional Derivatives

Many interesting time-invariant systems are described by an input-output convolutional relation, [10] $y(t) = h(t) * x(t)$, where $x(t)$ is the input, $y(t)$ is the output, and $h(t)$ is the *impulse response* of the system. Assume that the system is BIBO stable. Then, $h(t)$ is almost everywhere continuous, with bounded variation, and absolutely integrable, which implies the existence of its Fourier transform (FT), $H(\omega)$, that is called *frequency response* of the system, since, if the input of the system is $x(t) = e^{i\omega_0 t}$, $\omega_0, t \in \mathbb{R}$ with $i = \sqrt{-1}$, then the output is given by $y(t) = H(i\omega_0)e^{i\omega_0 t}$, provided that $H(i\omega_0)$ is finite [12], which happens always if the system is stable. Among the properties of the FT, there is one very interesting: the *Hermitean symmetry* that we can state as

If $h(t) \in \mathbb{R}$, *then* $H(-\omega) = H^*(\omega)$, where $(.)^*$ means "conjugate". Write it in the form

$$H(\omega) = A(\omega)e^{i\phi(\omega)}, \text{ with } A(\omega) = |H(\omega)|, \ \phi(\omega) = \arg(H(\omega)) \tag{1}$$

where $A(\omega)$ and $\phi(\omega)$ are called *amplitude and phase* spectra. These functions are even and odd, respectively and will be the features characterizing the hermitian fractional derivatives. We will say that *a derivative is hermitian if its frequency response verifies the hermitian property.*

There are several different approaches to define fractional derivatives. In general, we can consider one-sided or two-sided derivatives [10]. As we are primarily interested in dealing with causal systems we will consider one-sided derivatives. On the other hand, we select those for which the derivative of a sinusoid is a sinusoid. The Grünwald-Letnikov, Liouville (regularised or not), and Liouville-Caputo derivatives, [1,10,12] defined over $\mathbb{R}$ have such property. Each derivative can be treated as an elemental system with a corresponding frequency response. Assume that $y(t)$ is an absolutely integrable function defined on $\mathbb{R}$. The forward Grünwald-Letnikov derivative (GL), is given by

$${}^{GL}D^{\alpha}y(t) = \lim_{h\to 0^+} h^{-\alpha}\sum_{k=0}^{\infty}\frac{(-\alpha)_k}{k!}y(t-kh), \tag{2}$$

where $(a)_k = a(a+1)(a+2)\cdots(a+k-1)$, $(a)_0 = 1$, represents the Pochhammer symbol for the raising factorial. Applying the FT to derivative (2) we obtain, [11,12],

$$\mathcal{F}\left[D^{\alpha}y(t)\right] = (i\omega)^{\alpha}Y(\omega) \tag{3}$$

It is not hard to show that

$$(i\omega)^{\alpha} = |\omega|^{\alpha}\, e^{i\alpha\frac{\pi}{2}\operatorname{sgn}(\omega)}, \tag{4}$$

where sgn(.) is the signum function. The relation (4) shows that the GL derivative is hermitian, provided that the derivative order is real. The same happens with the Liouville, regularised Liouville and Liouville-Caputo derivatives, since they verify (3) [10,12]. With these results we can conclude that the two-sided derivatives, [13], namely, Riesz' and Feller's, of real order are also hermitian. Concerning the Riemann–Liouville and Caputo derivatives the procedure is not so simple because the RL or C derivative of a sinusoid is not a sinusoid. However, in the RL case, we can show that, if $y(t) = e^{i\omega t}$, $t \geq 0$. We have

$$ {}_0^{RL}D_+^{\alpha} y(t) = \frac{\mathrm{d}^m}{\mathrm{d}t^m}\left[\frac{e^{i\omega t}}{\Gamma(\alpha)}\int_0^t \xi^{\alpha-1}e^{i\omega\xi}\mathrm{d}\xi\right] $$

The last integral is the FT of a real function. Therefore, it is hermitian. This can be repeated to the integer order derivatives. So, the real order RL derivative is hermitian. Similarly, we can prove that the C derivative is also hermitian, if the order is real and that (4) is valid approximately.

3 Complex Order Derivatives

3.1 Non Hermitian Character

The frequency response to the complex order derivative reads (4):

$$ H(\omega) = (i\omega)^{a+ib} = (i\omega)^a (i\omega)^{ib} \tag{5} $$

The first term corresponds to a real order derivative and then is hermitian. Therefore, the second term that corresponds to an imaginary order derivative has to be studied separately. Let $\Phi(\omega) = (i\omega)^{ib}$ be the corresponding frequency response. As $i\omega = |\omega|e^{i\frac{\pi}{2}\mathrm{sgn}(\omega)}$, then

$$ \Phi(\omega) = (i\omega)^{ib} = e^{-b\frac{\pi}{2}\mathrm{sgn}(\omega)}e^{ib\ln|\omega|}. \tag{6} $$

Thus, the amplitude (gain) spectrum of the complex order fractional derivative is given by

$$ A(\omega) = |\Phi(\omega)| = |\omega|^a e^{-b\frac{\pi}{2}\mathrm{sgn}(\omega)} \neq A(-\omega) \tag{7} $$

and the phase by

$$ \phi(\omega) = \arg\left(\Phi(\omega)\right) = \frac{a\pi}{2}\mathrm{sgn}(\omega) + b\ln|\omega| \neq -\phi(-\omega). \tag{8} $$

As the gain is not symmetric and the phase is not anti-symmetric we conclude that *the imaginary order derivatives are not hermitian and consequently the same happens to the complex order derivatives.*

3.2 Using the Hilbert Transform

Makris seemed to be the first to note the abonormality of the complex order derivatives. In [8], he supposedly found a way of avoiding the problem: instead of working with a real signal, $x(t)$, he proposed the use of a complex one

$$ f(t) = x(t) + i\hat{x}(t) \tag{9} $$

where $\hat{x}(t)$ was constructed in such a way that $f(t)$ could be expressed in terms of sisoids with only positive frequencies. In fact he was computing the so-called *analytical signal* [14]. In this case, $\hat{x}(t)$ is the Hilbert transform of $x(t)$. Apparently, this solves the problem, since we only involve positive or negative, not both, types of frequencies. However, we continue to have two different possibilities as the next example shows. In fact, we can define two different derivatives of for $\cos(\omega_0 t)$:

$$D^{a+ib}\cos(\pm\omega_0 t) = \begin{cases} |\omega_0|^a e^{-b\frac{\pi}{2}}\cos(\omega_0 t + \frac{a\pi}{2} + b\log|\omega_0|) & \textit{positive frequency } (+\omega_0) \\ |\omega_0|^a e^{b\frac{\pi}{2}}\cos(\omega_0 t + \frac{a\pi}{2} - b\log|\omega_0|) & \textit{negative frequency } (-\omega_0). \end{cases} \tag{10}$$

So, *we lost unicity.* With generality, given a signal $x(t)$ with FT $X(\omega)$, we proceed as follows:

1. Construct one of the complex signals with FT $F_\pm(\omega) = 2X(\omega)u(\pm\omega)$, where $u(.)$ is the Heaviside unit step. The signals $f_\pm(t) = \mathcal{F}^{-1}F_\pm(\omega)$ are called upper (+) and lower (-) single sideband signals.
2. As $u(\pm\omega) = 1 \pm \operatorname{sgn}(\omega)$, $F_\pm(\omega) = X(\omega) \pm \operatorname{sgn}(\omega)X(\omega)$
3. The inverse FT is $f_\pm(t) = x(t) \pm i\hat{x}(t)$, with

$$\hat{x}(t) = \mathcal{F}\left[-i\operatorname{sgn}(\omega)X(\omega)\right] = \frac{1}{\pi}\text{p.v.}\int_{-\infty}^{+\infty}\frac{x(\tau)}{t-\tau}\mathrm{d}\tau$$

that is the Hilbert transform of $x(t)$ (p.v. means "principal value").

The single sideband signals $f_\pm(t)$ agree with and generalize Makris' proposal. However, we have two different solutions for each problem.

4 The Hermitian Part

Any complex function of real variable can be considered as a sum of hermitian and anti-hermitian functions, allowing to write

$$\mathcal{F}f(t) = F(\omega) = \frac{F(\omega) + F^*(-\omega)}{2} + \frac{F(\omega) - F^*(-\omega)}{2} = R(\omega) + Q(\omega)$$

where $R(\omega) = R^*(-\omega)$ and $Q(\omega) = -Q^*(-\omega)$, meaning that $R(\omega)$ is hermitian and $Q(\omega)$ is anti-hermitian. Therefore, a correct procedure to obtain a hermitian operator is the following. Let $\bar{D}^\alpha$ be an operator with frequency response given by

$$R(\omega) = \frac{(i\omega)^a(i\omega)^{ib} + (i\omega)^a(i\omega)^{-ib}}{2} = (i\omega)^a\frac{(i\omega)^{ib} + (i\omega)^{-ib}}{2} \tag{11}$$

Then,

$$\bar{D}^\alpha f(t) = \frac{D^\alpha + D^{\alpha^*}}{2} \tag{12}$$

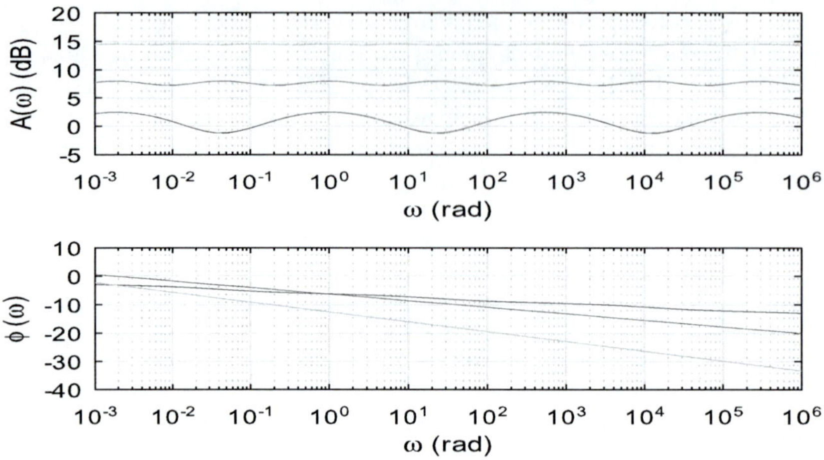

Fig. 1. Frequency response of system (13) for $b = 0.5, 1, 2$.

that coincides with the approach proposed in [6]. They suggested also the operator with frequency response $Q(\omega) = \frac{(i\omega)^a(i\omega)^{ib} - (i\omega)^a(i\omega)^{-ib}}{2}$, that is anti-hermitian. Its inverse FT is purely imaginary.

As $(i\omega)^a$ is hermitian, we will study the other factor only:

$$H(\omega) = \frac{(i\omega)^{ib} \pm (i\omega)^{-ib}}{2} \tag{13}$$

that is the frequency response of the operator $\bar{D}^{ib} = \frac{D^{ib} \pm D^{-ib}}{2}$. Noting that $(\pm i\omega)^{\pm ib} = e^{-v\frac{\pi}{2}\mathrm{sgn}(\omega)} e^{\pm ib\ln|\omega|}$, we obtain

$$H(\omega) = \frac{1}{2} e^{-b\frac{\pi}{2}\mathrm{sgn}(\omega)} e^{ib\ln|\omega|} \pm \frac{1}{2} e^{+b\frac{\pi}{2}\mathrm{sgn}(\omega)} e^{-ib\ln|\omega|}$$

Setting $H(\omega) = A(\omega)e^{i\theta(\omega)}$, we are led to

$$A(\omega) = \sqrt{\frac{\cosh(b\pi) \pm \cos(2b\ln|\omega|)}{2}} \tag{14}$$

and

$$\theta(\omega) = -\arctan\left\{\tanh(b\frac{\pi}{2})^{\pm 1}\tan(b\ln|\omega|)\right\}\mathrm{sgn}(\omega). \tag{15}$$

These relations allow us to conclude that

1. The operators with frequency responses given by (13) only one is hermitian: $\bar{D}^{ib} = \frac{D^{ib} + D^{-ib}}{2}$;
2. Its Bode amplitude spectrum is bounded and oscillating, but behaving almost as the amplitude of an allpass filter (see Fig. 1);
3. The corresponding Bode phase spectrum decreases almost linearly (see Fig. 1);

4. It cannot be considered as a fractional derivative, since it does not verify most of the required properties [15].

Besides, we can state that

1. $\bar{D}^{-ib} = \bar{D}^{ib}$.
2. $\bar{D}^{ib+iw} \neq \bar{D}^{ib}\bar{D}^{iw}$. It is the $w = -b$ case. We find $\bar{D}^{ib+iw} = \bar{D}^0 = 1$ and $\bar{D}^{ib}\bar{D}^{-ib} = \left[\frac{D^{ib}+D^{-ib}}{2}\right]^2 = \frac{D^{i2b}+D^{-i2b}}{4} + \frac{1}{2} = \frac{\bar{D}^{i2b}}{4} + \frac{1}{2}$.
3. The inverse of $\bar{D}^{ib}$ exists and is defined by $\frac{1}{A(\omega)}$ in (14) and by $-\theta(\omega)$ in (15), but there is no simple expression to define it in terms of $D^{\pm ib}$.

Conclusion. The complex order fractional derivatives are not hermitian, therefore not suitable for modeling real life phenomena.

References

1. Samko, S.G., Kilbas, A.A., Marichev, O.I.: Fractional Integrals and Derivatives. Gordon and Breach, Yverdon (1993)
2. Kilbas, A.A., Srivastava, H.M., Trujillo, J.J.: Theory and Applications of Fractional Differential Equations, vol. 204. Elsevier, Amsterdam (2006)
3. Oustaloup, A., Levron, F., Mathieu, B., Nanot, F.M.: Frequency-band complex non-integer differentiator: characterization and synthesis. IEEE Trans. Circuits Syst. I Fundam. Theory Appl. **47**(1), 25–39 (2000)
4. Cois, O., Levron, F., A. OUSTALOUP: complex-fractional systems: modal decomposition and stability condition. In: European Control Conference (ECC), pp. 1484–1489. IEEE 2001 (2001)
5. Jacob, J.A., Tare, A.V., Vyawahare, V.A, Pande, V.N.: A review of time domain, frequency domain and stability analysis of linear complex-order systems. In: 2016 IEEE International WIE Conference on Electrical and Computer Engineering (WIECON-ECE), pp. 164–169 (2016)
6. Adams, J.L., Hartley, T.T., Lorenzo, C.F.: Complex order-distributions using conjugated order differintegrals. In: Sabatier, J., Agrawal, O.P., Machado, J.A.T. (eds.) Advances in Fractional Calculus, pp. 347–360. Springer, Dordrecht (2007). https://doi.org/10.1007/978-1-4020-6042-7_24
7. Makris, N.: Complex-parameter Kelvin model for elastic foundations. Earthq. Eng. Struct. Dyn. **23**(3), 251–264 (1994)
8. Makris, N.: The imaginary counterpart of recorded motions. Earthq. Eng. Struct. Dyn. **23**(3), 265–273 (1994)
9. Makris, N.: On the physical meaning of time-domain constitutive models with complex parameters. Meccanica **55**(3), 453–467 (2020)
10. Ortigueira, M.D.: Fractional Calculus for Scientists and Engineers. Lecture Notes in Electrical Engineering, Springer, Dordrecht, Heidelberg (2011). https://doi.org/10.1007/978-94-007-0747-4
11. Ortigueira, M.D., Machado, J.T.: Fractional derivatives: the perspective of system theory. Mathematics **7**(2), 150 (2019)
12. Ortigueira, M.D., Valério, D.: Fractional Signals and Systems, vol. 7. Walter de Gruyter GmbH & Co KG, Berlin (2020)

13. Ortigueira, M.D.: Two-sided and regularised Riesz-Feller derivatives. Math. Methods Appl. Sci. **44**, 8057–8069 (2021)
14. Haykin, S.: Communication Systems. John Wiley & Sons, Hoboken (2008)
15. Ortigueira, M.D., Machado, J.T.: What is a fractional derivative? J. Comput. Phys. **293**, 4–13 (2015)

Transient Regime of Fractional *RLC* Circuit

Kristian Haška[1], Dušan Zorica[2,3](✉), and Stevan M. Cvetićanin[1]

[1] Department of Power, Electronic and Telecommunication Engineering, Faculty of Technical Sciences, University of Novi Sad, Trg D. Obradovića 6, 21000 Novi Sad, Serbia
{kristian.haska,stevan.cveticanin}@uns.ac.rs

[2] Department of Physics, Faculty of Sciences, University of Novi Sad, Trg D. Obradovića 4, 21000 Novi Sad, Serbia
dusan.zorica@df.uns.ac.rs

[3] Mathematical Institute, Serbian Academy of Arts and Sciences, Kneza Mihaila 36, 11000 Beograd, Serbia

Abstract. Fractional capacitor (inductor) is modeled by the sum of terms describing instantaneous and power type hereditary contributions, thus expressing charge (magnetic flux) in terms of voltage (current) history. Constitutive models were further used in transient regime analysis of the series *RLC* circuit comprised of generalized elements, and subject to electromotive force. Impulse response of the fractional *RLC* circuit proved to have three distinctive types of behavior: aperiodic, critically aperiodic, and damped oscillatory, depending on model parameters.

Keywords: Fractionally generalized electric elements · Memory type constitutive models · Circuit in transient regime

1 Introduction

Following the exposition of [1], classical constitutive equations of basic electrical elements - capacitor and inductor - are altered to account for memory effects, so that physical quantities characteristic for elements are connected through both hereditary and instantaneous terms, giving the charge-voltage (flux-current) constitutive equation of fractional capacitor (inductor)

$$q(t) = C\, u_C(t) + C_\alpha\, {}_0\mathrm{I}_t^{1-\alpha} u_C(t), \quad \alpha \in (0,1), \tag{1}$$

$$\phi(t) = L\, i_L(t) + L_\beta\, {}_0\mathrm{I}_t^{1-\beta} i_L(t), \quad \beta \in (0,1), \tag{2}$$

where the total charge q accumulated on generalized capacitor (total magnetic flux ϕ in generalized inductor) is history dependent on the capacitor voltage u_C (inductor current i_L), with $C[\mathrm{F}]$ and $C_\alpha[\frac{\mathrm{F}}{\mathrm{s}^{1-\alpha}}]$ being classical and fractional capacitance, while $L[\mathrm{H}]$ and $L_\beta[\frac{\mathrm{H}}{\mathrm{s}^{1-\beta}}]$ are classical and fractional inductance,

A. Dzielinski et al. (Eds.): ICFDA 2021, LNNS 452, pp. 45–51, 2022.
https://doi.org/10.1007/978-3-031-04383-3_4

where s stands for second, F stands for farad, while H stands for henry. Element's long-tail memory is modeled by utilizing Riemann-Liouville fractional integral

$$ {}_{0}\mathrm{I}_{t}^{\xi} f(t) = \frac{t^{\xi-1}}{\Gamma(\xi)} * f(t) = \frac{1}{\Gamma(\xi)} \int_{0}^{t} \frac{f(t')}{(t-t')^{1-\xi}} \mathrm{d}t', \tag{3} $$

of order $\xi \geq 0$, where $*$ denotes the convolution. This approach enabled accounting not only instantaneous contribution present in classical element, but also hereditary contribution with the power type hereditary kernel.

Charge voltage (1) and flux-current (2) constitutive relations of fractional capacitor and inductor were further utilized to model generalized series RLC circuit trough the governing Eq. (7), whose solution in the complex domain is obtained by the Laplace transform method as the product of electromotive force image and transfer function (9), corresponding to capacitor voltage. Further, inverse Laplace transform was applied to transfer function in order to obtain impulse responses and thus model transient response of fractional RLC circuit. It would be interesting to check whether the proposed constitutive equations when used in modeling electrical devices such as supercapacitors provide good agreement with experimental data.

The governing equations of classical RLC and RC circuits were generalized to include memory effects by replacing the ordinary time derivatives with the fractional ones, and transient regime analysis was performed in [2,3]. Physical meaning of the governing equation obtained in this manner is not clear, and therefore it is more reasonable to perform generalization on the level of electric element, either phenomenologically employing the fractional derivative, as done in numerous papers, or by modeling interactions of the electro-magnetic field with the material, as displayed in [4–6].

Transient regime analysis of the series RC_α circuit is performed in [7], where generalized capacitor was modeled by the series connection of resistor and fractional capacitor, while in [8,9] the transient response of the series and parallel $RL_\beta C_\alpha$, where the inductor is generalized in analogous manner, is considered. The transient response of RC and more complex circuit containing fractional capacitor and inductor was examined in [10].

The aim of this short communication is to summarize and review the results presented in [1].

2 Model Formulation and Analytical Solution

Examined generalized series RLC circuit consists of resistor of resistance R, generalized capacitor modeled by charge-voltage relation (1) and generalized inductor modeled by flux-current relation (2). The circuit is subjected to electromotive forcing $\mathcal{E}$. In order to analyze circuit's transient response, the second Kirchhoff's law

$$ \mathcal{E}(t) = R\,i(t) + u_L(t) + u_C(t) \tag{4} $$

is coupled with the time derivatives of elements' constitutive Eqs. (1) and (2), given by

$$i(t) = C\,\frac{\mathrm{d}}{\mathrm{d}t}u_C(t) + C_\alpha\,{}_0\mathrm{D}_t^{\alpha}u_C(t), \tag{5}$$

$$u_L(t) = L\,\frac{\mathrm{d}}{\mathrm{d}t}i(t) + L_\beta\,{}_0\mathrm{D}_t^{\beta}i(t), \tag{6}$$

where $i(t) = \frac{\mathrm{d}}{\mathrm{d}t}q(t)$ is the electric current, $u_L(t) = \frac{\mathrm{d}}{\mathrm{d}t}\phi(t)$ is the inductor voltage and where ${}_0\mathrm{D}_t^{n+\xi}$, with $n \in N_0$, $\xi \in (0,1)$, denotes the operator of Riemann-Liouville fractional differentiation, defined as

$${}_0\mathrm{D}_t^{n+\xi}f(t) = \frac{\mathrm{d}^{n+1}}{\mathrm{d}t^{n+1}}\,{}_0\mathrm{I}_t^{1-\xi}f(t) = \frac{\mathrm{d}^{n+1}}{\mathrm{d}t^{n+1}}\left(\frac{t^{-\xi}}{\Gamma(1-\xi)} * f(t)\right).$$

Subsequent reduction of the system of equations (4)–(6) to a single equation expressed in terms of u_C, yields the governing equation of the observed *RLC* circuit

$$\tau_L\tau_C\,\frac{\mathrm{d}^2}{\mathrm{d}t^2}u_C(t) + \tau_L\tau_\alpha\,{}_0\mathrm{D}_t^{1+\alpha}u_C(t) + \tau_C\tau_\beta\,{}_0\mathrm{D}_t^{1+\beta}u_C(t) + \tau_\alpha\tau_\beta\,{}_0\mathrm{D}_t^{\alpha+\beta}u_C(t)$$
$$+\,\tau_C\,\frac{\mathrm{d}}{\mathrm{d}t}u_C(t) + \tau_\alpha\,{}_0\mathrm{D}_t^{\alpha}u_C(t) + u_C(t) = \mathcal{E}(t), \tag{7}$$

with $\tau_C = RC\,[\mathrm{s}]$, $\tau_L = \frac{L}{R}\,[\mathrm{s}]$, $\tau_\alpha = RC_\alpha\,[\mathrm{s}^\alpha]$, $\tau_\beta = \frac{L_\beta}{R}\,[\mathrm{s}^\beta]$, being classical and fractional time-constants and $\alpha, \beta \in (0,1)$. During derivation of the governing Eq. (7) the property of Riemann-Liouville fractional derivative

$${}_0\mathrm{D}_t^{\xi}\,{}_0\mathrm{D}_t^{\zeta}f(t) = {}_0\mathrm{D}_t^{\xi+\zeta}f(t), \quad \text{for} \quad \xi, \zeta \in (0,1),$$

where function f is bounded at zero, is additionally exploited.

Laplace transform, defined as

$$\hat{f}(s) = \mathcal{L}[f(t)](s) = \int_0^\infty f(t)\,\mathrm{e}^{-st}\mathrm{d}t, \quad \text{for} \quad \mathrm{Re}s > 0, \tag{8}$$

is applied to governing Eq. (7) and transfer function relating capacitor voltage with electromotive force is obtained as

$$\hat{g}_C(s) = \frac{\hat{u}_C(s)}{\hat{\mathcal{E}}(s)} = \frac{1}{\Phi(s)}, \quad \text{with} \tag{9}$$

$$\Phi(s) = \tau_L\tau_C s^2 + \tau_L\tau_\alpha s^{1+\alpha} + \tau_C\tau_\beta s^{1+\beta} + \tau_\alpha\tau_\beta s^{\alpha+\beta} + \tau_C s + \tau_\alpha s^\alpha + 1. \tag{10}$$

Transfer functions connecting current running through *RLC* circuit and inductor voltage with electromotive force

$$\hat{g}_i(s) = \frac{\hat{i}(s)}{\hat{\mathcal{E}}(s)} = \frac{1}{R}\frac{\tau_C s + \tau_\alpha s^\alpha}{\Phi(s)} \quad \text{and} \tag{11}$$

$$\hat{g}_L(s) = \frac{\hat{u}_L(s)}{\hat{\mathcal{E}}(s)} = 1 - (\hat{g}_C(s) + R\,\hat{g}_i(s)) = 1 - \frac{\tau_C s + \tau_\alpha s^\alpha + 1}{\Phi(s)} \tag{12}$$

are obtained analogously from (5) and (6), respectively, utilizing the Laplace transform of Riemann-Liouville fractional derivative

$$\mathcal{L}\left[{}_0\mathrm{D}_t^{n+\xi} f(t)\right](s) = s^{n+\xi}\hat{f}(s) - \sum_{k=0}^{n}\left[\frac{\mathrm{d}^k}{\mathrm{d}t^k}{}_0\mathrm{I}_t^{1-\xi} f(t)\right]_{t=0} s^{n-k} = s^{n+\xi}\hat{f}(s),$$

holding for f bounded at zero.

By the means of the inverse Laplace transform, defined as

$$f(t) = \mathcal{L}^{-1}[\hat{f}(s)](t) = \frac{1}{2\pi\mathrm{j}}\int_{\Gamma_{\mathrm{Br}}} \hat{f}(s)\,\mathrm{e}^{st}\,\mathrm{d}s, \tag{13}$$

where Γ_{Br} is the Bromwich path, the impulse responses g_C, g_i, and g_L, i.e. the capacitor voltage, circuit's current, and inductor voltage, resulting from electromotive forcing assumed in the form of Dirac delta distribution, are acquired from transfer functions (9), (11), and (12) in the following forms

$$g_C(t) = \frac{1}{\pi}\int_0^\infty \frac{K_C(\rho)}{\left|\Phi\left(\rho\mathrm{e}^{\mathrm{j}\pi}\right)\right|^2}\mathrm{e}^{-\rho t}\,\mathrm{d}\rho + \begin{cases} 0, & \text{in Case I,} \\ -\frac{\mathrm{Re}P(\rho^*\mathrm{e}^{\mathrm{j}\pi})}{|P(\rho^*\mathrm{e}^{\mathrm{j}\pi})|^2}\rho^*\mathrm{e}^{-\rho^* t}, & \text{in Case II,} \\ 2\mathrm{Re}\left(\frac{1}{\frac{\mathrm{d}}{\mathrm{d}s}\Phi(s)\Big|_{s=s_0}}\mathrm{e}^{\mathrm{j}t\mathrm{Im}s_0}\right)\mathrm{e}^{-|\mathrm{Re}s_0|t}, & \text{in Case III,} \end{cases} \tag{14}$$

$$g_i(t) = \frac{1}{\pi R}\int_0^\infty \frac{K_i(\rho)}{\left|\Phi\left(\rho\mathrm{e}^{\mathrm{j}\pi}\right)\right|^2}\mathrm{e}^{-\rho t}\,\mathrm{d}\rho + \begin{cases} 0, & \text{in Case I,} \\ -\frac{1}{R}\frac{P_i(\rho^*)}{|P(\rho^*\mathrm{e}^{\mathrm{j}\pi})|^2}\rho^*\mathrm{e}^{-\rho^* t}, & \text{in Case II,} \\ \frac{2}{R}\mathrm{Re}\left(\frac{\tau_C s_0 + \tau_\alpha s_0^\alpha}{\frac{\mathrm{d}}{\mathrm{d}s}\Phi(s)\Big|_{s=s_0}}\mathrm{e}^{\mathrm{j}t\mathrm{Im}s_0}\right)\mathrm{e}^{-|\mathrm{Re}s_0|t}, & \text{in Case III,} \end{cases} \tag{15}$$

$$g_L(t) = \delta(t) - \left(g_C(t) + R\,g_i(t)\right), \tag{16}$$

where functions K_C, P, K_i and P_i are respectively given by

$$\begin{aligned} K_C(\rho) &= -\tau_L\tau_\alpha\,\rho^{1+\alpha}\sin(\alpha\pi) - \tau_C\tau_\beta\,\rho^{1+\beta}\sin(\beta\pi) \\ &\quad + \tau_\alpha\tau_\beta\,\rho^{\alpha+\beta}\sin((\alpha+\beta)\pi) + \tau_\alpha\,\rho^\alpha\sin(\alpha\pi), \end{aligned} \tag{17}$$

$$\begin{aligned} P(\rho^*\mathrm{e}^{\mathrm{j}\pi}) &= 2\tau_L\tau_C\left(\rho^*\right)^2 - (1+\alpha)\,\tau_L\tau_\alpha\left(\rho^*\right)^{1+\alpha}\mathrm{e}^{\mathrm{j}\alpha\pi} - (1+\beta)\,\tau_C\tau_\beta\left(\rho^*\right)^{1+\beta}\mathrm{e}^{\mathrm{j}\beta\pi} \\ &\quad + (\alpha+\beta)\,\tau_\alpha\tau_\beta\left(\rho^*\right)^{\alpha+\beta}\mathrm{e}^{\mathrm{j}(\alpha+\beta)\pi} - \tau_C\rho^* + \alpha\,\tau_\alpha\left(\rho^*\right)^\alpha\mathrm{e}^{\mathrm{j}\alpha\pi}, \end{aligned} \tag{18}$$

$$K_i(\rho) = \tau_\alpha\,\rho^\alpha\sin(\alpha\pi)\left(\tau_\alpha\tau_\beta\,\rho^{\alpha+\beta}\frac{\sin(\beta\pi)}{\sin(\alpha\pi)}\left|\frac{\tau_C}{\tau_\alpha}\rho^{1-\alpha}\mathrm{e}^{-\mathrm{j}\alpha\pi} - 1\right|^2 - 1\right) \tag{19}$$

$$\begin{aligned}P_i(\rho^*) &= -2\tau_L\tau_C^2\,(\rho^*)^3 + (3+\alpha)\tau_L\tau_C\tau_\alpha\,(\rho^*)^{2+\alpha}\cos(\alpha\pi)\\ &+ (1+\beta)\tau_C^2\tau_\beta\,(\rho^*)^{2+\beta}\cos(\beta\pi) + \tau_C^2\,(\rho^*)^2\\ &- \big((1+\beta)\cos((\alpha-\beta)\pi)+(\alpha+\beta)\cos((\alpha+\beta)\pi)\big)\tau_C\tau_\alpha\tau_\beta\,(\rho^*)^{1+\alpha+\beta}\\ &- (1+\alpha)\tau_L\tau_\alpha^2\,(\rho^*)^{1+2\alpha} - (1+\alpha)\tau_C\tau_\alpha\,(\rho^*)^{1+\alpha}\cos(\alpha\pi)\\ &+ (\alpha+\beta)\tau_\alpha^2\tau_\beta\,(\rho^*)^{2\alpha+\beta}\cos(\beta\pi) + \alpha\tau_\alpha^2\,(\rho^*)^{2\alpha},\end{aligned}$$

with cases in (14) and (15) differentiated according to the nature of zeros of function Φ, see (10), as given in Table 1.

Table 1. Nature of zeros of function Φ.

Case I:	If $\operatorname{Re}\Phi(-\rho^*) < 0$,	then Φ has no zeros in the complex plane.
Case II:	If $\operatorname{Re}\Phi(-\rho^*) = 0$,	then Φ has one negative real zero $s_0 = -\rho^*$.
Case III:	If $\operatorname{Re}\Phi(-\rho^*) > 0$,	then Φ has a pair of complex conjugated zeros s_0 and $\bar{s}_0$ having negative real part.

Solution of the equation $\operatorname{Im}\Phi\left(\rho e^{i\pi}\right) = 0$ is denoted by ρ^*.

It is evident from Fig. 1 that nature of zeros of function Φ determines behavior of the impulse response, so in Cases I and II, obtained for $\tau_C = 5$ and $\tau_C = 0.972$, non-oscillatory behavior is the consequence of the fact that function Φ either has no zeros or has one negative real zero. Additionally, the impulse response in Cases I and II is nonmonotonic due to the fact that function K_i, given by (19), changes its sign, while Case II differs from Case I by an additional term in (15) exponentially decaying to zero. Since in Case III, obtained for $\tau_C = 0.02$, function Φ has a pair of complex conjugated zeros, corresponding impulse response exhibits exponentially damped oscillatory character with the frequency determined by $\operatorname{Im} s_0$ and damping factor by $\operatorname{Re} s_0$.

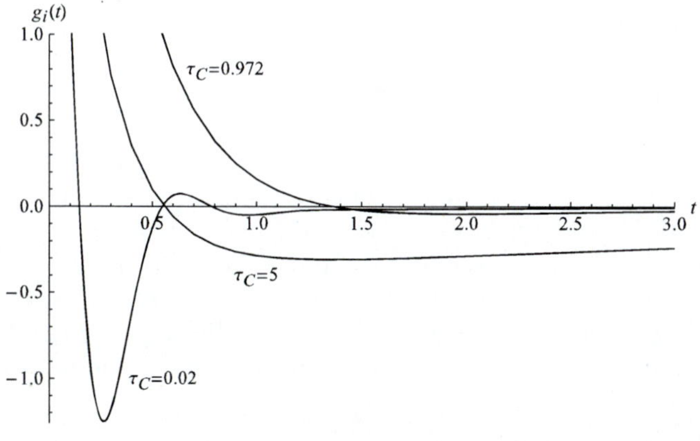

Fig. 1. Time profiles of current in *RLC* circuit as a response to electromotive force assumed as Dirac delta distribution, obtained analytically for model parameters: $R = 1$, $\tau_L = 0.2$, $\tau_\alpha = 0.1$, $\tau_\beta = 0.05$, $\alpha = 0.55$, and $\beta = 0.55$.

Generalized capacitor voltage, circuit's current, and inductor voltage resulting from arbitrary electromotive forcing is obtained as a convolution of corresponding impulse responses, given by (14), (15), and (16), and electromotive force as

$$u_C(t) = g_C(t) * \mathcal{E}(t), \quad i(t) = g_i(t) * \mathcal{E}(t), \quad \text{and} \quad u_L(t) = g_L(t) * \mathcal{E}(t).$$

Figure 2 presents the current as a response of *RLC* circuit to a harmonic electromotive force, that is initially shaped depending on the model parameters, while later is shaped as a harmonic function indicating that after a certain period of time the circuit enters the steady state regime.

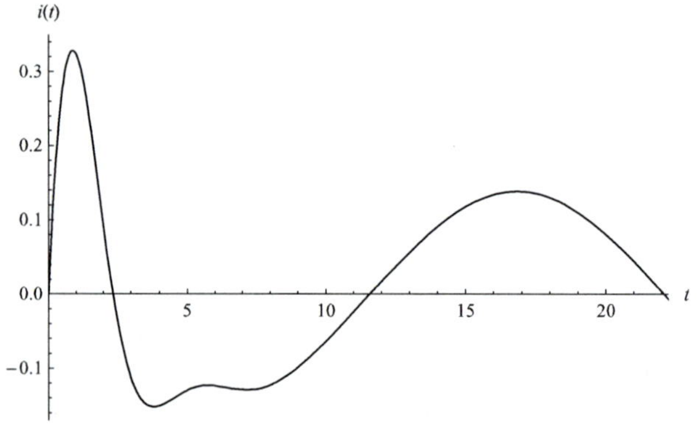

Fig. 2. Time profiles of current in *RLC* circuit as a response to electromotive force assumed as a cosine function of angular frequency $\omega = 0.3$, obtained analytically for model parameters: $R = 1$, $\tau_L = 1$, $\tau_C = 0.2$, $\tau_\alpha = 0.2$, $\tau_\beta = 0.5$, $\alpha = 0.8$, and $\beta = 0.65$.

3 Conclusion

Constitutive equations, describing behavior of generalized capacitor and inductor as elements appearing in fractionally generalized electric circuits, are proposed in the form that include element's classical, i.e., instantaneous response, as well as long memory hereditary response, modeled by the fractional integral. Constitutive models are further used in obtaining impulse response of fractional *RLC* circuit, so as circuit's response to an arbitrary electromotive force. Using the method of the Laplace transform it is obtained that the impulse response in transient regime has three distinctive types of behavior: aperiodic, critically aperiodic, and damped oscillatory, related to the number and nature of transfer function's poles.

References

1. Haška, K., Zorica, D., Cvetićanin, S.M.: Fractional RLC circuit in transient and steady state regimes. Commun. Nonlinear Sci. Numer. Simul. **96**, 105670 (2021)
2. Gómez, F., Rosales, J., Guía, M.: RLC electrical circuit of non-integer order. Cent. Eur. J. Phys. **11**, 1361–1365 (2013)
3. Guía, M., Rosales, J., Gómez, F.: Analysis on the time and frequency domain for the *RC* electric circuit of fractional order. Cent. Eur. J. Phys. **11**, 1366–1371 (2013)
4. Moreles, M.A., Lainez, R.: Mathematical modelling of fractional order circuit elements and bioimpedance applications. Commun. Nonlinear Sci. Numer. Simul. **46**, 81–88 (2017)
5. Stefański, T.P., Gulgowski, J.: Electromagnetic-based derivation of fractional-order circuit theory. Commun. Nonlinear Sci. Numer. Simul. **79**, 104897 (2019)
6. Stefański, T.P., Gulgowski, J.: Signal propagation in electromagnetic media described by fractional-order models. Commun. Nonlinear Sci. Numer. Simul. **82**, 105029 (2020)
7. Jakubowska, A., Walczak, J.: Analysis of the transient state in a circuit with supercapacitor. Poznan Univ. Technol. Acad. J. Elect. Eng. **81**, 71–77 (2015)
8. Jakubowska, A., Walczak, J.: Analysis of the transient state in a series circuit of the class $RL_\beta C_\alpha$. Circuits Syst. Signal Process. **35**, 1831–1853 (2016)
9. Jakubowska-Ciszek, A., Walczak, J.: Analysis of the transient state in a parallel circuit of the class $RL_\beta C_\alpha$. Appl. Math. Comput. **319**, 287–300 (2018)
10. Bošković, M.C., Šekara, T.B., Lutovac, B., Daković, M., Mandić, P.D., Lazarević, M.P.: Analysis of electrical circuits including fractional order elements. In: 6th Mediterranean Conference on Embedded Computing (MECO), Bar, Montenegro (2017)

Modelling Cardiovascular Diseases with Fractional-Order Derivatives

Cristina Nuevo-Gallardo(✉), José Emilio Traver, Inés Tejado, and Blas M. Vinagre

Industrial Engineering School, University of Extremadura, 06006 Badajoz, Spain
{cnuevog,jetraverb,itejbal,bvinagre}@unex.es

Abstract. Mathematical modelling of the cardiovascular system (CVS) helps to predict cardiovascular events in patients with heart failure, myocardial and valvular heart diseases. Most of the models found in the literature are based on electrical components with numerous parameters, being difficult to simulate pathological situations. This work presents an equivalent model of the CVS based on that of the Windkessel model that allows to simulate different pathologies in cardiac valves using a fractional-order behaviour for one of the model elements, by means of the variation of a single parameter, α. This, unlike other models, allows to modify the system dynamics and, consequently, offers more flexibility to emulate cardiovascular diseases. The results show that it is possible to simulate different valve conditions within the range $\alpha \in [0, 1]$.

Keywords: Cardiovascular system · Electrical model · Fractional model · Heart diseases · Pressure-volume loops

1 Introduction

The cardiovascular system (CVS) is a relatively complex system, which leads to the definition of models or simulators to get an overall understanding of its functioning and pathophysiological conditions. Plenty of information can be found in the literature about its modelling (see e.g. [4,10,14]), being most of these models based on equivalent electrical systems. Although several models can be found, even of fractional-order as in [1], the modelling of cardiac and heart diseases is still tedious and difficult due to the complexity of the models, the numerous parameters involved, and the estimation of these parameters to simulate a certain pathology.

This work proposes a simple method to simulate pathologies of the cardiac valves through the fractional-order modelling of some elements of the electrical model equivalent to the CVS presented in [12], which is based on the Windkessel model proposed in [13] and the modifications suggested in [10,14]. Starting from that model, and substituting the integer-order behaviour in one of the elements by a fractional-order behaviour of order α, changes in only this parameter within the range [0,1] allow to simulate different heart conditions which could be

A. Dzielinski et al. (Eds.): ICFDA 2021, LNNS 452, pp. 52–57, 2022.
https://doi.org/10.1007/978-3-031-04383-3_5

associated with already known diseases. The assessment is carried out through pressure-volume loops (PV loops, henceforth), which is a reference method for the study of cardiac mechanics and allows to detect pathologies related to cardiac valves and heart failure (HF).

The remainder of the paper is organised as follows. Section 2 describes the main characteristics of the CVS and PV loops. The electrical model used in this work is provided in Sect. 3, while Sect. 4 addresses the modelling of pathologies based on the non-integer-order model. Finally, the conclusions are drawn in Sect. 5.

2 Description of the Cardiovascular System

The CVS can be simply described as a distribution network of blood vessels that supplies blood to all the parts of a body thanks to the heart, which performs as a pump. Two stages can be distinguised: systemic circulation, which transports the oxygen and substances; and pulmonary circulation, being responsible for oxygenation of the blood [6]. The heart is composed of a double chamber atria-ventricle, where the ventricle is the pump and the atria is a pre-loaded chamber.

The contraction of the heart and the sucession of electrical and mechanical phenomena that occur during a heartbeat, known as the cardiac cycle [6], generate the necessary gradient pressure to power the blood through the vessels.

The cardiac cycle is usually characterized by PV loops, which represent changes in preload, afterload and contractility of the heart during a cardiac cycle by the plot of left ventricle volume (LVV) and left ventricle pressure (LVP) [3]. Usually, the form of these loops is rectangular or trapezoidal, and represent the differents phases of a cardiac cycle [2]. The main advantage of these diagrams is the ease of detecting cardiovascular diseases as HF, myocardial and valvular heart diseases, as well as sensing and monitoring (see [2] and the references therein) through several measurements as shown in Fig. 1.

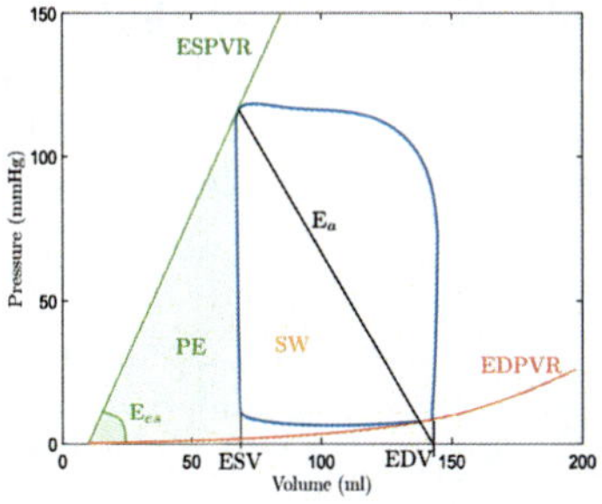

Abbreviation	Parameter	Meaning
ESPVR	End-systolic PV relationship	Maximal pressure of left ventricle
EDPVR	End-diastolic PV relationship	Left ventricle pressure in diastole.
E_{es}	End-systolic elastance	Peak chamber elastance during a beat
E_a	Effective arterial elastance	Relates EDP and EDV to ESV
SV	Stroke volume	The difference between ESV and EDV
SW	Stroke work	The area within the loop.
PE	Potential energy	The area within the loop and ESPVR.
PVA	Pressure-volume area	Sum of SW and potential energy PE.
ME	Mechanical efficiency	The ratio between SW and PVA.

Fig. 1. Parameters measured from PV loops (extracted from [2]).

3 Fractional-Order Model of the Cardiovascular System

This section summarises the electrical model equivalent to the CVS developed in [12] in which the current work focuses on, whose scheme is shown in Fig. 2.

The model was built based on the Windkessel electrical model [13] extended to the left common carotid artery in accordance with [10, 14]. The dynamics of the heart was reduced to its left side, being the left ventricle modelled by a time-varying capacitance ($C(t)$), while the left atria was interpreted as a passive element of accumulation (C_R). In respect to the valves, they were considered as non-ideal diodes. The model comprises a representation of the aortic arch and the sistemic circulation, this last included by means of a feedback in order to obtain a cyclic behaviour. The effects of pulmonary circulation were neglected since healthy conditions were assumed. The reader is referred to [12] for better comprehension of the remaining of the electrical components of the model.

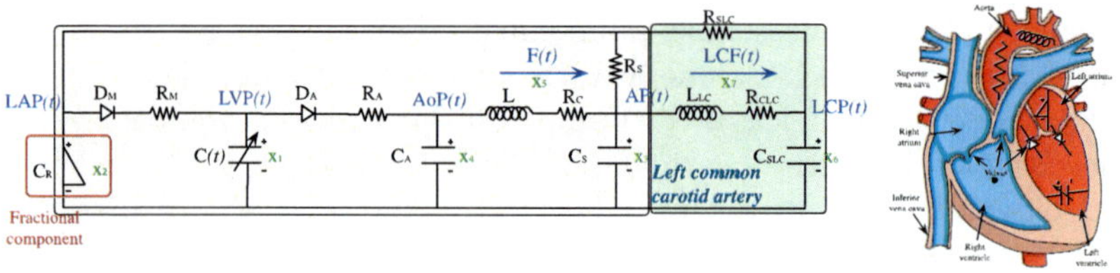

Fig. 2. Electrical model equivalent to the CVS with fractional-order component.

The application of Kirchhoff's laws to the electrical circuit leads to the definition of the differential equations of the model dynamics, choosing as state variables the following: x_1(t) as the left ventricle pressure (LVP); x_2(t) as the left atrial pressure (LAP); x_3(t) is the arterial pressure (AP); x_4(t) is the aortic pressure (AoP); and x_5(t) is the total blood flow (F).

Most CVS models allow to simulate patient conditions through the variation of model parameters such as the elastance, heart rate, and others, which implies a high challenge in what estimating the values of these parameters concerns. Experimental data is required to approximate their values and obtain an adequate simulation of patient conditions, being not always possible to acquire this data.

To solve this difficulty, a fractional-order approach is considered, being the fractional-order the only parameter to be modified while the rest are maintained. In this work, the capacitance C_R associated to the left atria is assumed to have a fractional-order behaviour, so the state variable x_2 is redefined to model the new fractional component. Therefore, the derivatives of the states are:

$$\frac{dx_1}{dt} = \frac{1}{C(t)}\left(-\dot{C}(t)x_1 + \frac{1}{\mathrm{R}_M}r(x_2 - x_1) - \frac{1}{\mathrm{R}_A}r(x_1 - x_4)\right) \tag{1}$$

$$\mathcal{D}^{\alpha}x_2 = \frac{1}{\mathrm{C}_R}\left(-\left(\frac{1}{\mathrm{R}_S} + \frac{1}{\mathrm{R}_{SLC}}\right)x_2 + \frac{1}{\mathrm{R}_S}x_3 + \frac{1}{\mathrm{R}_{SLC}}x_6 - \frac{1}{\mathrm{R}_M}r(x_2 - x_1)\right) \tag{2}$$

$$\frac{dx_3}{dt} = \frac{1}{\mathrm{C}_S}\left(\frac{1}{\mathrm{R}_S}x_2 - \frac{1}{\mathrm{R}_S}x_3 + x_5 - x_7\right) \tag{3}$$

$$\frac{dx_4}{dt} = \frac{1}{\mathrm{C}_A}\left(-x_5 + \frac{1}{\mathrm{R}_A}r(x_1 - x_4)\right) \tag{4}$$

$$\frac{dx_5}{dt} = \frac{1}{\mathrm{L}}\left(-x_3 + x_4 - \mathrm{R}_C x_5\right) \tag{5}$$

$$\frac{dx_6}{dt} = \frac{1}{\mathrm{C}_{SLC}}\left(\frac{1}{\mathrm{R}_{SLC}}x_2 - \frac{1}{\mathrm{R}_{SLC}}x_6 + x_7\right) \tag{6}$$

$$\frac{dx_7}{dt} = \frac{1}{\mathrm{L}_{LC}}\left(x_3 - x_6 - \mathrm{R}_{LC}x_7\right) \tag{7}$$

where $\mathcal{D}^\alpha$ is the fractional operator of order $\alpha \in \mathbb{R}^+$ defined according to Grünwald-Letnikov definition [7]; V_0 is a reference volume (the theoretical volume in the ventricle at zero pressure); r is a function derived from diodes operation; and $\mathrm{C}(t)$ is defined as the inverse of the elastance $E(t)$ [10–12].

4 Modelling of Pathologies

This section proves as the fractional-order model of the CVS described above allows to simulate different pathologies related to aortic valves, specifically aortic valve stenosis and aortic valve regurgitation.

Stenosis in aortic valve refers to the narrowing of the valve during systole, while a leakage of the aortic valve backward into the left ventricle during diastole results in aortic valve regurgitation, with similar causes as aortic valve stenosis [8]. In both pathologies, the left ventricle muscle thickens to try to overcome the stress, which leads to a hypertrophy in the left ventricle. The stiffness of the muscle can increase and cause a high pressure in the left ventricle. Given this context, patients with these pathologies present PV loops displaced to the right with respect to the healthy case, as well as an increased amplitude in terms of pressure [5,9]. Traduced to the parameters in Fig. 1, values of EDV and ESV increase due to the displacement of the loop to the right. On account of a wider and higher loop, the width of the loop (SV) and the area within the loop (SW) raise with respect to the healthy case. As SW is higher, PVA also increases (as it is the sum of SW and PE, two areas of the loop).

In order to find analogies between the pathologies described and the fractional-order model of the CVS presented above, this model was implemented and simulated in the MATLAB®/Simulink environment, considering different orders in the range $\alpha \in [0, 1]$. The results show an increase of the afterload and preload, as the PV loop is higher and wider, respectively, whereas the ESPVR line (and thus the contractility) remains constant independently of the value of α, as it can be observed in Fig. 3. On the other hand, a more quantitative analysis of the PV loops, summarised in Table 1, allows to determine that the values of EDV and ESV increase as the order of the model decreases, as well as the SV. The indices SW, PE, PVA and ME increase in the same way as SV for all the models. As of the results, the behaviour of fractional-order model shows similarities with patients with aortic valve stenosis and aortic valve regurgitation.

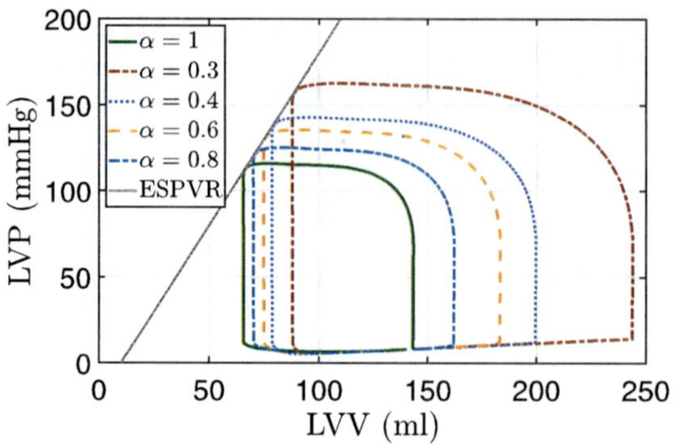

Fig. 3. PV loops resulting from different fractional-order ($\alpha = 1$ corresponds to a healthy patient).

Table 1. Measurements from PV loops of the electrical model with different fractional-order derivatives. The case $\alpha = 1$ corresponds to a healthy patient.

Model order	SV (ml)	SW (J/beat)	PE (J/beat)	PVA (J/beat)	ME (%)
$\alpha = 1$	77.69	1.03	0.38	1.42	72.92
$\alpha = 0.8$	92.01	1.34	0.45	1.79	74.78
$\alpha = 0.6$	108.62	1.70	0.52	2.23	76.30
$\alpha = 0.4$	121.08	2.00	0.59	2.59	77.21
$\alpha = 0.3$	156.01	2.89	0.76	3.66	79.09

5 Conclusions

The complexity of the cardiovascular system (CVS) leads to the use of equivalent models, most of them with electrical components, that allow to emulate the behaviour of cardiovascular diseases through the modification of numerous parameters. This work proposes a fractional-order model which simplifies the simulation of pathologies to the variation of a single parameter, the derivative order.

Four models were considered for the simulations, with fractional orders between $[0, 1]$, and its pressure-volume (PV) loops were compared. The results show that end-systolic volume (ESV), end-diastolic volume (EDV), stroke volume (SV), stroke work (SW), potential energy (PE), pressure-volume area (PVA) and mechanical energy (ME) increase while the order of the model decreases, showing similarities with patients with aortic valve stenosis and aortic valve regurgitation.

Our future works will focus on: 1) performing a more detailed study to identify other possible pathologies, 2) applying fractional-order derivatives to other state variables, and 3) comparing the results with experimental data.

Acknowledgments. This work has been supported in part by the Consejería de Economía, Ciencia y Agenda Digital (Junta de Extremadura) under the project IB18109 and the grant "Ayuda a Grupos de Investigación de Extremadura" (no. GR18159), by the Agencia Estatal de Investigación (Ministerio de Ciencia e Innovación) through the project PID2019-111278RB-C22/AEI/10.13039/501100011033, and in part by the European Regional Development Fund "A way to make Europe".

References

1. Bahloul, M.A., Laleg-Kirati, T.M.: Three-element fractional-order viscoelastic arterial windkessel model. In: 40th Annual International Conference of the IEEE Engineering in Medicine and Biology Society (EMBC), pp. 5261–5266 (2018). https://doi.org/10.1109/EMBC.2018.8513473
2. Bastos, M., et al.: Invasive left ventricle pressure-volume analysis: overview and practical clinical implications. Eur. Heart J. **41**, 1286–1297 (2019). https://doi.org/10.1093/eurheartj/ehz552
3. Goldstein, J.A., Kern, M.J.: Principles of Normal Physiology and Pathophysiology, chap. 1, pp. 1–34. John Wiley and Sons, Ltd., New Jersey (2018)
4. Gwak, K., Kim, H.D., Kim, C.: Feedback linearization control of a cardiovascular circulatory simulator. IEEE Trans. Control Syst. Technol. **23**(5), 1970–1977 (2015)
5. Itu, L.M., Sharma, P., Suciu, C. (eds.): Patient-specific Hemodynamic Computations: Application to Personalized Diagnosis of Cardiovascular Pathologies. Springer, Cham (2017). https://doi.org/10.1007/978-3-319-56853-9
6. Martini, F., Nath, J., Bartholomew, E.: Fundamentals of Anatomy & Physiology. Benjamin-Cummings Publishing Company, San Francisco (2015)
7. Monje, C., Chen, Y., Vinagre, B., Xue, D., Feliu, V.: Fractional Order Systems and Control - Fundamentals and Applications. Springer, London (2010). https://doi.org/10.1007/978-1-84996-335-0
8. Nishimura, R.A.: Aortic valve disease. Circulation **106**(7), 770–772 (2002). https://doi.org/10.1161/01.CIR.0000027621.26167.5E
9. Paul, A., Das, S.: Valvular heart disease and anaesthesia. Indian J. Anaesth. **61**(9), 721–727 (2017). https://doi.org/10.4103/ija.IJA_378_17
10. Simaan, M.A.: Rotary heart assist devices. In: Nof, S. (ed.) Springer Handbook of Automation. Springer Handbooks, pp. 1409–1422. Springer, Heidelberg (2009). https://doi.org/10.1007/978-3-540-78831-7_79
11. Traver, J.E., et al.: Simulador cardiovascular para ensayo de robots de navegación autónoma. Actas de las XXXVIII Jornadas de Automática, pp. 633–640 (2017)
12. Traver, J.E., Tejado, I., Prieto-Arranz, J., Vinagre, B.M.: Comparing classical and fractional order control strategies of a cardiovascular circulatory system simulator. IFAC-PapersOnLine **51**(4), 48–53 (2018)
13. Westerhof, N., Lankhaar, J., Westerhof, B.E.: The arterial windkessel. Med. Biol. Eng. Comput. **47**(2), 131–141 (2009)
14. Yu, Y., Boston, J.R., Simaan, M.A., Antaki, J.: Estimation of systemic vascular bed parameters for artificial heart control. IEEE Trans. Autom. Control **43**(6), 765–778 (1998)

Frequency Domain Modeling of an IPMC-Based Artificial Eukaryotic Flagellum Swimming Robot

José Emilio Traver(✉), Cristina Nuevo-Gallardo, Inés Tejado, and Blas M. Vinagre

Industrial Engineering School, University of Extremadura, 06006 Badajoz, Spain
{jetraverb,cnuevog,itejbal,bvinagre}@unex.es

Abstract. Ion polymer metal composite (IPMC) actuators promise a great advance in soft robotics, microrobotics and biomedical applications in a relatively near future, but this will require a deep knowledge in different aspects, such as manufacturing, material characterisation and control strategies. In this paper, frequency responses of a small-scale artificial Eukaryotic flagellum swimming robot manufactured in IPMC are obtained experimentally measuring the flagellum tip motion by means of an optical laser measurement system and actuating in a single point, the end of the flagellum attached to the head. For control purposes, two models consisting of a non-integer order integrator in series with a resonant system of both non-integer and integer order are considered and identified in the frequency range [0.5, 17] Hz. The results show that both models fit the experimental data.

Keywords: IPMC · Swimming · Microrobot · Fractional · Modeling · Frequency reponse

1 Introduction

Within a large number of current applications, ionic polymer metal composite (IPMC) has started a new revolution in soft and flexible robotics. Its attractive opportunities for flexible and underwater actuation make this material suitable for the development of swimming microrobots and biomedical applications [1,5]. However, working with such materials involves a great number of challenges, beginning from fabrication techniques, as well as behavior characterization and performance analysis in control, which requires a deeper comprehension to understand physical models as well as the influence of physical parameters [3,4]. In this respect, different models describing the behavior of IPMC actuators have been reported, based on integer order models [2] and non-integer order models described by two fractional poles and a fractional integrator [7].

In this context, this paper presents the frequency domain characterization of a small-scale 3-link artificial Eukaryotic flagellum (AEF) swimming robot

A. Dzielinski et al. (Eds.): ICFDA 2021, LNNS 452, pp. 58–64, 2022.
https://doi.org/10.1007/978-3-031-04383-3_6

manufactured in IPMC, where the first link is used as actuator and the other two perform as passive links. The aim of the work is to identify a model that describes the behavior of the swimmer robot for control purposes, on the basis of its frequency response. In particular, the models consist of a non-integer order integrator in series with a resonant system of non-integer or integer order. To obtain the frequency responses experimentally, the first IPMC link actuator was attached to a couple of gold electrodes, and the tip deflection of the last link was measured by means of a laser distance meter. It should be remarked that the final objective of this work is to obtain a swimming robot model in order to the generation of planar waves.

The remainder of this paper is organized as follows. Section 2 provides more details of IPMC actuators concerning technology and physical models. Section 3 describes the experiments carried out to obtain the frequency responses of the actuators. System identification for control applications is addressed in Sect. 4. Section 5 summarizes the main conclusions of this paper and future works.

2 Modeling of IPMC Actuators

IPMC actuators are electrically driven materials that undergo bending deformations in the presence of relatively low external voltages. A full-actuation model for IPMC can be decomposed in two dynamical models that recollect the electromechanical and mechanical behavior of the material [2]. The electromechanics model that relates the free tip displacement of an IPMC beam and the actuation voltage, when the mechanical dynamics of beams is ignored, can be defined as a function that mainly depends on the geometry and the electrical parameters of the material. In respect to the latter, some parameters can be only measured experimentally in a chemical laboratory. Taking into account this circumstance, and considering also the models proposed in [7] and in our previous work [6], the electromechanics model is approximated by a non-integer order integrator of order equal to 0.5, i.e. $H(s) \approx s^{-0.5}$, whose dynamic matches the analytical frequency response as it was also proved in [6] for other prototypes.

Regarding mechanical dynamics and hydrodynamic interactions if the actuator is submerged, it is incorporated by coupling a resonant system for the first vibration mode, being accurate enough if the application frequency range is relatively low. It may be given by a spring-mass-damper second order system or by a fractional order system of the forms:

$$G_1(s) = \frac{a_0}{b_2 s^2 + b_1 s + 1} \tag{1}$$

$$G_2(s) = \frac{a_0}{b_1 s^{\alpha} + 1} \tag{2}$$

with $\alpha \in (1, 2)$ and a_0, b_1 and b_2 constants. Notice that both models have the same number of parameters (i.e., three).

Then, the global dynamics of the system can be obtained by the product of both transfer functions, i.e., $M(s) = H(s)G_i(s)$, with $i = 1, 2$.

3 Experiments

3.1 Swimming Robot

Figure 1 shows a scheme and a picture and of the 3-link IPMC-based artificial Eukaryotic flagellum swimming robot under study. It consists in three independent and electrically isolated actuator-links manufactured over the same IPMC sheet. Likewise, if an actuator is not driven, it behaves as a flexible link. The geometry of swimming robot is given in Table 1. For this mechanical design, a micro laser etching machine was used to cut the swimming robot and isolate each one of the actuators.

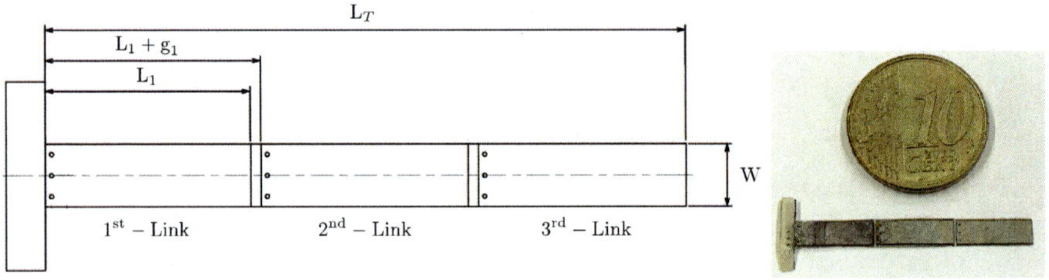

Fig. 1. 3-link IPMC-based AEF swimming robot

3.2 Set-up

A scheme of the experimental set-up used in this work to obtain the frequency responses of the manufactured swimming robot is illustrated in Fig. 2(a), which consists of:

- A water tank, used as environment, as the final objective of the swimming robot has to be able to swim in a fluid.
- A laser distance meter (OADM 20U2441), to measure the motion of the links.
- A clamp using gold electrodes, to transmit the voltage from the supplier to the IPMC surface.
- A USB data acquisition board of National Instruments (NI-USB6009), connected to a computer in which MATLAB® runs to collect the data.
- A power stage, to provide the enough power to actuator from a power DC supplier and the desired voltage indicated by the computer.

The voltage applied to the actuator and the output voltage of laser distance meter are measured with the USB data acquisition board, with a sampling time of 2 ms, and then sent to the computer in order to collect the data. Figure 2(b) shows a real image of the robot pointed by the laser during the experiments.

3.3 Measured Frequency Responses

The experiments were carried out performing a frequency sweep of sine waves between [0.1, 50] Hz, applied to the first actuator (A1) and measuring the deflection of the third link tip (L3). Likewise, every experiment was repeated ten times in order to test the reproducibility and to improve the final parameter estimation. Frequency spectra were obtained after an analysis of the time response data, estimating the amplitude and phase difference between the applied sine wave and the tip deflection of the third link of swimming robot. It should be mentioned that, in order to obtain the correct phase difference, a correction procedure was carried out. On the other hand, the frequency sweep was performed for different amplitudes with the objective of determining the importance of non-linearities of the material.

Table 1. Geometry of 3-link AEF swimming robot

Parameter	Description	Value	Unit	Parameter	Description	Value	Unit
L_T	Length of robot	31	mm	L_i	Length of link	10	mm
W	Width of robot	3	mm	N	Number of links	3	-
h	Thickness	250	μm	g_l	Space between links	500	μm

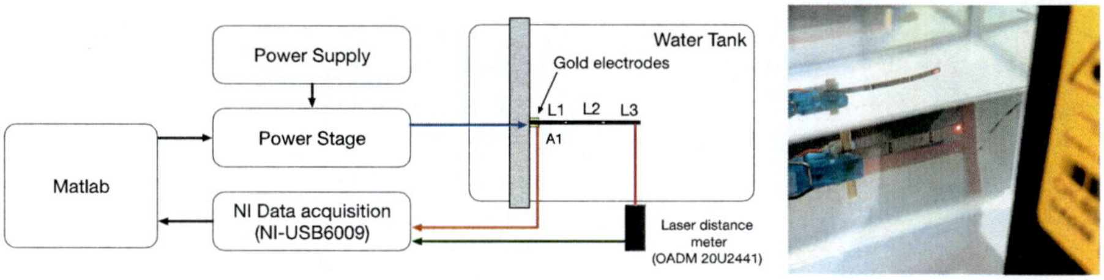

(a) (b)

Fig. 2. Experimental set-up: (a) scheme, (b) Swimming robot prototype pointed by the laser distance meter during an experiment.

Figure 3 shows the experimental frequency responses for different amplitudes and the ten measurements for the 3-V case. As it can be seen, the system has a resonant frequency around 62.83 rad/s and no differences are observed regarding the applied voltage. However, slight changes in behavior are noted for low frequencies and amplitudes.

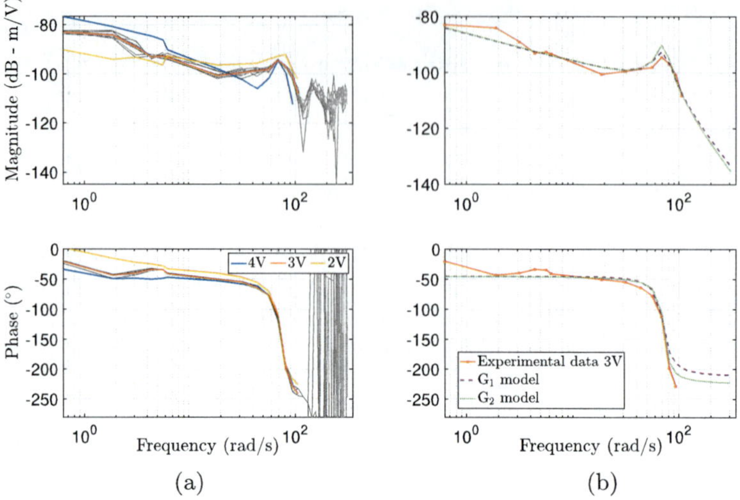

Fig. 3. Frequency responses of the 3-Link AEF swimming robot of IPMC for different input amplitudes and fitting results of experimental data for the 3-V case.

4 Identified Dynamic Models for Control

In this section, the dynamic models $M(s)$ for the first actuator of 3-link AEF swimming robot resulting from the identification are presented. It should be remarked that the model will be obtained for control purposes, i.e., their structure will be chosen as simplest as possible.

Previously to the identification process, several initial considerations must be stated. Firstly, after checking the measured frequency response shown in Fig. 3, it was noticed that the magnitude curves for high amplitude are not flat at low frequencies and fall out with a slope of -10 dB/dec, behavior that perfectly matches the approximation of $H(s) \approx s^{\beta} = s^{-0.5}$. In the case of low amplitudes, the frequency response does show a flat magnitude and the phase tends to 45°. Therefore, the swimming robot for a single point actuation shows a non-integer first order model dynamics, however, it will not be considered because the frequencies are too low for the application of interest. Hence, for identification purposes, the frequency response corresponding to this integrator ($H(j\omega)$) was removed from the measured responses ($M(j\omega)$). Consequently, identification procedure was reduced to only identify the part corresponding to mechanical behavior ($G(j\omega)$).

For the identification of $G(j\omega)$, models of the forms of (1) and (2) were identified in the frequency range between $[0.5, 17]$ Hz, as it is enough to capture the first resonant frequency of the swimming robot. Furthermore, the average frequency response of the ten measurements was used for identification. Concerning the identification process, the Levy's method is implemented in MATLAB® (with *levy* function [8]) for both integer and fractional order models. The adjustment is done minimizing $J = (G(j\omega)D(j\omega) - N(j\omega))^2$, i.e., the quadratic error per sampling frequency ($N(j\omega)$ and $D(j\omega)$ are the numerator and denominator of the transfer function of the model to be identified, respectively). Fractional models are found sweeping the $\alpha \in]1, 2[$ range, with a 0.05 step and keeping the best result.

Identification results are summed up in Table 2 including the cost function J, along with additional performance indices MSE, MAD, MD and R^2 calculated in the same way as in [6], in order to compare and evaluate the goodness of the models. Furthermore, fitting results are graphically represented in the right side of Fig. 3. From Table 2, where best fits are in bold, it can be stated that: i) two models are stable; ii) gains of transfer functions (coefficient a_0) are of the same order of magnitude. To this respect, the best results are obtained using the model $G_2(s)$ given by (2). Respect to the parameter β indicates the order of the integrator coupled to the models $G_1(s)$ and $G_2(s)$.

Table 2. Fitting results and performance indices.

Model	β	α	a_0 ($\times 10^{-5}$)	$b_1(\times 10^4)$	b_2
G_1	0.5	–	5.0062	24	1.8879×10^{-4}
G_2	0.5	1.84	4.9778	3.8143	–
	J ($\times 10^{-10}$)	MSE ($\times 10^{-11}$)	MAD ($\times 10^{-6}$)	MD ($\times 10^{-5}$)	R^2
G_1	14.0160	6.9199	4.6507	**2.5694**	**0.6737**
G_2	**9.9756**	**6.1250**	**3.9727**	2.6319	0.7112

5 Conclusions

This paper has analyzed the frequency responses of a 3-link AEF swimming robot manufactured in IPMC, in order to identify a model for control purposes. The frequency response was obtained by sweeping sine waves applied to the first link, through a couple of gold electrodes, and measuring the deflection of third link by means of a laser distance meter. In addition, the frequency response was obtained for different amplitudes of excitation, in order to determined if the non linearity of the material is reflected in the frequency range of interest.

Two models were considered for the identification process, consisting of a non-integer order integrator in series with a resonant system of non-integer or integer order in the frequency range $[0.5, 17]$ Hz. The results showed that those models can be adequate for control purposes.

Our future works will focus on: 1) designing robust control for tracking position of third link, 2) generating a desired waveform with a single point actuation, and 3) characterizing propulsion force.

Acknowledgment. This work was supported in part by the Consejería de Economía, Ciencia y Agenda Digital (Junta de Extremadura) under the project IB18109 and the grant "Ayuda a Grupos de Investigación de Extremadura" (no. GR18159), by the Agencia Estatal de Investigación (Ministerio de Ciencia e Innovación) through the project PID2019-111278RB-C22/AEI/10.13039/501100011033, and in part by the European Regional Development Fund "A way to make Europe". José Emilio Traver would like to thank the Ministerio de Educación, Cultura y Deporte its support through the scholarship no. FPU16/2045 of the FPU Program. The authors would like to thank S. Hassan HosseinNia and Andres Hunt at the Technical University of Delft, The Netherlands, where the robot was fabricated.

References

1. Chen, Z., Bart-Smith, H., Tan, X.: IPMC-actuated robotic fish. In: Du, R., Li, Z., Youcef-Toumi, K., Valdivia y Alvarado, P. (eds.) Robot Fish. STME, pp. 219–253. Springer, Heidelberg (2015). https://doi.org/10.1007/978-3-662-46870-8_8
2. Chen, Z., Tan, X.: A control-oriented and physics-based model for ionic polymer-metal composite actuators. IEEE/ASME Trans. Mechatron. **13**(5), 519–529 (2008)
3. Nemat-Nasser, S.: Micromechanics of actuation of ionic polymer-metal composites. J. Appl. Phys. **92**(5), 2899–2915 (2002)
4. Pugal, D.: Physics based model of ionic polymer-metal composite electromechanical and mechanoelectrical transduction. Ph.D. thesis, University of Nevada, Reno (2012)
5. Shahinpoor, M., Kim, K.J.: Ionic polymer metal composites: IV industrial and medical applications. Smart Mater. Struct. **14**(1), 197–214 (2004)
6. Tejado, I., Traver, J.E., Prieto-Arranz, J., López, M.Á., Vinagre, B.M.: Frequency domain based fractional order modeling of IPMC actuators for control. In: Proceedings of the 2019 18th European Control Conference (ECC), pp. 4112–4117. IEEE (2019)
7. Tenreiro Machado, J.A., Caponetto, R., Graziani, S., Sapuppo, F., Tomasello, V.: An enhanced fractional order model of ionic polymer-metal composites actuator. Adv. Math. Phys. **2013**, 717659 (2013)
8. Valério, D., Sá da Costa, J.: Ninteger: a fractional control toolbox for Matlab. In: Proceedings of the 1st IFAC Workshop on Fractional Differentiation and its Applications, Bordeaux (2004)

Accuracy Analysis of Approximated Fractional Order Transfer Function with Pole and Zero

Krzysztof Oprzedkiewicz(✉)

AGH University, al. A. Mickiewicza 30, 30-059 Kraków, Poland
kop@agh.edu.pl

Abstract. The paper deals with the accuracy analysis of a new approximation of fractional order transfer function containing both pole and zero. This class of fractional transfer functions can be employed e.g. to build model - based special control algorithms. The considered function uses Charef approximation. Results are illustrated by example.

Keywords: Fractional order systems · Fractional order transfer function · Charef approximation

1 Introduction

Fractional order transfer differential equations and consequently, fractional-order transfer functions describe a number of real physical phenomena see e.g. [2,3]. The possibility of effective modeling fractional order systems at the MATLAB/SIMULINK platfom is determined by possibility of their integer order approximation. Such approximations has been presented by many Authors. Fundamental results are given by Oustaloup (see e.g.: [1,4]. The approximation of elementary inertial trasfer function $\frac{k}{(Ts+1)^{\alpha}}$ was proposed by Charef [1]. Ideas of both approximations are close and they consist in fitting Bode magnitude plots of exact and approximated transfer functions.

However author does not know a simple approximation method dedicated to elementary fractional order transfer function containing both one pole and one zero $\frac{(T_{\beta}s+1)^{\beta}}{(T_{\alpha}s+1)^{\alpha}}$. Such a transfer function can describe a part of special, model - based controller e.g. cancellation controller.

In this paper a proposition of generalization the Charef approximation to describe an elementary fractional order inertial transfer function with pole and zero is presented.

The paper is organized as follows. At the beginning the Charef approximation is recalled. Next the proposed approximation is given and its accuracy is analysed. Finally illustrative example is given.

A. Dzielinski et al. (Eds.): ICFDA 2021, LNNS 452, pp. 65–73, 2022.
https://doi.org/10.1007/978-3-031-04383-3_7

2 Preliminaries-The Charef Approximation

In preliminaries the Charef approximation needs to be recalled. It allows to describe the FO inertial plant. It takes the following form [1]:

$$G(s) = \frac{1}{(Ts+1)^{\gamma}}. \tag{1}$$

where T denotes the time constant of the plant, $0 < \gamma < 1$ denotes the fractional order of the plant. The finite-dimensional approximation of the transfer function (1) is expressed as follows:

$$G_{ch}(s) = \frac{\prod_{n=0}^{N-1}(1+\frac{s}{z_i})}{\prod_{n=0}^{N}(1+\frac{s}{p_i})}. \tag{2}$$

where z_i and p_i are zeros and poles of approximation, N is its order. An idea of this approximation is to best fit the Bode magnitude plot of approximation to Bode magnitude plot of plant in given frequency band. Zeros and poles are calculated with the use of following recursive dependencies [1]:

$$\begin{cases} p = \frac{1}{T}, \\ p_0 = p\sqrt{b}, \\ z_0 = ap_0, \\ \dots \\ p_i = p_0 ab^i \quad i = 1...N, \\ z_i = ap_0 ab^i \quad i = 1...N. \end{cases} \tag{3}$$

where:

$$\begin{aligned} a &= 10^{\frac{\Delta}{10(1-\alpha)}}, \\ b &= 10^{\frac{\Delta}{10\alpha}}, \\ ab &= 10^{\frac{\Delta}{10\alpha(1-\alpha)}}. \end{aligned} \tag{4}$$

In (4) $\Delta > 0$ denotes maximum permissible error of Charef approximation, defined as the difference between Bode magnitude plot for model and plant, expressed in [dB].

3 The Considered Transfer Function with Zero and Pole

Consider the following, elementary transfer function:

$$G(s) = \frac{(T_\beta s+1)^{\beta}}{(T_\alpha s+1)^{\alpha}}. \tag{5}$$

where $0 < \alpha, \beta < 1$, $\alpha \neq \beta$ are fractional orders of numerator and denominator, $T_\alpha, T_\beta > 0$, $T_\alpha \neq T_\beta$ denote time constants of numerator and denominator respectively. We do not make any assumptions about commensurability of these coefficients. They can be commensurable or not.

The exact magnitude Bode plot for the considered plant can be obtained using MATLAB. The spectral transfer function of the plant described by (5) is the following function of angular frequency ω:

$$G(j\omega) = \frac{(j\omega T_\beta + 1)^\beta}{(j\omega T_\alpha + 1)^\alpha}. \tag{6}$$

The module $M(\omega)$ of the transfer function (6) as a function of ω is expressed as follows:

$$M(\omega) = \sqrt{\frac{(\omega T_\beta^2 + 1)^\beta}{(\omega T_\alpha^2 + 1)^\alpha}}. \tag{7}$$

The module (7) can be expressed in $[dB]$ as the following sum:

$$20 \log M(\omega) = 20 \log M_\beta(\omega) + 20 \log M_\alpha(\omega). \tag{8}$$

where:

$$20 \log M_\beta(\omega) = 10\beta \log(\omega T_\beta^2 + 1),\ 20 \log M_\alpha \omega = -10\alpha \log(\omega T_\alpha^2 + 1). \tag{9}$$

Equations (7)–(8) describe the exact Bode module plot for the transfer function we deal with. It can be easily obtained using MATLAB or Bode approximation. The shape of the diagram is determined by orders α, β and time constants T_α, T_β. The factors (9) can be directly approximated using Charef approximation.

4 The Proposed Approximation of the Transfer Function

The approach presented above can be also applied to approximate multimodal fractional order transfer functions (see [1]). The Bode magnitude plot of the function we deal with is a sum of plots described by (9), because the transfer function (5) can be presented as the following product:

$$G(s) = G_\alpha(s) G_\beta(s). \tag{10}$$

where:

$$G_\alpha(s) = \frac{1}{(T_\alpha s + 1)^\alpha}. \tag{11}$$

$$G_\beta(s) = (T_\beta s + 1)^\beta = \frac{1}{\frac{1}{(T_\beta s + 1)^\beta}}. \tag{12}$$

The transfer function (11) can be directly approximated using approximation (2). Denote this approximation by $G_{ch\alpha}(s)$:

$$G_{ch\alpha}(s) = \frac{\prod_{n_\alpha=0}^{N_\alpha-1} 1 + \frac{s}{z_{n_\alpha}}}{\prod_{n_\alpha=0}^{N_\alpha} 1 + \frac{s}{p_{n_\alpha}}} = \frac{L_{ch\alpha}(s)}{D_{ch\alpha}(s)}. \tag{13}$$

In (13) notation is the same, as in (2). The order of approximation is equal N_α.

More problematic is the approximation of the second factor of transfer function (5) , expressed by (12). This transfer function is the inverse of inertial transfer function (5). This implies that its approximation can be proposed as an inverse of approximation for inertial plant described by (13):

$$G_{ch\beta}(s) = \frac{\prod_{n_\beta=0}^{N_\beta-1} 1 + \frac{s}{p_{n_\beta}}}{\prod_{n_\beta=0}^{N_\beta} 1 + \frac{s}{z_{n_\beta}}} = \frac{D_{ch\beta}(s)}{L_{ch\beta}(s)}. \tag{14}$$

Consequently, the approximation $G_{ch}(s)$ of the whole considered transfer function can be expressed as follows:

$$G_{ch}(s) = G_{ch\alpha}(s)G_{ch\beta}(s) = \frac{L_{ch\alpha}(s)D_{ch\beta}(s)}{D_{ch\alpha}(s)L_{ch\beta}(s)}. \tag{15}$$

The form of the above approximation can casuse pole-zero cancellation problem, analyzed with details in paper [5].

The accuracy of the proposed approximation can be estimated using maximum perimissible errors of both partial approximations. Denote this error by $\Delta_s(\omega)$. It is described by the following proposition:

Proposition 1. *Consider the approximation of fractional order transfer function (5). Assume that errors of both partial approximations are the same and equal Δ.*

The maximum error of the proposed approximation can be estimated as follows:

$$max\Delta_s(\omega) = 2\Delta \tag{16}$$

where:

$$\Delta_s(\omega) = M(\omega) - M_{ch}(\omega). \tag{17}$$

The proof of the above proposition is given in the paper [5].

5 Examples

5.1 Example 1

As the beginning consider the following transfer function with "dominating" pole ($\alpha > \beta$):

$$Gs = \frac{(25s+1)^{0.15}}{(87s+1)^{0.7}}. \tag{18}$$

The approximation of the above transfer function was built for the maximum error $\Delta = 0.5$ [dB]. The accuracy of approximation was tested for equal orders of both parts: $N_\alpha = N_\beta = N$. Results of simulations are described by Table 1 and illustrated by Figs. 1, 2 and 3.

Table 1. The accuracy of approximation for different orders N.

N	ω_l	ω_h	$\max \Delta_s[dB]$
6	-3	-1	1.8546
12	-4	1	1.8512
24	-4	1	1.8512

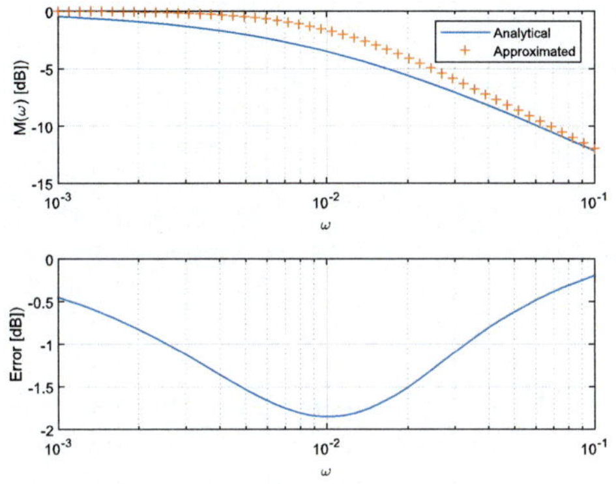

Fig. 1. Bode magnitude plots and approximation errors for transfer function (18) and $N = 6$.

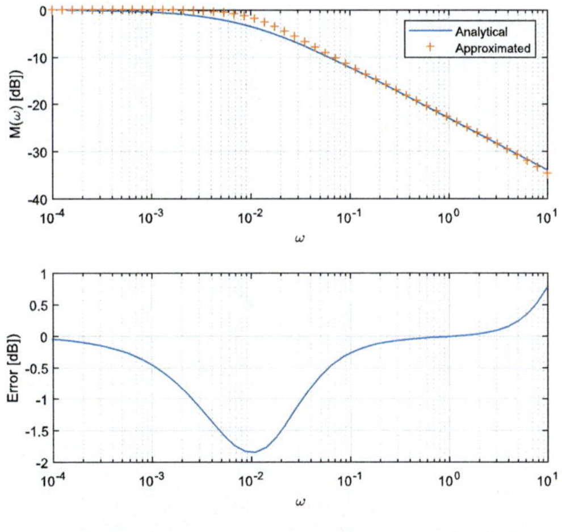

Fig. 2. Bode magnitude plots and approximation errors for transfer function (18) and $N = 12$.

5.2 Example 2

As the next example consider the following transfer function with "dominating" zero ($\alpha < \beta$):

$$Gs = \frac{(25s+1)^{0.75}}{(78s+1)^{0.29}}. \tag{19}$$

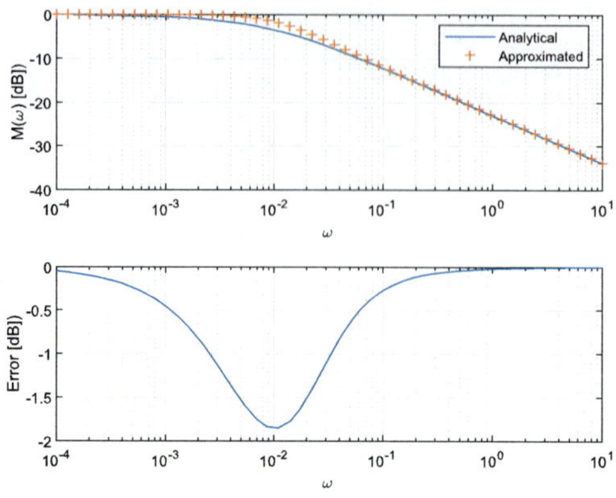

Fig. 3. Bode magnitude plots and approximation errors for transfer function (18) and $N = 24$.

The approximation of the above transfer function was built for the maximum error $\Delta = 0.5$ [dB]. The accuracy of approximation was tested for equal orders of both parts: $N_\alpha = N_\beta = N$. Results of simulations are described by Table 1 and illustrated by Figs. 4, 5 and 6. From analysis of plots presented in Figs. 1, 2, 3, 4, 5 and 6 and Tables 1, 2 it can be concluded that the approximation is correct and its maximum error does not exceed estimation, equal 2 [dB]. Next, the good accuracy can be achieved for relatively low order of Charef approximation N.

Table 2. The accuracy of approximation for different orders N.

N	ω_l	ω_h	$\max \Delta_s [dB]$
6	-4	1	1.7941
12	-4	1	1.7940
24	-4	1	1.7940

5.3 Example 3

As the last example consider the transfer function (18) for different frequency ranges ω_l and ω_h and orders N equal: $N = 6$, $N = 12$. Results of simulations for different frequency ranges are given in the Table 3 and illustrated by Figs. 7 and 8. Copmaring Figs. 1 vs 7 and 2 vs 8 and Tables 1 vs 3 it can be observed that the accuracy of the approximation descreases with increasing of the frequency range. This phenomenon is stronger for lower order of approximation. In addition, the accuracy decreases far from the corner of diagram. This is a little big surprising and it should be explained during further investigations.

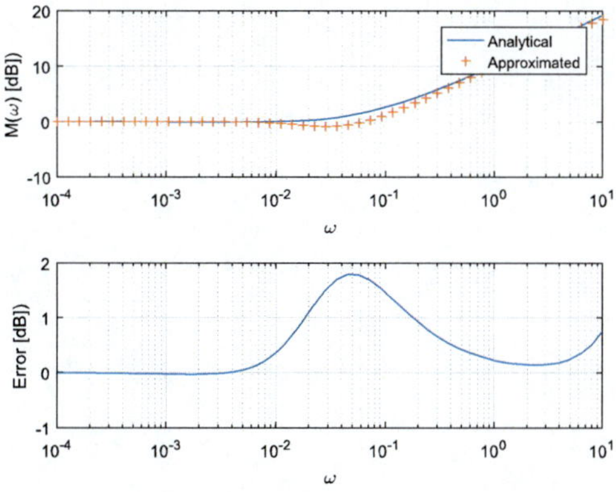

Fig. 4. Bode magnitude plots and approximation errors for transfer function (19) and $N = 6$.

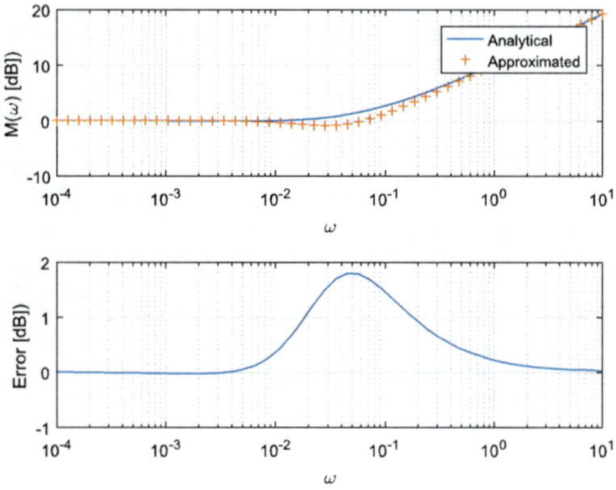

Fig. 5. Bode magnitude plots and approximation errors for transfer function (19) and $N = 12$.

Table 3. The accuracy of approximation for different orders N.

N	ω_l	ω_h	max $\Delta_s[dB]$
6	-4	-1	1.8546
6	-4	4	12.9881
12	-4	1	1.8512
12	-4	4	1.8973

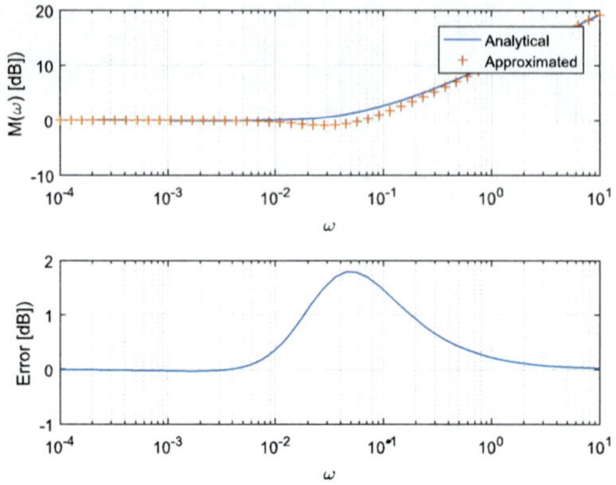

Fig. 6. Bode magnitude plots and approximation errors for transfer function (19) and $N = 24$.

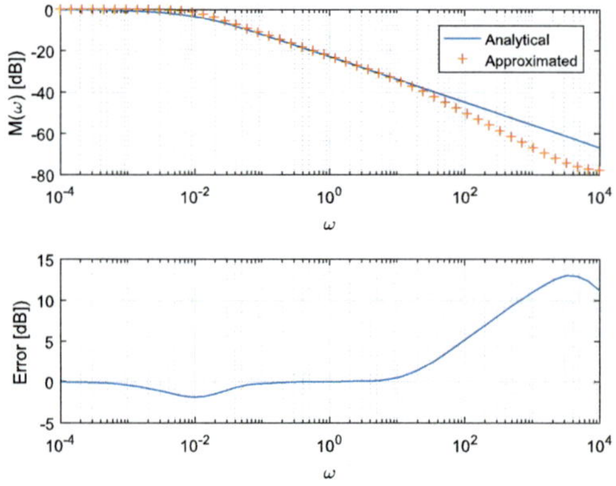

Fig. 7. Bode magnitude plots and approximation errors for transfer function (18), $N = 6$ and frequency range: $\omega_l = -4$, $\omega_h = 4$.

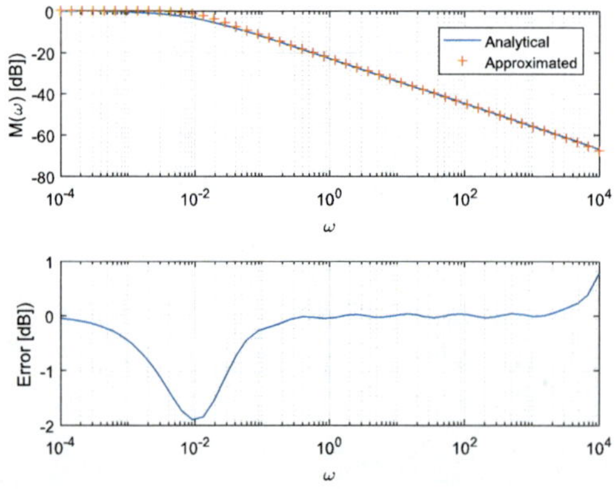

Fig. 8. Bode magnitude plots and approximation errors for transfer function (18), $N = 12$ and frequency range: $\omega_l = -4$, $\omega_h = 4$.

6 Final Conclusions

The main final conclusion from the paper is that the proposed approximation is correct and its maximum error is localized in expected places (corners of the Bode magnitude plot).

An interesting issue is the multimodal case, when the transfer function (5) contains more than one zero and pole. It will be presented with details as a sequel of this paper.

Acknowledgement. This paper was sponsored by AGH project no 16.16.120.773.

References

1. Charef, A., Sun, H.H., Tsao, Y.Y., Onaral, B.: Fractal system as represented by singularity function. IEEE Trans. Autom. Control **37**(9), 1465–1470 (1992)
2. Das, S.: Functional Fractional Calculus for System Identification and Controls. Springer, Heidelberg (2008). https://doi.org/10.1007/978-3-540-72703-3
3. Das, S., Pan, I.: Intelligent Fractional Order Systems and Control: An Introduction. Springer, Heidelberg (2013). https://doi.org/10.1007/978-3-642-31549-7
4. Merrikh-Bayat, F.: Rules for selecting the parameters of Oustaloup recursive approximation for the simulation of linear feedback systems containing PID_{μ} controller. Commun. Nonlinear Sci. Numer. Simul. **17**, 1852–1861 (2012)
5. Oprzedkiewicz, K.: Approximation method for a fractional order transfer function with zero and pole. Arch. Control Sci. **24**(4), 409–425 (2014)

Check for updates

Fractional Order Models Are Doubly Infinite Dimensional Models and Thus of Infinite Memory: Consequences on Initialization and Some Solutions

Jocelyn Sabatier(✉)

Bordeaux University, 15 Rue Naudet - CS 10 207, 33175 Gradignan Cedex, France
jocelyn.sabatier@u-bordeaux.fr

Abstract. Using a few mathematical transformations, this article sheds light on the nature of fractional models (described by fractional differential equations or pseudo state space descriptions). Computation of the impulse response of a fractional model using the Cauchy method shows that they exhibit infinitely small and high time constants. This impulse response can be rewritten as a diffusive representation whose Fourier transform permits a representation of a fractional model by a diffusion equation on an infinite space domain. Fractional models can thus be viewed as doubly infinite dimensional models (distributed and therefore infinite, on an infinite domain). This infinite domain or the infinitely large time constants of the impulse response reveal a property intrinsic to fractional models: their infinite memory. Solutions to generate fractional behaviours without infinite memory are finally proposed.

Keywords: Fractional models · Infinite memory · Initial conditions

1 Introduction

Nowadays there is a confusion or rather an implicit link between fractional behaviours (of physical, biological, thermal, etc. origin) and fractional models (a tool to model fractional behaviours). It is now evident that fractional behaviours are ubiquitous in numerous domains and that appropriate modelling tools are needed. Due to the dynamic behaviour similarity between fractional behaviours and fractional models, the latter have therefore naturally been used as modelling tools. But the connection between fractional behaviours and fractional models was not based on any physical justification, as recalled in the present paper. Many researchers have nevertheless contributed to the study of fractional calculus and fractional models, sometimes yielding to the ease of generalizing the results dedicated to classical integer systems (which are sometimes referred to as "fractionalisations"). This field of research has thus undergone numerous (perhaps not really fundamental) developments [1].

However, it is also well-known now that several drawbacks are associated to fractional models [2–4]. The most important one, from which some others derive, is they

A. Dzielinski et al. (Eds.): ICFDA 2021, LNNS 452, pp. 74–89, 2022.
https://doi.org/10.1007/978-3-031-04383-3_8

were not deduced from physical considerations but are only generalizations of existing (integer) tools. Consequently, fractional models are fitted with an intrinsic and unrealistic property: they are doubly infinite dimensional models:

- infinite as they are distributed
- and infinite also as they are defined (the parameter distribution) on an infinite spatial domain.

It is only by trying to analyze the physical meaning of fractional models that this property is revealed.

Giving meaning to fractional models (a posteriori) has nevertheless been a concern of many researchers. Mathematical [5, 6], physical [7, 8], geometrical [9, 10], statistical [11] meanings were therefore sought but these approaches do not really help to understand their nature and especially their advantages and disadvantages. It was even attempted to deduce a model from a physical phenomenon [12, 13] but infinite spatial dimensions were considered in order to do so, which tends on the contrary to demonstrate the physical inconsistency of a fractional model.

The infinite spatial dimension is the conclusion also reached in this work after some mathematical transformations that lead to a representation of a fractional model by a diffusion equation. This representation is obtained through the Fourier transform of the diffusive representation of a fractional model. It is also shown that the diffusive representation can be deduced from the impulse response of a fractional model computed using the Cauchy method. The diffusive representation obtained exhibits infinitely large time constants that are a consequence of the definition of fractional models on an infinite spatial domain. This infinite dimension or these infinitely large time constants are responsible for the infinite memory of fractional models. All these concepts are detailed and demonstrated in this paper. Finally, a list of alternative solutions to fractional models to produce fractional behaviours is given.

2 A Physical and Systemic Analysis of Fractional Models

2.1 Models Definition

Even if the terms used to designate them are very similar, fractional behaviour and fractional models are two different concepts. Fractional (dynamical) behaviour qualifies the nature of some real systems that exhibit a power law behaviour in a limited time or frequency band (the system impulse response behaves as $t^{-\nu}$ on a given time band or the system frequency response behaves as $(j\omega)^{-1+\nu}$ on a given frequency band). Fractional models are a class of models that was introduced to capture fractional behaviours. Without considering physical justifications, fractional models were introduced by simply generalizing existing modelling tools, by simply replacing the classical differentiation operators by fractional differentiation operators. These fractional derivation operators had long been mathematically defined and were in fact known to exhibit fractional behaviours (but not only over a limited time or frequency band). For this reason, fractional differential equations appeared as possible models to capture some fractional behaviours.

With $u(t) \in \mathbb{R}$ and $y(t) \in \mathbb{R}$ respectively as input and output, a fractional differential equation is found in the literature under the form.

$$\sum_{k=0}^{N_a} a_k \boldsymbol{D}_{t_0}^{\nu_{a_k}}(y(t)) = \sum_{k=0}^{N_b} b_k \boldsymbol{D}_{t_0}^{\nu_{b_k}}(u(t)). \tag{1}$$

In relation (1) $a_k \in \mathbb{R}$ and $b_k \in \mathbb{R}$. $\boldsymbol{D}_{t_0}^{\nu_{a_k}}$ and $\boldsymbol{D}_{t_0}^{\nu_{b_k}}$ denote respectively fractional differentiation operators of orders $\nu_{a_k} \in \mathbb{R}$ and $\nu_{b_k} \in \mathbb{R}$ [14]. Various definitions can be used to describe these operators [15], which raises the question of choice. These definitions are in certain respects equivalent, but differences exist notably regarding how initial conditions are taken into account.

The "fractionalisation" of another well-known modelling tool was also proposed in the literature if the differentiation orders ν_{a_k} and ν_{b_k} met the following conditions:

$$\nu_{a_k} = k/q \quad \text{and} \quad \nu_{b_k} = k/q, q \in \mathbb{N}^*. \tag{2}$$

These conditions, known as commensurability conditions, enable relation (1) to be rewritten as (under null initial conditions):

$$\begin{cases} \boldsymbol{D}_{t_0}^{\nu} X(t) = AX(t) + Bu(t) \\ y(t) = CX(t) + Du(t) \end{cases}. \tag{3}$$

This representation is known in the literature as a pseudo state space description. "Pseudo" because the vector $X(t) \in \mathbb{R}^n$ cannot be viewed as the model state since this is not the minimal amount of information required to predict the future of the model. In relation (3), $\nu = 1/q$ is the fractional order of the model and $A \in \mathbb{R}^{nxn}$, $B \in \mathbb{R}^{nx1}$, $C \in \mathbb{R}^{1xn}$ and $D \in \mathbb{R}$ are constant matrices.

Laplace transform applied to relations (1) or (3) (with null initial conditions) permits a transfer function representation of the form:

$$H(s) = \frac{Y(s)}{U(s)} = \frac{\mathcal{L}\{y(t)\}}{\mathcal{L}\{u(t)\}}. \tag{4}$$

Models (1) and (3) are very widely used in the literature when talking about fractional models.

Example: The following fractional model, which will be used repeatedly throughout the article is now considered:

$$D_0^{1.8} y(t) + y(t) = 8\boldsymbol{D}_0^{1.2} u(t) - 8D_0^{0.6} u(t) + 10u(t). \tag{5}$$

As all the fractional orders are multiples of 0.6, such a model admits the following pseudo state space description:

$$\begin{cases} \boldsymbol{D}_0^{0.6} X(t) = AX(t) + Bu(t) \\ y(t) = CX(t) \end{cases}. \tag{6}$$

with

$$X(t) = \begin{bmatrix} x(t) \\ \boldsymbol{D}_0^{0.6} x(t) \\ \boldsymbol{D}_0^{1.2} x(t) \end{bmatrix} \quad A = \begin{bmatrix} 0 & 1 & 0 \\ 0 & 0 & 1 \\ -1 & 0 & 0 \end{bmatrix} \quad B = \begin{bmatrix} 0 \\ 0 \\ 1 \end{bmatrix} \quad C = \begin{bmatrix} 10 & -8 & 8 \end{bmatrix}. \tag{7}$$

This model transfer function is thus given by:

$$H(s) = \frac{8s^{1.2} - 8s^{0.6} + 10}{s^{1.8} + 1}. \tag{8}$$

■

As previously mentioned, the definition of these models did not result from physical analyses; they are "fractionalizations" of classical models. The objective of the following sections is therefore to propose an analysis of the physical meaning of these models.

2.2 Poles and Time Constants Distribution and Infinite Memory

The first physical analysis proposed is based on the impulse response of the models (1) and (3). This impulse response is given by the inverse Laplace transformation relation:

$$h(t) = \mathcal{L}^{-1}\{H(s)\} = \frac{1}{2j\pi}\int_{c-j\infty}^{c+j\infty} H(s)e^{st}ds. \tag{9}$$

In relation (9), the value of c is chosen greater than the $H(s)$ singular point abscissa and the integral computation is done on the Bromwich-Wagner path Γ of Fig. 1 that bypasses the negative axis around the branching point $z = 0$ for $t > 0$. It thus avoids the complex plane domain for which the transfer function $s^{\nu} = 0$ is not defined, i.e. the segment $]-\infty, 0]$. As shown by Fig. 1, the path is the union of six sub-paths: $\Gamma = \gamma_1 \cup \ldots \cup \gamma_6$.

Using the residue theorem, relation (9) can be rewritten as:

$$h(t) = -\frac{1}{2\pi j}\int_{\Gamma-\gamma_1} H(s)e^{ts}ds + \sum\nolimits_{\substack{poles \\ in\,\Gamma}} \mathrm{Res}\left[H(s)e^{ts}\right] \tag{10}$$

with, in the case of a simple pole p for $H(s)e^{ts}$

$$Res_p\left[H(s)e^{ts}\right] = \lim_{z\to p}(z-p)H(z)e^{tz}. \tag{11}$$

Relation (10) thus imposes the computation of the poles of $H(s)$ and shows that the impulse responses of models (1) and (3) can be split into two parts:

$$h(t) = h_p(t) + h_d(t). \tag{12}$$

The function $h_p(t)$ is computed from the poles of models (1) and (3) (residues of the Cauchy method). The function $h_d(t)$ is linked to the diffusive part of the model. The name "diffusive" is explained in the sequel (see Sect. 2.2). The function $h_d(t)$ is defined by [16]:

$$h_d(t) = \int_0^{\infty} \mu(z)e^{-tz}dz, \tag{13}$$

in which the function $\mu(x)$ is given in [17] by:

$$\mu(z) = \frac{1}{2i\pi}\left[H\left((-z)^-\right) - H\left((-z)^+\right)\right] \text{ with } \mu(z) \in \mathbb{R}. \tag{14}$$

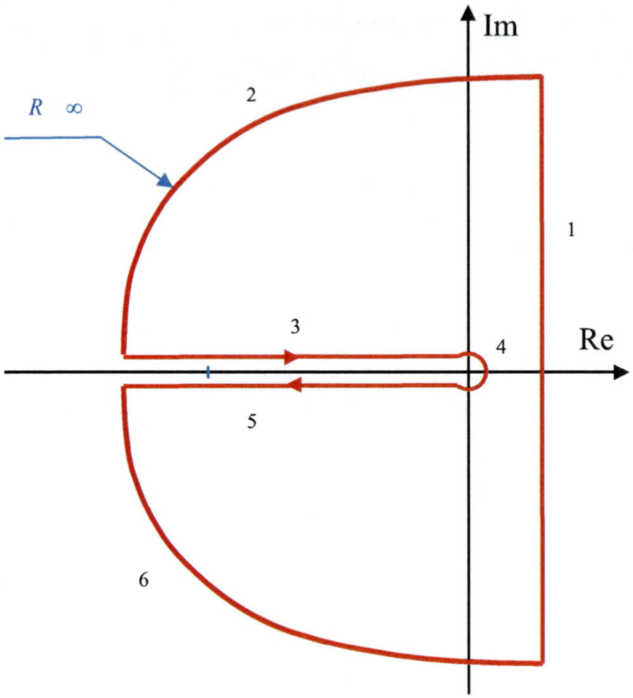

Fig. 1. The Γ path considered for impulse response computation of $h(t)$

The Laplace transform of function $h_d(t)$ is given by:

$$h_d(s) = \int_0^{\infty} \frac{\mu(z)}{s+z} dz. \tag{15}$$

Example: The model defined by relation (5) is again considered. The corresponding transfer function (relation (8)) can be rewritten as:

$$H(s) = \frac{2}{s^{1.8}+1} + \frac{8}{s^{0.6}+1} \quad \text{as} \quad s^{1.8}+1 = \left(s^{0.6}+1\right)\left(s^{1.2}-s^{0.6}+1\right). \tag{16}$$

The poles of the model, denoted $s_k = \rho_k e^{j\theta_k}$, are the values of s such that $\theta_k \in]-\pi, \pi[$ and:

$$s_k^{1.8}+1=0 \quad \text{or} \quad s_k^{0.6}+1=0 \tag{17}$$

or

$$s_k^{1.8} = e^{j(\pi+2k\pi)} \quad \text{or} \quad s_k^{0.6} = e^{j(\pi+2k\pi)} \tag{18}$$

and thus

$$s_k = e^{j\frac{(\pi+2k\pi)}{1.8}} \quad \text{or} \quad s_k = e^{j\frac{(\pi+2k\pi)}{0.6}} \tag{19}$$

if there exist values of k such that

$$\frac{(\pi + 2k\pi)}{1.8} \in]-\pi, \pi[\quad \text{or} \quad \frac{(\pi + 2k\pi)}{0.6} \in]-\pi, \pi[. \tag{20}$$

Two poles are produced by relation (20):

$$s_0 = e^{j\frac{\pi}{1.8}} \quad \text{or} \quad s_1 = e^{-j\frac{\pi}{1.8}} \tag{21}$$

that generate the function

$$h_p(t) = 2\left(e^{e^{j\frac{\pi}{1.8}t}} + e^{e^{-j\frac{\pi}{1.8}t}}\right) = 4e^{cos(\frac{\pi}{1.8})t} cos\left(sin\left(\frac{\pi}{1.8}\right)t\right). \tag{22}$$

According to relations (13) and (14), the function $h_d(t)$ is given by:

$$h_d(t) = \int_0^{\infty} \left(\frac{2sin(1.8\pi)}{\pi} \frac{z^{1.8}}{1 - 2z^{1.8}cos(1.8\pi) + z^{3.6}} + \frac{8sin(0.6\pi)}{\pi} \frac{z^{0.6}}{1 - 2z^{0.6}cos(0.6\pi) + z^{1.2}} \right) e^{-tz} dz \tag{23}$$

which corresponds to relation (13) with

$$\mu(z) = \frac{2sin(1.8\pi)}{\pi} \frac{z^{1.8}}{1 - 2z^{1.8}cos(1.8\pi) + z^{3.6}} + \frac{8sin(0.6\pi)}{\pi} \frac{z^{0.6}}{1 - 2z^{0.6}cos(0.6\pi) + z^{1.2}}. \tag{24}$$

that is represented by Fig. 2.

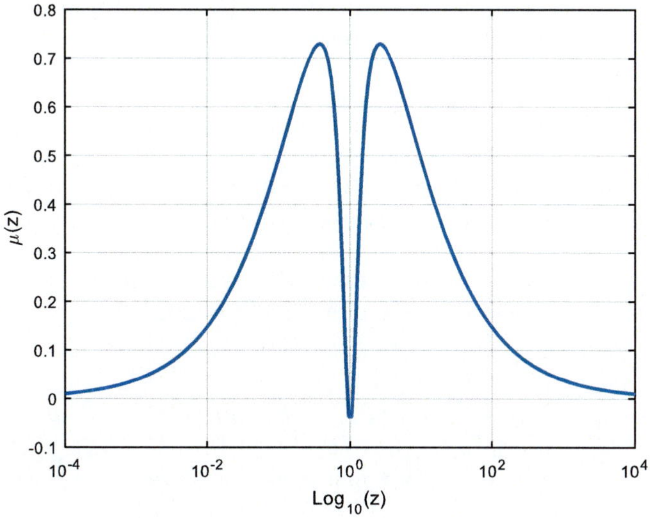

Fig. 2. Function $\mu(z)$

■

If an input $u(t)$ is applied to the diffusive part with an impulse response $h_d(t)$, the output $y_d(t)$ is the convolution product:

$$y_d(t) = \int_0^t \int_0^\infty \mu(z) e^{-(t-\tau)z} u(\tau) dz d\tau = \int_0^\infty \mu(z) \int_0^t e^{-(t-\tau)z} u(\tau) d\tau dz. \quad (25)$$

Introducing the function $w(t, z)$ such that:

$$w(t, z) = \int_0^t e^{-(t-\tau)z} u(\tau) d\tau \quad (26)$$

relation (26) can be rewritten as:

$$\begin{cases} \frac{\partial w(t,z)}{\partial t} = -zw(t, z) + u(t) \\ y_d(t) = \int_0^\infty \mu(z) w(t, z) dz \end{cases} \quad z \in \mathbb{R}_+. \quad (27)$$

This representation is known is the literature as the diffusive representation [16, 17].

Representations (15) and (27) of fractional models are particularly interesting because they make it possible to reveal properties that are poorly known and yet fundamental because they provide information on the internal nature of fractional models and how to initialize them. Relation (15) highlights that models (1) and (3) have poles (values of z) distributed from 0 to $-\infty$. Models (1) and (3) are thus fitted with infinitely fast and infinitely slow time constants (they are attenuated by the function $\mu()$, but they exist). Relation (27) reveals the impact of these poles (or time constants) on the initialisation of fractional models. All the poles must be initialised as the function $w(0, z)$ must be initialised with $z \in \mathbb{R}_+$. This provides fractional models with an infinite memory.

2.3 Spatial Definition and Infinite Memory

In this section, we try to obtain a spatial representation of fractional models. To this end, model (27) is rewritten as:

$$\begin{cases} \frac{\partial w(t,z)}{\partial t} = -zw(t, z) + u(t) \\ y_d(t) = \frac{1}{2} \int_0^\infty \mu(z) w(t, z) dz + \frac{1}{2} \int_0^\infty \mu(z) w(t, z) dz \end{cases} \quad z \in \mathbb{R}_+. \quad (28)$$

The changes of variable $\rho = -\sqrt{z}$ and $\rho = \sqrt{z}$ are respectively used in the first and second integral of relation (28). The resulting model is thus

$$\begin{cases} \frac{\partial \psi(t,\rho)}{\partial t} = -\rho^2 \psi(t, \rho) + u(t) \\ y_d(t) = \frac{1}{2} \int_0^{-\infty} 2\rho \mu(\rho^2) \psi(t, \rho) d\rho + \frac{1}{2} \int_0^\infty 2\rho \mu(\rho^2) \psi(t, \rho) d\rho \end{cases} \quad \rho \in \mathbb{R} \quad (29)$$

with $\psi(t, \rho) = w(t, \rho^2)$. After simplifications, relations (29) become

$$\begin{cases} \frac{\partial \psi(t,\rho)}{\partial t} = -\rho^2 \psi(t, \rho) + u(t) \\ y_d(t) = \int_{-\infty}^\infty |\rho| \mu(\rho^2) \psi(t, \rho) d\rho \end{cases} \quad \rho \in \mathbb{R}. \quad (30)$$

If the inverse Fourier transform is applied to the first equation of relation (30) (on variable ρ), it can be rewritten as:

$$\frac{\partial \phi(t,\zeta)}{\partial t} = \frac{\partial^2 \phi(t,\zeta)}{\partial \zeta^2} + u(t)\delta(\zeta) \quad \zeta \in \mathbb{R} \tag{31}$$

where

$$\psi(t,\rho) = \mathcal{F}\{\phi(\mathrm{t},\zeta)\} = \int_{-\infty}^{\infty} \phi(\mathrm{t},\zeta)\mathrm{e}^{-\mathrm{j}\rho\zeta}\mathrm{d}\zeta$$

and

$$\phi(t,\zeta) = \mathcal{F}^{-1}\{\psi(\mathrm{t},\rho)\} = \frac{1}{2\pi}\int_{-\infty}^{\infty} \psi(\mathrm{t},\rho)\mathrm{e}^{\mathrm{j}\rho\zeta}\mathrm{d}\rho \tag{32}$$

As $\psi(t,\rho)$ is a real and even function, according to the Fourier transform properties, the function $\phi(t,\zeta)$ is also a real and even function. Model (30) can be rewritten as:

$$\begin{cases} \frac{\partial \phi(t,\zeta)}{\partial t} = \frac{\partial^2 \phi(t,\zeta)}{\partial \zeta^2} + u(t)\delta(\zeta) \\ y_d(t) = \int_{-\infty}^{\infty} |\rho|\mu(\rho^2)\mathcal{F}\{\phi(\mathrm{t},\zeta)\}\mathrm{d}\rho \end{cases}. \tag{33}$$

As:

$$\begin{aligned} y_d(t) &= \int_{-\infty}^{\infty} |\rho|\mu\left(\rho^2\right)\left(\int_{-\infty}^{\infty} \phi(t,\zeta)e^{-j\rho\zeta}d\zeta\right)d\rho \\ &= \int_{-\infty}^{\infty} \phi(t,\zeta)\int_{-\infty}^{\infty} |\rho|\mu\left(\rho^2\right)e^{-j\rho\zeta}d\rho d\zeta \end{aligned} \tag{34}$$

relation (33) becomes

$$\begin{cases} \frac{\partial \phi(t,\zeta)}{\partial t} = \frac{\partial^2 \phi(t,\zeta)}{\partial \zeta^2} + u(t)\delta(\zeta) \\ y_d(t) = \int_{-\infty}^{\infty} m(\zeta)\phi(t,\zeta)d\zeta \end{cases} \quad \zeta \in \mathbb{R} \tag{35}$$

with $m(\zeta) = \mathcal{F}\{|\rho|\mu(\rho^2)\}$.

As $\mu\left(\rho^2\right)$ is a real and even function, the function $m(\zeta)$ is also a real and even function. The integral defining $y_d(t)$ in relation (35) can thus be rewritten as follows:

$$y_d(t) = 2\int_0^{\infty} m(\zeta)\phi(t,\zeta)d\zeta = \int_0^{\infty} m(\zeta)2\phi(t,\zeta)d\zeta \tag{36}$$

and model (35) can be rewritten as:

$$\begin{cases} \frac{\partial \phi(t,\zeta)}{\partial t} = \frac{\partial^2 \phi(t,\zeta)}{\partial \zeta^2} + 2u(t)\delta(\zeta) \\ y_d(t) = \int_0^{\infty} m(\zeta)\phi(t,\zeta)d\zeta \end{cases} \quad \zeta \in \mathbb{R}_+. \tag{37}$$

Relation (37) provides information on the nature of fractional models. As illustrated by Fig. 3, they can be described by a diffusion equation on an infinite dimensional domain

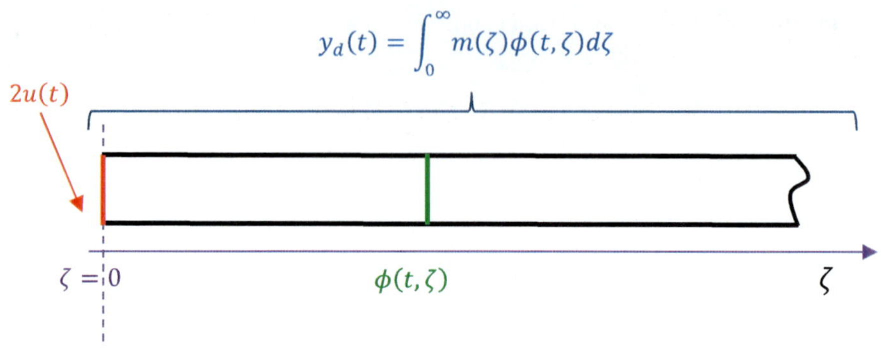

Fig. 3. Spatial representation of the diffusive part of a fractional model

($\zeta \in \mathbb{R}_+$). They can thus be seen as doubly infinite models (infinite because distributed and infinite as defined on an infinite spatial domain). This infinite spatial domain imposes an initialisation with an infinite amount of information: $\phi(0, \zeta), \zeta \in \mathbb{R}_+$.

Example: The model defined by relation (5) is once again considered. The response of the diffusive part of the impulse response $h_d(t)$ given by relation (23) to an input $u(t)$ is provided by the diffusion Eq. (37) defined on an infinite spatial domain, where the function $m(\zeta)$ is defined by:

$$m(\zeta) = \int_{-\infty}^{\infty} |\rho| \left(\frac{2sin(1.8\pi)}{\pi} \frac{\left(\rho^2\right)^{1.8}}{1 - 2\left(\rho^2\right)^{1.8} cos(1.8\pi) + \left(\rho^2\right)^{3.6}} + \frac{8sin(0.6\pi)}{\pi} \frac{\left(\rho^2\right)^{0.6}}{1 - 2\left(\rho^2\right)^{0.6} cos(0.6\pi) + \left(\rho^2\right)^{1.2}} \right) e^{-j\rho\zeta} d\rho \tag{38}$$

which is also equal to

$$m(\zeta) = 2\int_0^\infty \left(\frac{2sin(1.8\pi)}{\pi} \frac{\rho^{4.6}}{1 - 2\rho^{3.6} cos(1.2\pi) + \rho^{7.2}} + \frac{8sin(0.6\pi)}{\pi} \frac{\rho^{2.2}}{1 - 2\rho^{1.2} cos(0.4\pi) + \rho^{2.4}} \right) \cos(\rho\zeta) d\rho \tag{39}$$

■

3 Other Modelling Solutions

The need is for modelling tools that can capture fractional behaviours. Fractional models are only one class of models possible, but they are associated to several drawbacks, mainly resulting from their intrinsic property revealed in the previous sections: their infinite memory. As briefly described in the next sections and also discussed in [3] other solutions are possible, while solving the drawbacks associated with fractional models.

3.1 Kernels with Limited Memory

The Riemann-Liouville or the Caputo definitions are the ones most widely used in the literature for fractional differentiation operators in relations (1) and (3). These operators are based on the fractional integrator operator in their definitions. The fractional integral of a function $f(t)$ is given by [14]:

$$I_{t_0}^{\nu}[f(t)] = \frac{1}{\Gamma(\nu)} \int_{t_0}^{t} \frac{f(\tau)}{(t-\tau)^{1-\nu}} d\tau. \tag{40}$$

This operator is in fact the convolution product of the function to integrate with the singular kernel:

$$\eta(t) = \frac{1}{\Gamma(\nu) t^{1-\nu}}. \tag{41}$$

To solve the infinite memory problem, a solution is to modify this kernel. This solution is often used in the literature to introduce the concept of tempered fractional integration [18]

$$\eta(t) = \frac{t^{\nu-1}}{\Gamma(\nu)} e^{-\omega_{min} t}. \tag{42}$$

The Laplace transform of this kernel is given by

$$\chi(s) = \frac{\omega_{min}^{\nu}}{(s+\omega_{min})^{\nu}}. \tag{43}$$

In the frequency domain, this kernel no longer exhibits a fractional behaviour at low frequency. Computation of its impulse response using the method described in Sect. 2.1 leads to:

$$\eta(t) = \frac{sin(\nu\pi)}{\pi} \int_{\omega_{min}}^{\infty} \frac{1}{\left(\frac{x}{\omega_{min}} - 1\right)^{\nu}} e^{-tx} dx. \tag{44}$$

This relation shows that the kernel no longer exhibits infinitely large time constants, but still has infinitely small time constants, which is questionable from a physical point of view.

We must therefore go further, and introduce kernels with no fractional behaviour at low and high frequency. The following kernel [19] whose impulse response is

$$\eta(t) = C_0 \frac{(\omega_{max})^{\nu}}{(\omega_{min})^{\nu}} \phi_2(\nu-1, 1-\nu, 1; \omega_{min} t, \omega_{max} t) \quad \text{with} \quad C_0 = \frac{\left(\frac{1}{\omega_{min}} + 1\right)^{\frac{1-\nu}{2}}}{\left(\frac{1}{\omega_{max}} + 1\right)^{\frac{1-\nu}{2}}} \tag{45}$$

where $\phi_2(\beta, \beta', \gamma; x, y)$ is a two-variable hypergeometric function defined by [20],

$$\phi_2(\beta, \beta', \gamma; x, y) = \sum_{m,n} \frac{(\beta)_m (\beta')_n}{(\gamma)_{m+n} m! n!} x^m y^n \tag{46}$$

is a possible candidate. The Laplace transform of this kernel is defined by [20] (p. 238):

$$\eta(s) = C_4 \frac{\left(\frac{s}{\omega_{min}} + 1\right)^{1-\nu}}{s\left(\frac{s}{\omega_{max}} + 1\right)^{1-\nu}} \tag{47}$$

Computation of its impulse response using the method described in Sect. 2.1 leads to:

$$\eta(t) = C_4 \left(H_e(t) + \left(\frac{\omega_{max}}{\omega_{min}}\right)^{1-\nu} \frac{\sin((1-\nu)\pi)}{\pi} \int_{\omega_{min}}^{\omega_{max}} \frac{(x-\omega_{min})^{1-\nu}}{x(\omega_{max}-x)^{1-\nu}} e^{-xt} dx \right) \tag{48}$$

where $H_e(t)$ is the Heaviside function. Relation (48) shows that this kernel has time constants distributed in a limited domain.

3.2 Volterra Integro-Differential Equations

It can be easily demonstrated [21] that representation (3) is a special case of a Volterra equation of the form:

$$\int_0^t \eta(t-\tau)x(\tau)d\tau = v(t) \quad v(t) = \int_0^t Bu(\tau)d\tau \quad y(t) = \mathrm{C}x(t) \tag{49}$$

with

$$\eta(t) = \left(\frac{t^{-\nu}}{\Gamma(1-\nu)} I_n - A\right). \tag{50}$$

If the constraint (50) is not imposed, it is possible to generate more varied behaviours with the relation (49) while leaving free the choice of the kernel $\eta(t)$, in particular those mentioned in the previous section. Thus many kinds of fractional behaviours can be generated by relation (53) and integro-differential equations in general, while overcoming the drawbacks associated with the models (1) and (3) with an adequate choice of the kernels.

3.3 Time Delay Models

It is shown in [22, 23] that fractional behaviours can be generated using distributed time delay models of the form:

$$\frac{d}{dt}x(t) = A_0 x(t) + A_1 \int_0^{T_f} \eta(\tau)x(t-\tau)d\tau + Bu(t). \tag{51}$$

The conditions on parameters A_0, A_1, B and on the kernel $\eta(\tau)$ to reach a fractional behaviour are also defined in [3, 22]. The latter paper shows that the dynamic behaviour of a lithium-ion battery can be modelled with this class of equation.

3.4 Nonlinear Models

In [24] and [25] Random Sequential Adsorption (RSA) was studied. RSA exhibits a fractional behaviour that a fractional model cannot correctly capture. In the above cited paper this behaviour is captured using a drift free control affine nonlinear model of the form

$$\frac{dx(t)}{dt} = f(x(t))u(t), \tag{52}$$

where the function $f(x(t))$ is a polynomial to be identified in order to fit the analysed behaviour. Relation (52) is thus a potential model class for fractional behaviour modelling that was used in [24] to model hydrogen adsorption.

3.5 Time Varying Models

thout referring to fractional models, the Avrami model is often used to model the kinetics of crystallization, as well as other phase changes or chemical reactions [26, 27]. This model is described by the relation

$$x(t) = K\left(1 - e^{-(at)^{\nu}}\right). \tag{53}$$

The Laplace transform of relation (52) is given by:

$$x(s) = K\left(\frac{1}{s} - \frac{1}{a}\sum_{r=0}^{\infty} \frac{(-\tau)^r}{r!} \frac{\Gamma(1+\nu r)}{\left(\frac{s}{a}\right)^{1+\nu r}}\right). \tag{54}$$

Figure 4 shows the frequency response of $x(s)$ and demonstrates a fractional behaviour.

Function $x(t)$ in relation (53) is also a solution of the differential equation:

$$\frac{dx(t)}{dt} = -\nu a^{\nu} t^{\nu-1} x(t) + K\nu a^{\nu} t^{\nu-1} \tag{55}$$

This equation involves singular coefficients. Such a matter can be solved by a differential equation of the form

$$\frac{dx(t)}{dt} = -\nu a^{\nu} (t+c)^{\nu-1} x(t) + K\nu a^{\nu} (t+c)^{\nu-1} \tag{56}$$

which also exhibits a fractional behaviour and thus shows that time varying models are possible models to capture fractional behaviours.

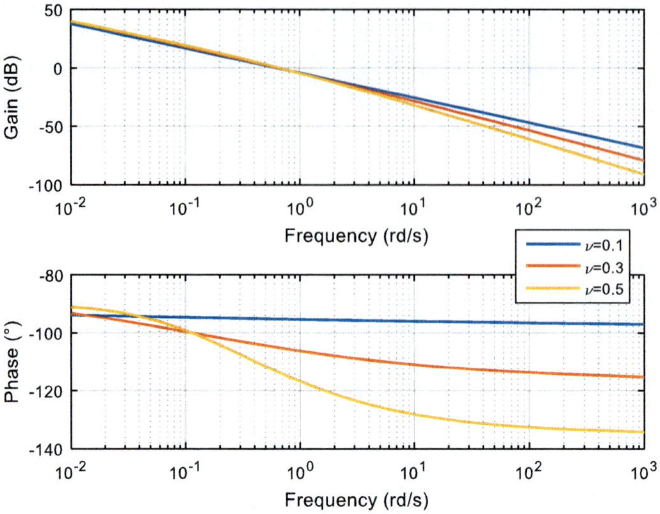

Fig. 4. Frequency response of $x(s)$ for $a = 1$, $\nu = 0.1$, $\nu = 0.3$, $\nu = 0.5$

3.6 Diffusion Equation with Spatially Variable Coefficients

With a judicious choice of the spatial functions $\gamma(z)$ and $\beta(z)$, it is shown in [28] that the following heat equation

$$\frac{\partial T(z,t)}{\partial t} = \gamma(z)\frac{\partial}{\partial z}\left(\beta(z)\frac{\partial T(z,t)}{\partial z}\right) \quad y(t) = \left.\frac{\partial T(z,t)}{\partial z}\right|_{z=0} \quad u(t) = T(0,t) \tag{57}$$

generates fractional behaviours. An infinite number of combinations are possible for the functions $\gamma(z)$ and $\beta(z)$. This leaves considerable room for theoretical investigations in the search for these functions, then for the analysis of the properties of Eq. (57), and also for the development of methods to identify the functions $\gamma(z)$ and $\beta(z)$ on real data.

4 Conclusion

Fractional operators and the resulting fractional models are known for their memory property. This is an interesting property that makes it possible to capture fractional behaviours that are often encountered in modelling problems. For this class of behaviours, the need is indeed models with memory, but not with an infinite memory. And this is unfortunately the case with fractional models. Their infinite memory is demonstrated in this paper by applying mathematical transformations that are described by Fig. 5.

These transformations highlight that fractional models (described by a fractional differential equation or a pseudo state space description) are fitted with infinite small and infinitely large time constants. They also describe a fractional model with a diffusion equation defined on an infinite spatial domain. And it is precisely these infinitely large time constants or the infinite spatial domain that are the reasons for an infinite memory.

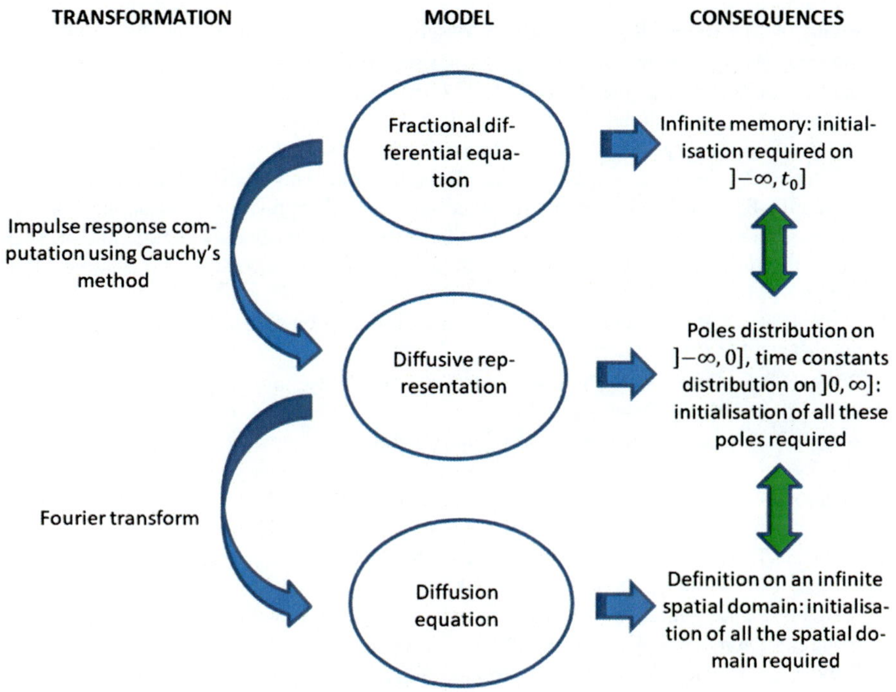

Fig. 5. Various representations of fractional models exhibiting the same problem: infinite memory

Regrettably, many studies dedicated to fractional models seem to ignore this way of considering fractional models, which leads to errors in how these models are initialised. Their infinite memory imposes considering all the model past and not only an initial condition as with the Caputo definition. This is also described by Fig. 5.

Fractional models are, however, just one class of models among many that are able to capture fractional behaviours without the drawbacks of fractional models. Some other solutions are presented in the paper: operators with limited memory kernels, Volterra equations, distributed time delay models, time varying models, nonlinear models, and diffusion equations with spatially variable coefficients. Considering fractional behaviours without being limited to fractional models opens up countless avenues of research in the field of model analysis and identification.

References

1. Machado Tenreiro, J.A., Kiryakova, V., Mainardi, F.: Recent history of fractional calculus. Commun. Nonlinear Sci. Numer. Simul. **16**(3), 1140–1153 (2011)
2. Dokoumetzidis, A., Magin, R., Macheras, P.: A commentary on fractionalization of multi-compartmental models. Pharmacokinet Pharmacodyn **37**, 203–207 (2010)
3. Sabatier, J., Farges, C., Tartaglione, V.: Some alternative solutions to fractional models for modelling power law type long memory behaviours. Mathematics **8**(2), 196 (2020)

4. Balint, A.M., Balint, S.: Mathematical description of the groundwater flow and that of the impurity spread, which use temporal Caputo or Riemann–Liouville fractional partial derivatives, is non-objective. Fractal Fract. **4**(36) (2020)
5. Gorenflo, R.: Afterthoughts on interpretation of fractional derivatives and integrals. In: Rusev, P., Dimovski, I., Kiryakova, V. (eds.) Transform Methods and Special Functions, Varna 1996. Institute of Mathematics and Informatics, Bulgarian Academy of Sciences, Sofia (1998)
6. Mainardi, F.: Considerations on fractional calculus: interpretations and applications. In: Rusev, P., Dimovski, I., Kiryakova, V. (eds.) Transform Methods and Special Functions, Varna'96, Institute of Mathematics and Informatics, Bulgarian Academy of Sciences, Sofia (1998)
7. Nigmatullin, R.R.: A fractional integral and its physical interpretation. Theor. Math. Phys. **90**(3), 242–251 (1992)
8. Rutman, R.S.: On physical interpretations of fractional integration and differentiation. Theor. Math. Phys. **105**(3), 393–404 (1995)
9. Ben, A.F.: Geometric interpretation of the fractional derivative. J. Fract. Calc. **11**, 21–52 (1997)
10. Podlubny, I.: Geometric and physical interpretation of fractional integration and fractional differentiation. J. Fract. Calc. Appl. Anal. **5**(4), 357–366 (2002)
11. Machado Tenreiro, J.A.: A probabilistic interpretation of the fractional-order differentiation. J. Fract. Calc. Appl. Anal. **6**(1), 73–80 (2003)
12. Metzler, R., Joseph, K.J.: The random walk's guide to anomalous diffusion: a fractional dynamics approach. Phys. Rep. **339**(1), 1–77 (2000)
13. Barkai, E., Metzler, R., Klafter, J.: From continuous time random walks to the fractional Fokker-Planck equation. Phys. Rev. E **61**(1), 132–138 (2000)
14. Samko, S.G., Kilbas, A.A., Marichev, O.I.: Fractional Integrals and Derivatives. Gordon and Breach Science Publishers, Langhorne (1993)
15. Capelas de Oliveira E., Tenreiro Machado J.A.: A review of definitions for fractional derivatives and integral. Math. Phys. Eng. **2014**, 1–6 (2014). Article ID 238459
16. Montseny, G.: Diffusive representation of pseudo-differential time-operators. ESAIM Proc. **5**, 159–175 (1998)
17. Matignon, D.: Stability properties for generalized fractional differential systems. ESAIM Proc. **5**, 145–158 (1998)
18. Meerschaert, M.M., Sabzikar, F., Chen, J.: Tempered fractional calculus. J. Comput. Phys. **293**, 14–28 (2015)
19. Sabatier, J.: Non-singular kernels for modelling power law type long memory behaviours and beyond. Cybern. Syst. **51**(4), 383–401 (2020). https://doi.org/10.1080/01969722.2020.1758470
20. Erdelyi, A.: Tables of Integral Transforms, vol. 1, McGraw-Hill, New York City (1954)
21. Sabatier, J.: Fractional state space description: a particular case of the Volterra equation. Fractal Fract. **4**(23) (2020). https://doi.org/10.3390/fractalfract4020023
22. Sabatier, J.: Distributed time delay systems for power law type long memory behaviors modelling. In: 58th IEEE Conference on Decision and Control, CDC 2019, 11th–13th December, Nice, France (2019)
23. Sabatier, J.: Power law type long memory behaviors modeled with distributed time delay systems. Fractal Fract. **4**(1) (2020). https://doi.org/10.3390/fractalfract4010001
24. Tartaglione, V., Farges, C., Sabatier, J.: Nonlinear dynamical modeling of adsorption and desorption processes with power-law kinetics: application to CO_2 capture. Phys. Rev. E **102**, 052102 (2020)
25. Tartaglione, V., Sabatier, J., Farges, C.: Adsorption on fractal surfaces: a non linear modeling approach of a fractional behavior. Fractal Fract. **5**(65) (2021)

26. Avrami, M.: Kinetics of phase change. I. General theory. J. Chem. Phys. **7**(12), 1103–1112 (1939)
27. Fanfoni, M., Tomellini, M.: M. The Johnson-Mehl- Avrami-Kohnogorov model: a brief review. Nouv. Cim. D **20**, 1171–1182 (1998)
28. Sabatier, J.: Beyond the particular case of circuits with geometrically distributed components for approximation of fractional order models: application to a new class of model for power law type long memory behaviour modelling, J. Adv. Res. **25**, 243–255 (2020)

Check for updates

Stability Analysis of COVID-19 via a Fractional Order Mathematical Model

Sadia Arshad[1], Mubashara Wali[1], Ozlem Defterli[2], and Dumitru Baleanu[2,3](✉)

[1] COMSATS University Islamabad, Lahore Campus, Lahore 54000, Pakistan
[2] Department of Mathematics, Cankaya University, Ankara 06790, Turkey
dumitru@cankaya.edu.tr
[3] Institute of Space Sciences, Magurele-Bucharest 077125, Romania

Abstract. In this work, a four compartmental SEIR model is constructed for the transmission of the Novel Coronavirus infectious disease using Caputo fractional derivative. The disease-free equilibrium and endemic equilibrium are investigated with the stability analysis correspondingly. The solution at different fractional orders is obtained using the Laplace Adomian Decomposition method. Furthermore, the dynamics of the proposed fractional order model are interpreted graphically to observe the behaviour of the spread of disease by altering the values of initially exposed individuals and transmission rate.

1 Introduction and Preliminaries

A novel deadly virus, which the scientists named corona virus or Covid-19 hit the globe two years back and declared as the biggest global threat of the present era. Due to the critical situation emerged by the severity of this pandemic, the researchers throughout the globe joined hands in observing and analyzing the behaviour of this virus in order to predict its future effect. Most of them adopted the idea of choosing fractional differential equations in analyzing the dynamical behaviour of the epidemic systems by modeling the natural phenomena as fractional calculus made a significant impact in better explaining the dynamics of various processes in the fields biology, chemistry, medicine, nanotechnology, thermodynamics, mechanics, control theory, etc. over the past two decades. We here considered a simple SEIR epidemic model [1,2] by dividing the total population in different compartments to explain the spread of the virus considering Caputo fractional order derivative.

In this section, we will recall some of the fundamental definitions of fractional calculus [4].

Definition 1. *For an integrable function f on $[0, T]$, the fractional order Caputo derivative is defined as* ${}^{C}D_0^{\beta} f(t) = \frac{1}{\Gamma(m-\beta)} \int_0^t (t-r)^{m-\beta-1} f^{(m)}(r)dr$, *where* $m = [\beta] + 1$ *and* $[\beta]$ *represents the integer part of* β.

A. Dzielinski et al. (Eds.): ICFDA 2021, LNNS 452, pp. 90–95, 2022.
https://doi.org/10.1007/978-3-031-04383-3_9

Lemma 1 [5]. *For $g(t) \in C([0,T])$, the solution of the ${}^{C}D_0^{\beta} f(t) = g(t), \quad t \in [0,T], \quad m-1 < t < m, \quad f(0) = g_0$ is given by*
$f(t) = \sum_{i=0}^{m-1} M_i t^i + \frac{1}{\Gamma(\beta)} \int_0^t (t-r)^{\beta-1} g(r)dr$, *for* $M_i \in R$, $i = 0,1,2,...,(m-1)$.

Theorem 1 [5]. *Let $\daleth : X \to X$ be a compact and continuous operator where X is a Banach space, if the set $E = \{f \in X : f = n\daleth f, n \in (0,1)\}$ is bounded, then $\daleth$ has a unique fixed point.*

2 Description of the SEIR Covid-19 Model

For investigating the spread of Covid-19, Caputo derivative of fractional order ρ of a simple SEIR epidemic disease model in [1,2] is considered.

$$\begin{aligned} {}^{C}D_0^{\rho} S(t) &= \Lambda - \mu S - \beta S \frac{I}{N}, & {}^{C}D_0^{\rho} E(t) &= \beta S \frac{I}{N} - (\mu + \epsilon)E, \\ {}^{C}D_0^{\rho} I(t) &= \epsilon E - (\gamma + \mu + \alpha)I, & {}^{C}D_0^{\rho} R(t) &= \gamma I - \mu R, \end{aligned} \tag{1}$$

along with the initial conditions $S(0) = S_0, E(0) = E_0, I(0) = I_0, R(0) = R_0$. N is equals to the sum of the four classes S, E, I and R. The total population is divided into four different classes as susceptible, exposed, infected and recovered denoted by $S(t)$, $E(t)$, $I(t)$ and $R(t)$ respectively to describe the model equations. The parameters are interpreted as follows:
μ : the natural death rate per-capita, $1/\gamma$: infectious period,
γ : the rate of recovery of infectious individuals,
α : virus infected average fatality rate, Λ: the birth rate per-capita,
β : disease transmission rate(dimensionless), ϵ: progression rate from E to I,
$1/\epsilon$: incubation period, having units per unit time T. From the system of equations (1), the basic reproduction number R_0 is derived correspondingly as $R_0 = \frac{\beta\epsilon}{(\mu+\epsilon)(\gamma+\mu+\alpha)}$.

3 Stability Analysis of the Model

Disease-Free Equilibrium

Since in disease-free equilibrium, there is no spread of the disease, so $E = I = R = 0$. Now, equating (1) to zero and simplifying we get $S = \frac{\Lambda}{\mu}$, so the disease-free equilibrium point is $\mathbb{E}^0 = \left(\frac{\Lambda}{\mu}, 0, 0, 0\right)$.

Endemic Equilibrium

As endemic equilibrium point is used to describe the possibility of disease spread, so each of the classes S, E, I and R is not equals to zero. Equating (1) to zero and simplifying, we get the following endemic equilibrium point, $\mathbb{E}^* = (S^*, E^*, I^*, R^*)$, where

$$S^* = \frac{N(\gamma+\mu+\alpha)(\mu+\epsilon)}{\beta\epsilon}, \quad E^* = \frac{\Lambda}{\mu+\epsilon} - \frac{\mu N(\gamma+\mu+\alpha)}{\beta\epsilon},$$

$$I^* = \frac{\Lambda\epsilon}{(\gamma+\mu+\alpha)(\mu+\epsilon)} - \frac{\mu N}{\beta}, \quad R^* = \frac{\gamma\Lambda\epsilon}{\mu(\gamma+\mu+\alpha)(\mu+\epsilon)} - \frac{\gamma N}{\beta}.$$

3.1 Local Stability Analysis

Theorem 2. *If $R_0 < 1$ then the disease free equilibrium point of Covid-model* (1) *is locally asymptotically stable.*

Proof. The Jacobian of the system (1) is given as

$$J = \begin{bmatrix} -\mu - \frac{\beta I}{N} & 0 & -\frac{\beta S}{N} & 0 \\ \frac{\beta I}{N} & -(\mu+\epsilon) & \frac{\beta S}{N} & 0 \\ 0 & \epsilon & -(\gamma+\mu+\alpha) & 0 \\ 0 & 0 & \gamma & -\mu \end{bmatrix}.$$

Considering this Jacobian matrix around disease free equilibrium point $\mathbb{E}^0\left(\frac{\Lambda}{\mu}, 0, 0, 0\right)$ and putting $\frac{S^0}{N} = 1$ for this equilibrium, the eigenvalues of this matrix are $\lambda_1 = \lambda_2 = -\mu$, $\lambda_3 = -\frac{(2\mu+\epsilon+\gamma+\alpha)}{2} - \frac{\sqrt{(\epsilon-\gamma-\alpha)^2+4\beta\epsilon}}{2}$ and $\lambda_4 = -\frac{(2\mu+\epsilon+\gamma+\alpha)}{2} + \frac{\sqrt{(\epsilon-\gamma-\alpha)^2+4\beta\epsilon}}{2}$. It is clear that λ_1, λ_2 and λ_3 are negative. Now if $R_0 < 1$, we can see clearly that $(\epsilon-\gamma-\alpha)^2+4\beta\epsilon < (2\mu+\epsilon+\gamma+\alpha)^2$ and thus λ_4 is also negative. Hence, for $R_0 < 1$ the disease free equilibrium point is locally asymptotically stable.

Theorem 3. *If $R_0 > 1$ then the disease endemic equilibrium point $\mathbb{E}^*$ of Covid-model* (1) *is locally asymptotically stable.*

Proof. The Jacobian of the system (1) at the endemic equilibrium point $\mathbb{E}^*(S^*, E^*, I^*, R^*)$ is found as

$$J^* = \begin{bmatrix} -\mu - \frac{\beta}{N}\left(\frac{\Lambda\epsilon}{(\gamma+\mu+\alpha)(\mu+\epsilon)} - \frac{\mu N}{\beta}\right) & 0 & -\frac{\beta}{N}\left(\frac{N(\gamma+\mu+\alpha)(\mu+\epsilon)}{\beta\epsilon}\right) & 0 \\ \frac{\beta}{N}\left(\frac{\Lambda\epsilon}{(\gamma+\mu+\alpha)(\mu+\epsilon)} - \frac{\mu N}{\beta}\right) & -(\mu+\epsilon) & \frac{\beta}{N}\left(\frac{N(\gamma+\mu+\alpha)(\mu+\epsilon)}{\beta\epsilon}\right) & 0 \\ 0 & \epsilon & -(\gamma+\mu+\alpha) & 0 \\ 0 & 0 & \gamma & -\mu \end{bmatrix}.$$

We can see clearly from above matrix that, one of its eigenvalues is negative i.e. $\lambda_1 = -\mu$. Now after applying some matrix operations the reduced matrix becomes

$$J^* = \begin{bmatrix} -\frac{\beta\Lambda\epsilon}{N(\gamma+\mu+\alpha)(\mu+\epsilon)} & 0 & -\frac{(\gamma+\mu+\alpha)(\mu+\epsilon)}{\epsilon} \\ 0 & -\frac{\beta\Lambda\epsilon(\mu+\epsilon)}{N(\gamma+\mu+\alpha)(\mu+\epsilon)} & \frac{\mu(\gamma+\mu+\alpha)(\mu+\epsilon)}{\epsilon} \\ 0 & \epsilon & -(\gamma+\mu+\alpha) \end{bmatrix}.$$

Now from above matrix, the 2nd eigenvalue is $\lambda_2 = -(\mu+\epsilon)$, which is negative and the reduced matrix becomes, $J^* = \begin{bmatrix} -\frac{\beta\Lambda\epsilon(\mu+\epsilon)}{N(\gamma+\mu+\alpha)(\mu+\epsilon)} & \frac{\mu(\gamma+\mu+\alpha)(\mu+\epsilon)}{\epsilon} \\ \epsilon & -(\gamma+\mu+\alpha) \end{bmatrix}$, applying the row operation $\frac{\beta\Lambda\epsilon(\mu+\epsilon)}{N(\gamma+\mu+\alpha)(\mu+\epsilon)} R_2 + \epsilon R_1$, we get,

$$J^* = \begin{bmatrix} -\frac{\beta\Lambda\epsilon(\mu+\epsilon)}{N(\gamma+\mu+\alpha)(\mu+\epsilon)} & \frac{\mu(\gamma+\mu+\alpha)(\mu+\epsilon)}{\epsilon} \\ 0 & -\frac{\beta\Lambda\epsilon(\mu+\epsilon)}{N(\gamma+\mu+\alpha)(\mu+\epsilon)}(\gamma+\mu+\alpha) + \mu(\gamma+\mu+\alpha)(\mu+\epsilon) \end{bmatrix}.$$

Now from above matrix, the last two eigenvalues are $\lambda_3 = -\frac{\beta\Lambda\epsilon(\mu+\epsilon)}{N(\gamma+\mu+\alpha)(\mu+\epsilon)}$, which is negative and $\lambda_4 = -\frac{\beta\Lambda\epsilon}{N} + \mu(\gamma+\mu+\alpha)(\mu+\epsilon)$. For $\lambda_4 < 0$, we obtain $-\frac{\beta\Lambda\epsilon}{N} + \mu(\gamma+\mu+\alpha)(\mu+\epsilon) < 0$, $\frac{\beta\Lambda\epsilon}{N\mu(\gamma+\mu+\alpha)(\mu+\epsilon)} > 1$, hence for $R_0 > 1$ the endemic equilibrium point is locally asymptotically stable.

4 Numerical Simulation and Discussion

In this section, numerical simulation of the fractional model (1) is presented graphically. Using Laplace transform and inverse Laplace of Caputo fractional derivative on (1) within the initial conditions, and applying Laplace Adomian Decomposition method the numerical solutions are obtained. The initial values and estimated model parameters for COVID-19 epidemic are presented in the following table [3].

Table 1. Parameter values of the proposed COVID-19 model

Parameter	Value	Parameter	Value	Model variables at initial time	Value
N	10000000	ϵ	1/4.25	S (0)	9988539
μ	0.000033	Λ	$\mu \times N$	E (0)	11460
γ	1/4.02	α	0.00144	I (0)	1
β	0.75			R (0)	0

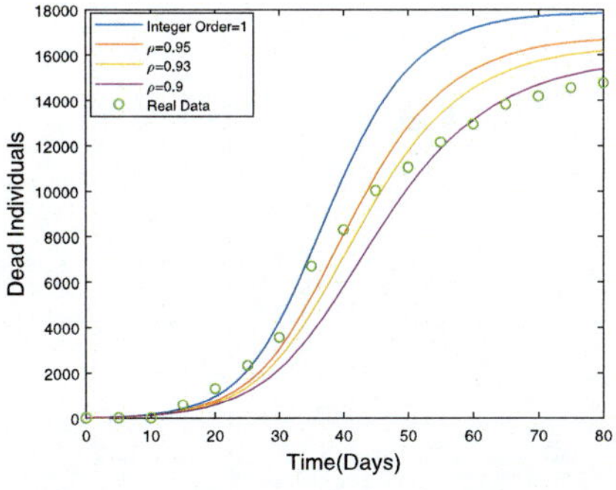

Fig. 1. Comparison of number of dead individuals for different fractional order vs. real data.

Figure 1 shows the comparison of integer order model, fractional model for $\rho = 0.95, 0.93, 0.9$ and real data of total number of deaths $D = \alpha I$ due to COVID-19 epidemic in the region Lombardy (Italy) [3] for first 80 days starting

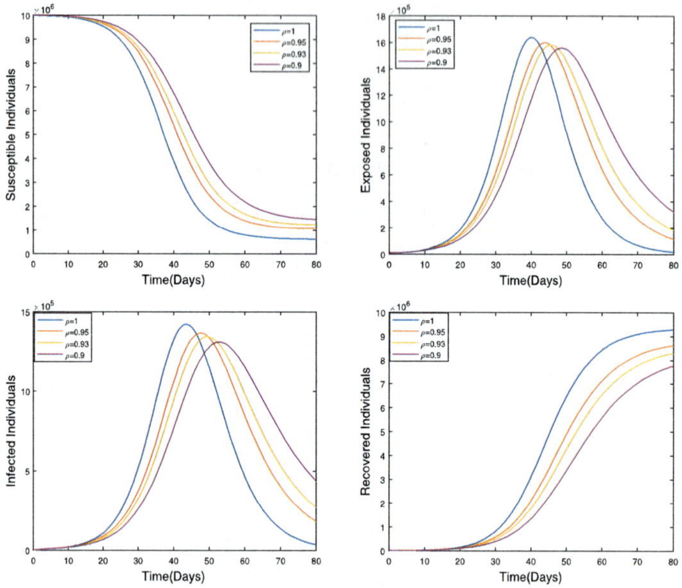

Fig. 2. The behaviour of all the four classes for different fractional order.

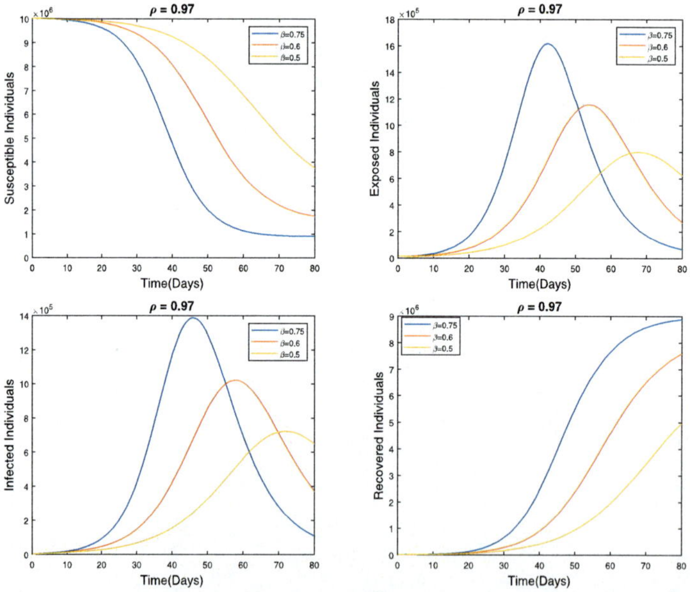

Fig. 3. The number of individuals for different values of β (disease transmission rate).

from 23rd February to 13th May. The respective data of COVID-19 is available at https://github.com/pcmdpc/COVID-19. In Fig. 2, we have presented the simulations for different compartments by varying the fractional order. Initially the people are considered susceptible, but with the start in the virus outbreak, decrease in the susceptible class started resulting an increase in the exposed and infected classes. This increase also effect the recovered class consisting of recovered and dead individuals as more people are being infected. We can see from the figure that smaller the fractional order, respective number of exposed and then infected individuals is decreasing for the initial transient. The concerned rate of increase is different at different fractional orders. Figure 3 clearly shows the effect of decrease in the disease transmission rate. By decreasing the transmission rate, the number of exposed and infected individuals are also decreased. The results presented in the Fig. 3 give us the idea that by keeping the social distance and adopting the quarantine strategy, exposed class and probability of disease transmission rate can be minimized and hence the spread of the virus can be controlled.

5 Concluding Remarks

In this article, we investigated the different conditions for the qualitative analysis and the dynamics of the proposed model within Caputo fractional operator. The conditions for the local stability are discussed by calculating disease-free and endemic equilibrium points along with the reproduction number. At the end, the solution of the concerned fractional order model is obtained using the Laplace Adomian decomposition method and graphical interpretation of the dynamics of the concerned system is made within different fractional orders to explain the behaviour of the Covid-19 virus.

References

1. Hethcote, H.W.: The mathematics of infectious diseases. SIAM Rev. **42**, 599–653 (2000)
2. Keeling, M.J., Rohani, P.: Modeling Infectious Diseases in Humans and Animals. Princeton, NJ, Oxford, Princeton University Press (2008)
3. Carcion, J.M., Santos, J.E., Bagaini, C., Ba, J.: A simulation of the COVID-19 epidemic based on a determinstic SEIR model, Frontiers in Public Health (2020)
4. Samko, S.G., Kilbas, A.A., Marichev, O.I.: Fractional Integrals and Derivatives: Theory and Applications. CRC Press, London (1993)
5. Granas, A., Dugundji, J.: Fixed Point Theory. Springer-Verlag, New York (2003)

Adsorption on Fractal Surfaces: A Non Linear Modeling Approach of a Fractional Behavior

Vincent Tartaglione(✉), Jocelyn Sabatier, and Christophe Farges

IMS Laboratory, Bordeaux University, UMR CNRS 5218,
351, Cours de la Libération, 33405 Talence Cedex, France
{vincent.tartaglione,jocelyn.sabatier,christophe.farges}@u-bordeaux.fr

Abstract. This article deals with the Random Sequential Adsorption (RSA) of 2D disks of same size on fractal surfaces with Hausdorff dimension $1 < d < 2$. According to literature and confirmed by numerical simulations in the paper, the high coverage regime exhibits fractional dynamics, i.e. dynamics in $t^{-1/d}$ where d is the fractal dimension of the surface. As main contribution, this paper proposes to capture this behavior with a particular class of nonlinear model: a driftless control affine model.

1 Introduction

Adsorption is a widely encountered phenomena, especially in physics, chemistry or biology, e.g. to capture pollutants [1,11], for water purification [3], in gas sensors [9,17]. From a mathematical point of view, the idealised adsorption process is known as the Random Sequential Adsorption (RSA) which produces fractional behaviors.

RSA has been widely studied in the literature especially in 2 dimensions. There are results on the final jamming value [6,19,24], and on the kinetic of free places density which exhibits a $t^{-1/2}$ behavior [7,19,21]. Numerous models have been proposed to capture this kinetic. An interesting analysis of these models is proposed in [2]. If $q(t)$ denotes the amount of adsorbed particles, most of the discussed models express the adsorption rate $\frac{\mathrm{d}\,q(t)}{\mathrm{d}\,t}$ as a function of the difference $q_e - q(t)$, where q_e denotes the amount of adsorbed particles at the equilibrium. The following models have been proposed: by Lagergren in 1898 [13]: $\frac{\mathrm{d}\,q(t)}{\mathrm{d}\,t} = k_1(q_e - q(t))$; by Kopelman in 1988 [12]: $\frac{\mathrm{d}\,q(t)}{\mathrm{d}\,t} = k_2 t^{-h}(q_e - q(t))$ with $0 \leq h \leq 1$; by Ho and Mckay in 2000 [10]: $\frac{\mathrm{d}\,q(t)}{\mathrm{d}\,t} = k_3(q_e - q(t))^2$; by Brouers and Sotolongo-Costain in 2006 [4]: $\frac{\mathrm{d}^\alpha\,q(t)}{\mathrm{d}\,t^\alpha} = k_4(q_e - q(t))^n$.

If $\theta(t) = \frac{q(t)}{q_m}$ denotes the relative surface coverage, where q_m is the maximal amount of adsorbed particles and if c denotes the concentration of particles close to the surface (bulk), the following model was also proposed by Haerifar and Azizian in 2012 [8] and Bashiri and Shajari in 2014 [2]:

$$\frac{\mathrm{d}\,\theta(t)}{\mathrm{d}\,t} = k_5 t^{-h} c(1 - \theta(t)) \text{ with } 0 \leq h \leq 1.$$

A. Dzielinski et al. (Eds.): ICFDA 2021, LNNS 452, pp. 96–105, 2022.
https://doi.org/10.1007/978-3-031-04383-3_10

However, the RSA process is non-linear. If the flow of particles is doubled, the filling dynamic of the surface is almost doubled but the final value is the same as for a single flow. Moreover, with the RSA process if the flow is stopped then the filling is obviously stop. And if the flow restarts, the filling will restart at the same point. Thus, none of the time varying and fractional models previously cited permit a consistent modeling of the RSA phenomenon. That is why a non linear model is proposed by the authors of the paper in [20].

In practice, the surface on which the adsorption occurs is not absolutely flat, which leads some authors to consider RSA on fractal surfaces [5]. Analysis of RSA on fractal surfaces has produced a major result: the high coverage regime exhibits a fractional behavior depending on the surface dimension d, i.e. dynamics in $t^{-1/d}$ [5,7,19]. Physical examples of adsorption on fractal surfaces are also considered in numerous works [15,16,18,22,23].

In line with reference [20], as a main contribution the paper proposes to capture the RSA kinetics on fractal surfaces with driftless control affine non linear models. Two models are proposed: a simple one which is proved analytically to produce a fractional dynamic behavior and a more complexe and accurate one. The accuracy of the last model is assesed via numerical simulations on three types of fractal surfaces.

2 Evidence of the Fractionnal Asymptotic Behavior for Some Fractal Surfaces

This article deals with Random Sequential Adsorption (RSA) of 2D disks on a fractal surface. A definition of RSA is now given.

Definition 1 (RSA). *Random Sequential Adsorption is a process in which particles are constantly trying to attach themselves to a random location on a surface. If the incoming particle does not overlap any previously attached particles, then it binds irreversibly.*

In order to model the physical adsorption phenomena, the Random Sequential Adsorption model is used. For any surfaces, it is suggested in the literature [5,19] that the long time coverage density of the RSA follows a power law:

$$\theta_\infty - \theta(t) \sim t^{-1/d}, \tag{1}$$

where d is the Hausdorff dimension of the surface on which the 2D disks arise. The considered fractal surfaces are a set of points contained in the square $[0,1] \times [0,1]$.

In order to simulate the RSA process of 2D disks on a given fractal surface, the RSA algorithm used is the same as in [20] and is now reminded:

Algorithm 1 (RSA). *A random point $c = (p, q)$ of the fractal is selected at each iteration of the process. A disc of radius R and center c will fix on the surface if:*

- *a part of the disc does not lie outside the surface* $[0,1] \times [0,1]$ *containing the fractal, which is true if the following two conditions are satisfied:* $p, q \geq R$ *and* $p, q \leq 1 - R$*;*
- *there is no overlap between the current disc and a previously fixed disc, that is if* $d(c, c_k) \geq 2R$*, where for all* $k \in \mathbb{N}$*,* $c_k = (p_k, q_k)$ *is the center of a disc previously fixed on* S *and where* $d(c, c_k) = \sqrt{(p - p_k)^2 + (q - q_k)^2}$ *is the distance between discs.*

The considered fractal surfaces are illustrated in Fig. 1.

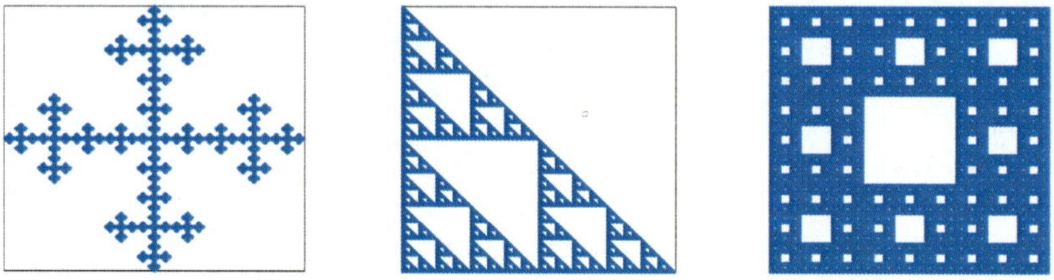

Fig. 1. Vicsek fractal (left), Sierpinski triangle (middle) and Sierpinski carpet (right).

For each of the fractal surfaces previously cited, a RSA process is performed. As shown in [5], the asymptotic value of the density depends of the number of iterations in the fractal building. The number of iterations needed to have relevant results seems to be greater than 6. The final values given in Table 1 are those proposed in [5] and are close to the one that appear in Fig. 2. Differences can be explained by the simulation conditions differences (the number of iterations in the fractal building, the number of trials, the ratio between the size of the disks and the size of the fractal).

Table 1. Final values of the density of adsorbed disks

Fractal	Final value
Vicsek	~0.68
Sierpinski triangle	~0.62
Sierpinski carpet	~0.58

The densities of adsorbed disks as a function of time (number of trials) for each of the considered fractals are shown in Fig. 2.

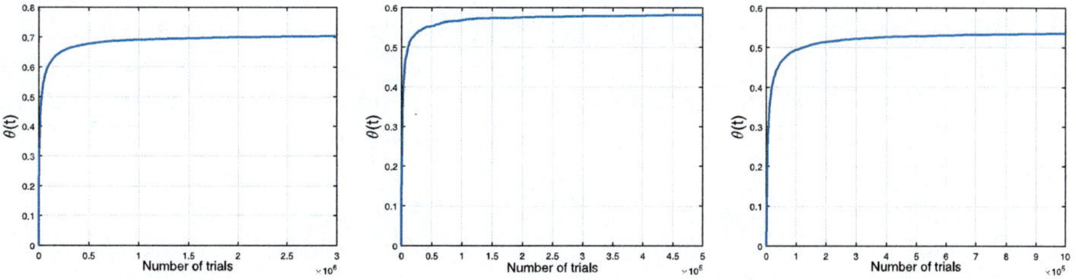

Fig. 2. Densities of adsorbed disks on Vicsek fractal (left), Sierpinski triangle (middle) and Sierpinski carpet (right).

The Fig. 3 shows $\theta_\infty - \theta$ as a function of $t^{-1/d}$, where d is the dimension of the fractal. This figure support that the density of adsorbed disks has a power-law asymptotic behavior of the form

$$\theta_\infty - \theta(t) \sim K t^{-1/d}, \tag{2}$$

where d is the fractal surface dimension (Hausdorff dimension). The Hausdorff dimension of the Vicsek fractal is $d_V = \frac{\log(5)}{\log(3)} \approx 1.4649$, $d_{ST} = \frac{\log(3)}{\log(2)} \approx 1.585$ for the Sierpinski triangle, $d_{SC} = \frac{\log(8)}{\log(3)} \approx 1.8928$ for the Sierpinski carpet. The straight line on each subfigure of Fig. 3 is the plot of $Kt^{-1/d}$.

Figure 3 shows that the results of the simulations are consistent with the expected power-law behaviors.

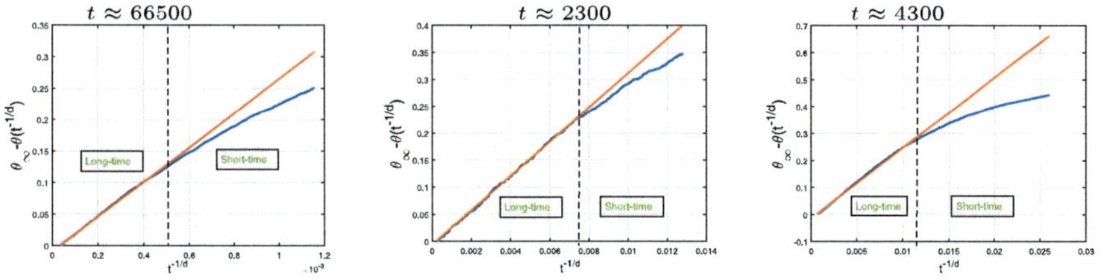

Fig. 3. High regime coverage behavior of RSA on Vicsek fractal (left), Sierpinski triangle (middle) and Sierpinski carpet (right).

3 Power-Law Non Linear Dynamical Modeling

Due to paper size limitation, the modeling approach is done only on the Vicsek fractal. The other fractals produce similar behaviors that can be captured by the same method.

In this section, the considered class of model is called a driftless non linear affine in control model and the first one is defined by the relation

$$\dot{x}(t) = A(x(t) + C)^{1-\frac{1}{\alpha}} \cdot u(t). \tag{3}$$

$u(t)$ can be viewed as an input and x(t) is the model state and output. $A \in \mathbb{R}$, $C \in \mathbb{R}$, $K \in \mathbb{R}$ and $-1 < \alpha < 1$ ($\alpha \neq 0$). This model produce trajectories of the form

$$x(t) = \frac{1}{\alpha}\left(\int Au(t)\mathrm{d}t + K\right)^{\alpha} - C. \tag{4}$$

For $u(t) = 1$, $A = 1$, $K = 0$ and $C = 0$ Eq. (4) becomes

$$x(t) = \frac{1}{\alpha}t^{\alpha}, \tag{5}$$

and the corresponding trajectories are given in Fig. 4 for various values of α.

The trajectories in Fig. 4 are very close to the densities of fixed particles considered above in Fig. 2.

This model is thus used in a first approach to model the RSA density on the fractals considered in the previous section. An optimization is done on parameter A and C.

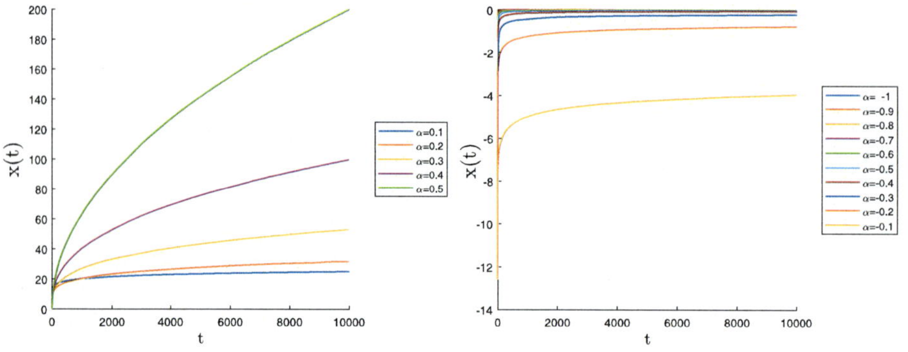

Fig. 4. Solution $x(t)$ with $A = 1$, $u(t) = 1$, $K = 0$ and $C = 1$ for $\alpha \in \{0.1, 0.2, \ldots, 0.5\}$ (left) and $\alpha \in \{-1, -0.9, \ldots, -0.1\}$ (right).

Since the power-law behavior only holds for the high coverage times (long times) of the RSA process, two optimization criterions are used:

$$\text{criterion 1:} \qquad \varepsilon = \sum_{t=0}^{t_{max}} (\theta(t) - \theta_{model}(t))^2 \tag{6}$$

$$\text{criterion 2:} \qquad \varepsilon_{TL} = \sum_{t=t_1}^{t_{max}} (\theta(t) - \theta_{model}(t))^2. \tag{7}$$

In (6) and (7), t_{max} denotes the last time point in the considered RSA data and t_1 denotes the first time value considered for the computation of the criterion.

For the Vicsek fractal, parameter α is imposed equal to $-\frac{1}{d_V}$ and the result of this optimization with criterion ε is given in Fig. 5. The following parameters are obtained: $A = 4.5743 \times 10^{-4}$ and $C = -0.7137$.

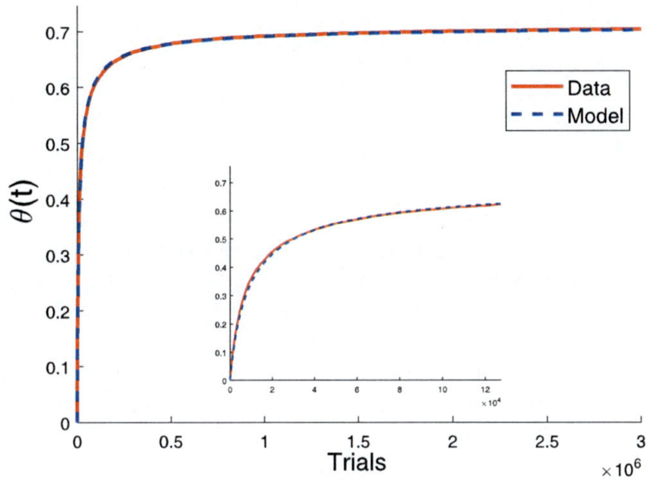

Fig. 5. Comparison of RSA density data with model (3) response with criterion ε.

The same optimization was also done using criterion ε_{TL} and $t_1 = 1 \times 10^5$. A comparison of the RSA density data and of the models obtained with criterion ε and ε_{LT} is done in Fig. 6.

Even if model parameters are computed using criterion ε_{LT}, criterion ε was computed for the resulting model, and conversely, criterion ε_{LT} is computed with the model obtained throught the minimization of criterion ε. The results are gathered in Table 3. This table shows that ε_{LT} is lower when the parameters are computed with ε_{LT}. Model (3) is thus a very good candidate to model fractional dynamics (high coverage times) and realize a compromise when computed with criterion ε.

In order to reduce this error, the model

$$\dot{x}(t) = f(x(t)) \cdot u(t), \tag{8}$$

is now considered, where f is a polynomial function:

$$f(x) = a_0 + a_1 x + a_2 x^2 + a_3 x^3 + a_4 x^4. \tag{9}$$

If $u(t) = 1$, and according to relation (8), this polynomial evaluated in $\theta(t)$ is equal to the derivative of $\theta(t)$. An optimisation is done on the parameters a_i in order to minimize the quadratic error ε.

The function $\dot{\theta}(t)$ computed numerically from data of Fig. 2 and the optimized function f are plotted in Fig. 7 as functions of $\theta(t)$.

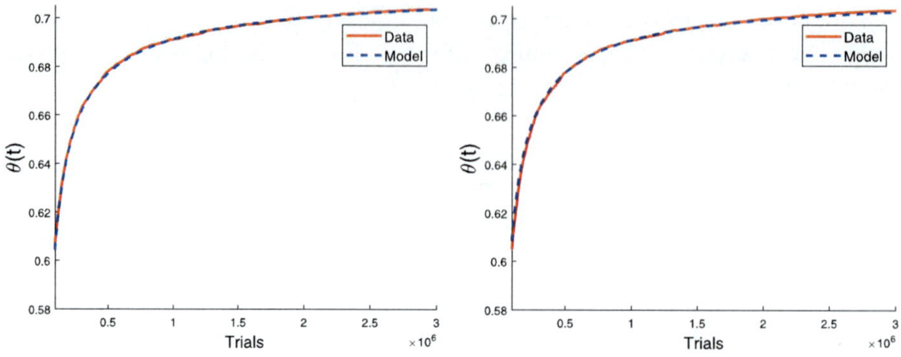

Fig. 6. Comparison of RSA density data with model (3) response optimized with criterion ε_{TL} (left) and with criterion ε (right).

The diagram in Fig. 8 is an implementation of relation (8) where function f is given by relation (9). Such a diagram is used for simulation of the density of adsorbed disks as a function of time.

A comparison of RSA density data and the response of model (8) is proposed by Fig. 9.

The parameters a_0, a_1, a_2, a_3, a_4 given in Table 2 were obtained using MATLAB function `fmincon`.

Table 2. Parameters of f from relation (9)

a_0	a_1	a_2	a_3	a_4
8.2105×10^{-5}	-2.4909×10^{-4}	9.6858×10^{-5}	2.9337×10^{-4}	-2.3250×10^{-4}

Both on the long time and on all the data, the errors are lower than the first approach with ε. Even if the error on the long time is greater than for the optimization done only on the long time (see Table 3), the compromise is better. Considering this, model (8) is better than model (3).

Table 3. Error criterion for the three modeling approaches.

	Model (3) with ε	Model (3) with ε_{LT}	Model (8) with ε
ε_{TL}	2.0669	0.3012	1.6765
ε	5.5433	11.9854	2.2896

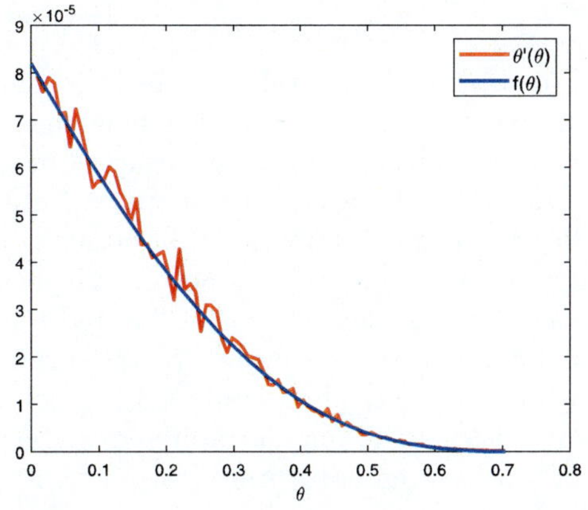

Fig. 7. Approximation of the derivative of $\theta(t)$ (red) and function f evaluate in θ after optimization (blue)

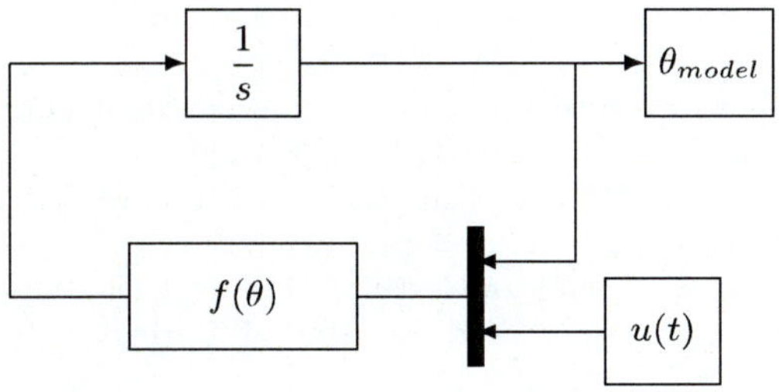

Fig. 8. Block diagram for simulation of model (8)

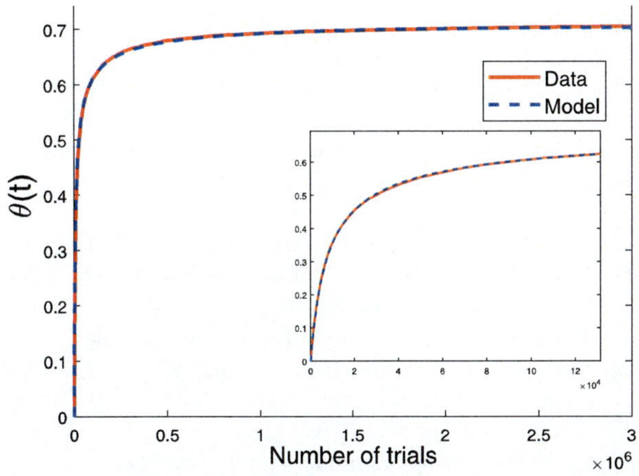

Fig. 9. Comparison between the RSA density data (red) and the model (8) (blue)

4 Conclusion

Random Sequential Adsorption on fractal surfaces is considered in this paper. Numerical simulations confirm that the studied phenomenon exhibits a fractional dynamic linked to the considered fractal surfaces dimension. Two driftless affine in control non linear model are proposed to capture this dynamic. For the first one, its ability to produce a fractional behavior is proved analytically. It is accurate on high coverage regime but not on the beginning of the process. For this reason a more complex but with a limited number of parameters is proposed. Numerical simulations on several fractal surfaces allow to confirm its efficiency. The authors now intend to apply this modeling approach to real adsorption phenomena such as the one encountered in Love wave sensors [9], nitrogen adsorption [14] or oxygen adsorption [15].

Acknowledgments. The authors are thankful to Michal Ciesla for the exchanges which allow us to progress faster.

References

1. Awad, A.M., et al.: Adsorption of organic pollutants by nanomaterial-based adsorbents: an overview. J. Mol. Liq. **301**, 112335 (2020)
2. Bashiri, H., Shajari, A.: Theoretical study of fractal-like kinetics of adsorption. Adsorpt. Sci. Technol. **32**(8), 623–634 (2014)
3. Bonilla-Petriciolet, A., Mendoza-Castillo, D.I., Reynel-Ávila, H.E.: Adsorption Processes for Water Treatment and Purification. Springer, Cham (2017). https://doi.org/10.1007/978-3-319-58136-1
4. Brouers, F., Sotolongo-Costa, O.: Generalized fractal kinetics in complex systems (application to biophysics and biotechnology). Phys. A **368**(1), 165–175 (2006)
5. Ciesla, M., Barbasz, J.: Random sequential adsorption on fractals. J. Chem. Phys. **137**, 044706 (2012)
6. Cieśla, M., Ziff, R.M.: Boundary conditions in random sequential adsorption. J. Stat. Mech. Theory Exp. **2018**(4), 043302 (2018)
7. Feder, J., Giaever, I.: Adsorption of ferritin. J. Colloid Interface Sci. **78**(1), 144–154 (1980)
8. Haerifar, M., Azizian, S.: Fractal-like adsorption kinetics at the solid/solution interface. J. Phys. Chem. C **116**, 13111–13119 (2012)
9. Halil, H., Menini, P., Aubert, H.: Novel microwave gas sensor using dielectric resonator with SnO_2 sensitive layer. Procedia Chem. **1**, 935–938 (2009)
10. Ho, Y.-S., Mckay, G.: The kinetics of sorption of divalent metal ions onto sphagnum moss peat. Water Res. **34**, 735–742 (2000)
11. Ighalo, J.O., Adeniyi, A.G.: Adsorption of pollutants by plant bark derived adsorbents: an empirical review. J. Water Process Eng. **35**, 101228 (2020)
12. Kopelman, R.: Fractal reaction kinetics. Science **241**(4873), 1620–1626 (1988)
13. Lagergren, S.: About the theory of so-called adsorption of soluble substances. Kungliga Svenska Vetenskapsakademiens Handlingar **24**(4), 1–39 (1898)
14. Liu, K., Ostadhassan, M., Jang, H.W., Zakharova, N.V., Shokouhimehr, M.: Comparison of fractal dimensions from nitrogen adsorption data in shale via different models. Roy. Soc. Chem. Adv. **11**(4), 2298–2306 (2021)

15. Lv, X., Liang, X., Xu, P., Chen, L.: A numerical study on oxygen adsorption in porous media of coal rock based on fractal geometry. Roy. Soc. Open Sci. **7**(2), 191337 (2020)
16. Marques, C., Joanny, J.: Adsorption of semi-dilute polymer solutions on fractal colloidal grains. J. Phys. France **49**(7), 1103–1109 (1988)
17. Nikolaou, I., Hallil, H., Conédéra, V., Deligeorgis, G., Dejous, C., Rebiere, D.: Inkjet-printed graphene oxide thin layers on love wave devices for humidity and vapor detection. IEEE Sens. J. **16**(21), 7620–7627 (2016)
18. Qi, H., Ma, J., Wong, P.-Z.: Adsorption isotherms of fractal surfaces. Colloids Surf. A **206**(1), 401–407 (2002)
19. Swendsen, R.H.: Dynamics of random sequential adsorption. Phys. Rev. A **24**, 504–508 (1981)
20. Tartaglione, V., Farges, C., Sabatier, J.: Non linear dynamical modeling of adsorption and desorption processes with power-law kinetics: application to CO_2 capture. Phys. Rev. E **102**, 052102 (2020)
21. Viot, P., Tarjus, G., Ricci, S., Talbot, J.: Random sequential adsorption of anisotropic particles. I. Jamming limit and asymptotic behavior. J. Chem. Phys. **97**, 5212–5218 (1992)
22. Watt-Smith, M., Edler, K., Rigby, S.: An experimental study of gas adsorption on fractal surfaces. Langmuir **21**(6), 2281–2292 (2005)
23. Wu, M.K.: The roughness of aerosol particles: surface fractal dimension measured using nitrogen adsorption. Aerosol Sci. Technol. **25**(4), 392–398 (1996)
24. Zhang, G., Torquato, S.: Precise algorithm to generate random sequential addition of hard hyperspheres at saturation. Phys. Rev. E **88**, 053312 (2013)

Vectorization Calculation Method of the Fractional-Order Mem-Elements

Xiang Xu, Gangquan Si(✉), Zhang Guo, and Babajide Oluwatosin

State Key Laboratory of Electrical Insulation and Power Equipment, Shaanxi Key Laboratory of Smart Grid, School of Electrical Engineering, Xi'an Jiaotong University, Xi'an 710049, Shaanxi, China
sigangquan@mail.xjtu.edu.cn

Abstract. Mem-elements have recently received widespread attention in various cross-cutting fields, but many aspects of the research are still in their infancy, especially fractional-order memory devices. As a promising research field, there are multiple difficulties in the study of fractional-order mem-elements. The calculation process is often very complicated, especially when solving nonlinear and fractional mixed problems. This article proposes a vectorized solution for the superposition problem of fractional-order HP mem-elements. Furthermore, submit a three-dimensional coordinate system calculation method to solve the calculation problem when multiple devices are connected.

Keywords: Fractional-order · Memristor · Vectorization

1 Introduction

Memristor is a passive two-terminal nonlinear circuit element. Its theoretical model was first proposed by Professor Chua of the University of California in 1971 based on the relationship between circuit variables. Theoretically, circuit elements reflect the relationship between magnetic flux and electricity, called memristor [1]. Then the concept of a memristive element is extended to a memristive system that is not limited to the relationship between magnetic flux and electric quantity [2]. Since no corresponding physical memristor has been found, the related theoretical research of memristor did not receive attention. Until 2008, Strukov et al. found physical devices with memristive characteristics [3, 4], which has set off a global research boom on memristors and circuit systems containing memristors. For example, Pershin et al. discovered the spin memory effect in semiconductor spintronic devices and proposed spintronic memristive devices [3]. In 2009, Wang et al. elaborated on three possible examples of magnetoresistive devices and invented one electron spin memristors based on electronic, magnetic properties [5]. Then, Williams et al. declared in 2009 that memristors could be stacked, which can increase the memory capacity of memristive chips by 4–8 times [6]. In 2010, They also proved that the memristor has the ability of mathematical logic operations [7] and pointed out that the combination of the memristor and crossbar latch means that the arithmetic circuit and the memory circuit can coexist at the same time. It is possible to

A. Dzielinski et al. (Eds.): ICFDA 2021, LNNS 452, pp. 106–111, 2022.
https://doi.org/10.1007/978-3-031-04383-3_11

develop a new generation of intelligent computer with memristor. In 2012, Thomas took the memristor as a critical component of the artificial brain, that is, the new generation of computers which is memristive. And for the first time in the literature [8], explained how this kind of neural-like computer transforms natural phenomena into technical systems.

In Thomas' work, the memristor is combined with a transistor to create a more energy-efficient analogue computer to simulate the human brain, completely changing the modern computer information processing method and significantly improving the computer's ability to solve complex problems.

Fractional order mem-element is a new branch which is deriving from the above fields [9, 10]. Some researchers have already done the essential work about their characters. Considering that the memory effect of mem-elements, the technique of fractional calculus is used to analyze memristor characteristics [11]. A chaotic system, while the fractional-order memristor has been used as a critical component, is proposed recently [12]. At the same time, the dynamic characteristics of fractional order mem-elements are discussed [13].

However, there are still many problems that have not been solved. The most representative one is the series-parallel operation of fractional-order mem-elements.

2 Vectorization of Fractional-Order Memristor, Meminductor and Memcapacitor

2.1 HP Model and Fractional Order Mem-Elements

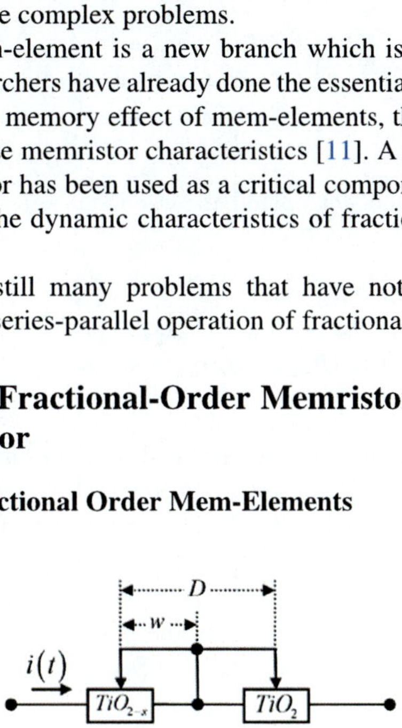

Fig. 1. Schematic diagram of HP memristor

The memristance R_M is calculated by:

$$\begin{cases} R_M = R_{\text{on}}\frac{w(t)}{D} + R_{\mathit{off}}(1 - \frac{w(t)}{D}) = R_{\mathit{off}} + Kq(R_{\text{on}} - R_{\mathit{off}}) = R_0 + R_d x \\ u(t) = R_M i(t) = R_M(q) \cdot dq/dt \end{cases} \tag{1}$$

The current i(t) flowing through the memristor, and its integral is charge q(t). R_{on} and R_{off} respectively represent the high and low resistance states of the memristor. D represents the thickness of the doped region of the device, and w(t) is the state variable. K means the rate factor of the memristor (Fig. 1).

All HP models can be expressed as the formula that initial value and a constant multiplied by the moving distance x of the doped region.

The flux controlled memristor model can be described as:

$$\begin{cases} i = R_M^{-1} \cdot u \\ R_M = R_M + R_d x \\ R_M^{-1} = R_M^{-1} + R_{Md}^{-1}\varphi \\ u = d\varphi/dt = U\cos\omega t \end{cases} \tag{2}$$

Here, u means the excitation voltage, and φ represents the flux.

Similar to the memristor, memcapacitor and meminductor have the same structure:

$$\begin{cases} q = C_M u \\ C_M = C_{\min} + C_d x \\ C_M = C_{M0} + C_{Md}\varphi \\ u = d\varphi/dt \end{cases} \quad (3)$$

$$\begin{cases} i = L_M^{-1} \cdot \varphi \\ L_M = L_{\min} + L_d x \\ L_M = L_{M0} + L_{Md} q \\ \varphi = d\rho/dt \end{cases} \quad (4)$$

When window function is not considered, the fractional-order models of mem-elements are as follows:

$$\begin{cases} B_M = B_0 + B_d x \\ B_M = B_{\min} + B_d(x_0 + kD_t^{-\alpha} v) \\ = (B_{\min} + B_d x_0) + kB_d D_t^{-\alpha} v \\ = B_0 + KD_t^{-\alpha} v \end{cases} \quad (5)$$

Here, B_M can represent any kinds of memristive devices like memristor, meminductor or memcapacitor. The current and voltage are differential of electric charge and magnetic flux respectively. According to the characteristics of the sine function, these parameters in the above formula can be expressed as sine or cosine wave. Therefore, the following derivation is valid.

2.2 The Vectorization of Fractional Order Mem-Elements

By considering the Riemann-Liouville definition and the mathematical method proposed by A.G. Radwan et al. in 2013 [13], the fractional-order sine and cosine function could be calculated as:

$$\begin{cases} \mathcal{D}_t^{-\alpha} \sin(2\pi nft) = \frac{2\pi nft_1^{\alpha+1} F_2\left[1; 1+\frac{\alpha}{2}, \frac{3+\alpha}{2}; -\pi^2 n^2 f^2 t^2\right]}{\Gamma(\alpha+2)} \\ \mathcal{D}_t^{-\alpha} \cos(2\pi nft) = \frac{t_1^{\alpha} F_2\left[1; \frac{\alpha+1}{2}, 1+\frac{\alpha}{2}; -\pi^2 n^2 f^2 t^2\right]}{\Gamma(\alpha+1)} \\ {}_1F_2[a; (b_1, b_2); z] = \sum_{n=0}^{\infty} \frac{(a)_n}{(b_1)_n (b_2)_n} \frac{z^n}{n!}; \left\langle (x)_n = \frac{\Gamma(x+n)}{\Gamma(x)} \right\rangle \end{cases} \quad (6)$$

When t tends to infinity, the whole system tends to stable, and a simplified formula of fractional-order is listed as:

$$\begin{cases} \mathcal{D}_{\mathrm{t}}^{-\alpha} \sin(\omega \mathrm{t}) \approx \omega^{-\alpha} \sin(\omega t - \alpha \frac{\pi}{2}), \\ \mathcal{D}_{\mathrm{t}}^{-\alpha} \cos(\omega \mathrm{t}) \approx \omega^{-\alpha} \cos(\omega t - \alpha \frac{\pi}{2}). \end{cases} \quad (7)$$

Then, for any fractional-order mem-elements, the model can be expressed as:

$$B_M = B_0 + K\omega^{-\alpha} \sin(\omega t - \alpha \frac{\pi}{2}) \quad (8)$$

In this expression, the parameter with a more significant impact on the device characteristics are (B_0, K, α). Assuming that the two devices are connected in series and parallel directly with (B_0, K, α), the parameter calculation will be complicated. The vectorized expression can significantly reduce the computational complexity of this process.

The specific method is to reconstruct the original parameters by the following form:

$$(\rho, \theta) = \left(K\omega^{-\alpha}, \alpha\frac{\pi}{2} + (\eta - 1)\frac{\pi}{2}\right) \tag{9}$$

Here, η is the positive direction of design in the entire system.

After this conversion, the device can directly perform vector superposition and can be converted into the corresponding original parameters (B_0, K, α) through the inverse operation.

In Fig. 2, the picture on the left represents the intrinsic representation of the mem-elements named the hysteresis loop. Three smaller hysteresis loops represent three superimposed devices which represent the memcapacitor in parallel or memristors in series. The same direction makes the memory effect enhanced.

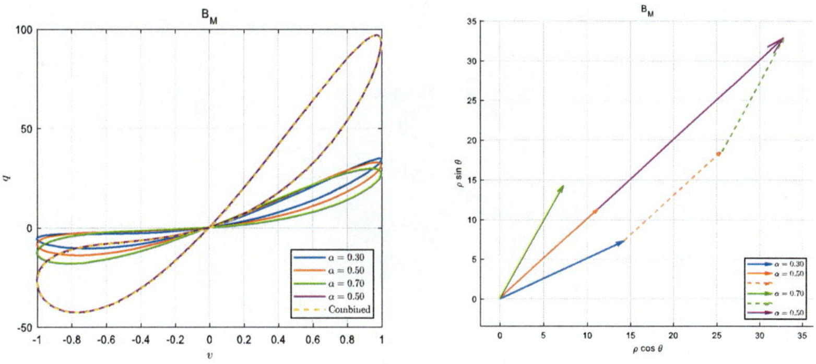

Fig. 2. Schematic diagram of memory devices and vectorization operations

The picture on the right side represents the mem-elements after conversion. Three shorter solid line vectors are superimposed to form a longer real vector, and the dashed line is the translation of the solid vector in the same colour. The result of the conversion calculation is consistent with the calculation result of the traditional superposition method.

For the same kind of mem-elements, this method is equivalent to the operation in the two-dimensional coordinate system, and then further promote this method to the three-dimensional coordinate system:

2.3 Different Kinds of Mem-Elements Superposition

Assuming that there are superposition operations of similar devices, two different devices operations can simultaneously perform in two planes. At the same time, three planes can simultaneously perform superposition operations of three other devices. Similar devices are superimposed. After the completion, only one operation of the device is required to solve the mixed connection operation of the three components of the fractional memristive.

As shown in Fig. 3, according to the calculation method in Sect. 2.2, the three memory devices are placed on different space planes for vector calculations. The results will eventually change into single memristor, meminductor, and memcapacitor calculations, which can significantly simplify fractional-order devices calculations.

3 Conclusion

The research of fractional-order memory devices is vital, but in a specific study, merging multiple fractional-order devices is very difficult and often requires very complicated calculations. The above method dramatically simplifies this process. It can accurately calculate the similar superposition problem of the fractional-order memory device under the condition without window function in the stable system.

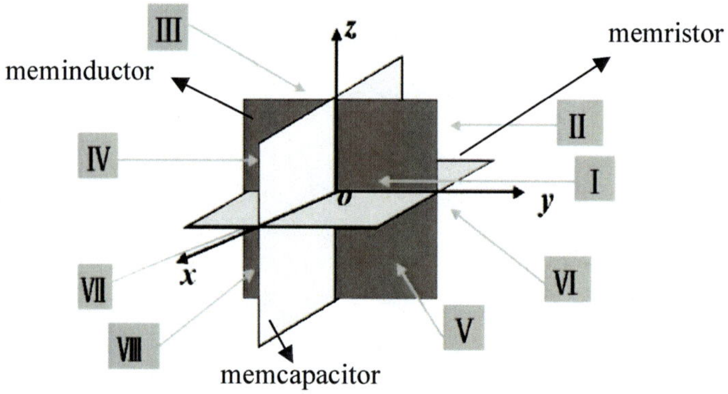

Fig. 3. Three-dimensional coordinate space and the mem-elements

References

1. Chua, L.: Memristor-the missing circuit element. IEEE Trans. Circuit Theory **18**(5), 507–519 (1971)
2. Chua, L.O., Kang, S.M.: Memristive devices and systems. Proc. IEEE **64**(2), 209–223 (1976)
3. Ventra, M.D., Pershin, Y.V., Chua, L.O.: Circuit elements with memory: memristors, memcapacitors, and meminductors. Proc. IEEE **97**(10), 1717–1724 (2009)
4. Strukov, D.B., Snider, G.S., Stewart, D.R., et al.: The missing memristor found. Nature
5. Wang, X.B., Chen, Y.R., Xi, H., Li, H., Dimitrov, D.: IEEE Electron Device Lett. **30**, 294 (2009)
6. Borghetti, J., et al.: Proc. Natl. Acad. Sci. USA **106**, 1699 (2009)
7. Borghetti, J., Snider, G.S., Kuekes, P.J., Yang, J.J., Stewart, D.R., Williams, R.S.: Nature **464**, 873 (2010)

8. Thomas, A.: J. Phys. D: Appl. Phys. **46**, 093001 (2013)
9. Monje, C.A., Chen, Y.Q., Vinagre, B.M., Xue, D.Y., Feliu, V.: Fractional-Order Systems and Controls: Fundamentals and Applications. Springer, London (2010). https://doi.org/10.1007/978-1-84996-335-0
10. Petrá, I., Chen, Y.Q.: Fractional-order circuit elements with memory. In: International Carpathian Control Conference (2012)
11. Zhang, G., Gangquan, S., Xiang, X., et al.: Generalized modeling and character analyzing of composite fractional-order memristors in series connection. Nonlinear Dyn. **95**, 101–115 (2018)
12. Rajagopal, K., Karthikeyan, A., Srinivasan, A.: Dynamical analysis and FPGA implementation of a chaotic oscillator with fractional-order memristor components. Nonlinear Dyn. **91**, 1491–1512 (2017)
13. Fouda, M.E., Radwan, A.G.: Memcapacitor response under step and sinusoidal voltage excitations. Microelectron. J. **45**(11), 1372–1379 (2014)

Which Kind of Fractional Partial Differential Equations Has Solution with Exponential Asymptotics?

Changpin Li(✉), Zhiqiang Li, and Chuntao Yin

Department of Mathematics, Shanghai University, Shanghai 200444, China
lcp@shu.edu.cn

Abstract. It is known that the solutions to the Cauchy problems with integer-order derivative or fractional Caputo derivative defined on $\mathbb{R}^+ \times \mathbb{R}^d$ can behave algebraically asymptotical, and the Cauchy problems with Caputo-Hadamard derivative defined on $\mathbb{R}_a^+ \times \mathbb{R}^d$ can behave logarithmically asymptotical, where $\mathbb{R}_a^+ = [a, +\infty), a > 0$. Does there exist a kind of fractional derivatives such that the corresponding Cauchy problems have exponentially asymptotical solution? The answer is positive. We can introduce a new derivative with exponential kernel such that the associate solution is exponentially asymptotic.

Keywords: Fractional derivative with exponential kernel · Fractional laplacian · Exponential asymptotics

1 Introduction

From the references available, the Cauchy problems with integer-order derivatives or fractional Caputo derivatives generally have algebraically asymptotical solutions [1,5], whilst those with Caputo-Hadamard fractional derivatives have logarithmically asymptotical solutions [2–4]. In this paper, we introduce a fractional derivative with exponential kernel such that the corresponding Cauchy problem defined on $\mathbb{R}_a^+ \times \mathbb{R}^d$ has exponentially asymptotical solution.

Definition 1. *Assume that the given function $f(t) \in L^1(a,b)$ $(-\infty \leq a < b \leq \infty)$. Then the left and right fractional integrals with exponential kernel are defined respectively by*

$$ {}_{e}\mathrm{D}_{a,t}^{-\alpha} f(t) = \frac{1}{\Gamma(\alpha)} \int_a^t (e^t - e^\tau)^{\alpha-1} f(\tau) e^\tau \mathrm{d}\tau,\ t > a, \tag{1}$$

and

$$ {}_{e}\mathrm{D}_{t,b}^{-\alpha} f(t) = \frac{1}{\Gamma(\alpha)} \int_t^b (e^\tau - e^t)^{\alpha-1} f(\tau) e^\tau \mathrm{d}\tau,\ t < b. \tag{2}$$

A. Dzielinski et al. (Eds.): ICFDA 2021, LNNS 452, pp. 112–117, 2022.
https://doi.org/10.1007/978-3-031-04383-3_12

Definition 2. *Let $n-1<\alpha<n\in\mathbb{N}$. Assume that the given function $f(t)\in AC^n_{\delta_e}[a,b]=\{f:[a,b]\to\mathbb{R}:\delta_e^{n-1}f(t)\in AC[a,b]\}$ $(-\infty\le a<b\le\infty)$. Then the left and right Riemann–Liouville-type fractional derivatives of order α with exponential kernel are defined respectively in the following way,*

$$ {}_e\mathrm{D}^{\alpha}_{a,t}f(t)=\frac{1}{\Gamma(n-\alpha)}\delta_e^n\int_a^t(e^t-e^{\tau})^{n-\alpha-1}f(\tau)e^{\tau}\mathrm{d}\tau,\ t>a, \tag{3}$$

and

$$ {}_e\mathrm{D}^{\alpha}_{t,b}f(t)=\frac{(-1)^n}{\Gamma(n-\alpha)}\delta_e^n\int_t^b(e^{\tau}-e^{t})^{n-\alpha-1}f(\tau)e^{\tau}\mathrm{d}\tau,\ t<b, \tag{4}$$

where $\delta_e^n f(t)=\left(e^{-t}\frac{\mathrm{d}}{\mathrm{d}t}\right)f(t)$, $\delta_e^n f(t)=\delta_e(\delta_e^{n-1}f(t))$, and $\delta_e^0 f(t)=f(t)$.

Definition 3. *Let $n-1<\alpha<n\in\mathbb{N}$. Assume that the given function $f(t)\in AC^n_{\delta_e}[a,b]$ $(-\infty\le a<b\le\infty)$. Then the left and right Caputo-type fractional derivatives of order α with exponential kernel are defined respectively as follows,*

$$ {}_{Ce}\mathrm{D}^{\alpha}_{a,t}f(t)=\frac{1}{\Gamma(n-\alpha)}\int_a^t(e^t-e^{\tau})^{n-\alpha-1}\delta_e^n f(\tau)e^{\tau}\mathrm{d}\tau, \tag{5}$$

and

$$ {}_{Ce}\mathrm{D}^{\alpha}_{t,b}f(t)=\frac{(-1)^n}{\Gamma(n-\alpha)}\int_t^b(e^{\tau}-e^{t})^{n-\alpha-1}\delta_e^n f(\tau)e^{\tau}\mathrm{d}\tau. \tag{6}$$

In the following, we introduce the amended Laplace transform and its inverse one for the left fractional integral and derivatives.

Definition 4. *Assume that a function $f(t)$ is defined on $[a,+\infty)$ $(a\in\mathbb{R})$. The amended Laplace transform of $f(t)$ is defined by*

$$\overline{f}(\lambda)=\mathscr{L}_a[f(t),\lambda]=\int_a^{\infty}e^{-(e^{\lambda}-e^{a})(e^{t}-e^{a})}f(t)e^{t}\mathrm{d}t,\ \lambda\in\mathbb{C}. \tag{7}$$

Then the inverse amended Laplace transform of $f(t)$ is given by

$$f(t)=\mathscr{L}_a^{-1}[\overline{f}(\lambda),t]=\frac{1}{2\pi i}\int_{c-i\infty}^{c+i\infty}e^{(e^{\lambda}-e^{a})(e^{t}-e^{a})}\overline{f}(\lambda)e^{\lambda}\mathrm{d}\lambda,\ c>0. \tag{8}$$

Definition 5. *Assume that functions $f(t)$ and $g(t)$ are defined on $[a,+\infty)$ $(a\in\mathbb{R})$. The convolution of amended Laplace transform of $f(t)$ and $g(t)$ is defined by*

$$f(t)*_a g(t)=\int_a^t f(\tau)g(\log[e^t+e^a-e^{\tau}])e^{\tau}\mathrm{d}\tau. \tag{9}$$

From the above definitions, we have following lemmas.

Lemma 1. *Let* $\mathscr{L}_a[f(t),\lambda]=\overline{f}(\lambda)$ *and* $\mathscr{L}_a[g(t),\lambda]=\overline{g}(\lambda)$. *Then*

$$\mathscr{L}_a[f(t)*_a g(t),\lambda]=\overline{f}(\lambda)\overline{g}(\lambda)=\mathscr{L}_a[f(t),\lambda]\mathscr{L}_a[g(t),\lambda],$$

or equivalently,

$$\mathscr{L}_a^{-1}[\overline{f}(\lambda)\overline{g}(\lambda),t]=f(t)*_a g(t).$$

Proof. The proof is straightforward so is omitted here.

Lemma 2. *Let* $n-1<\alpha<n\in\mathbb{N}$ *and* $a\in\mathbb{R}$. *Then one has:*

$$\mathscr{L}_a[{}_e\mathrm{D}_{a,t}^{-\alpha}f(t),\lambda]=(e^\lambda-e^a)^{-\alpha}\mathscr{L}_a[f(t),\lambda], \tag{10}$$

$$\begin{aligned}\mathscr{L}_a[{}_e\mathrm{D}_{a,t}^{\alpha}f(t),\lambda]=&(e^\lambda-e^a)^{\alpha}\mathscr{L}_a[f(t),\lambda]\\&-\sum_{k=0}^{n-1}(e^\lambda-e^a)^{n-k-1}\,{}_e\mathrm{D}_{a,t}^{\alpha+k-n}f(t)|_{t=a},\end{aligned} \tag{11}$$

$$\mathscr{L}_a[{}_{Ce}\mathrm{D}_{a,t}^{\alpha}f(t),\lambda]=(e^\lambda-e^a)^{\alpha}\mathscr{L}_a[f(t),\lambda]-\sum_{k=0}^{n-1}(e^\lambda-e^a)^{\alpha-k-1}\delta_e^k f(a). \tag{12}$$

Proof. The proof is quite similar to that of Lemma 2.1 in [4] so is omitted here.

The amended Laplace transform of Mittag-Leffler function is

$$\begin{aligned}&\int_a^\infty e^{-(e^\lambda-e^a)(e^t-e^a)}(e^t-e^a)^{\alpha k+\beta-1}E_{\alpha,\beta}^{(k)}(\pm\zeta(e^t-e^a)^\alpha)e^t\mathrm{d}t\\=&\frac{k!(e^\lambda-e^a)^{\alpha-\beta}}{((e^\lambda-e^a)^\alpha\mp\zeta)^{k+1}},\ \mathrm{Re}(\lambda)>|\zeta|^{\frac{1}{\alpha}},\end{aligned} \tag{13}$$

where $E_{\alpha,\beta}(y)$ denotes the Mittag-Leffler function and $E_{\alpha,\beta}^{(k)}(y)=\frac{\mathrm{d}^k}{\mathrm{d}y^k}E_{\alpha,\beta}(y)$.

Definition 6. *Assume that a function* $f(t)$ *is defined on* $[a,\infty)$ $(a\in\mathbb{R})$. *The amended Mellin transform of* $f(t)$ *is defined by*

$$\widetilde{f}(\xi)=\mathscr{M}_a[f(t),\xi]=\int_a^\infty(e^t-e^a)^{\xi-1}f(t)e^t\mathrm{d}t,\ \gamma_1<\mathrm{Re}(\xi)<\gamma_2. \tag{14}$$

The inverse amended Mellin transform is given by

$$f(t)=\mathscr{M}_a^{-1}[\widetilde{f}(\xi),t]=\frac{1}{2\pi i}\int_{c-i\infty}^{c+i\infty}(e^t-e^a)^{-\xi}\widetilde{f}(\xi)\mathrm{d}\xi,\ t>a,\ c=\mathrm{Re}(\xi). \tag{15}$$

It is east to verify that the following relation holds,

$$\mathscr{M}_a[f(t),\xi]=\frac{1}{\Gamma(1-\xi)}\mathscr{M}_a[\mathscr{L}_a[f(t),\lambda],1-\xi]. \tag{16}$$

Similarly, we can also define the amended Laplace and Mellin transforms for the right fractional integral and derivatives.

In the following, we consider the Cauchy problem

$$\begin{cases} {}_{Ce}\mathrm{D}_{a,t}^{\alpha}u(\mathrm{x},t) = -(-\Delta)^s u(\mathrm{x},t),\ \mathrm{x} \in \mathbb{R}^d,\ t > a(\geq 0), \\ u(\mathrm{x},a) = u_a(\mathrm{x}),\ \mathrm{x} \in \mathbb{R}^d, \end{cases} \tag{17}$$

where $\alpha \in (0,1)$, $s \in (0,1)$, and the initial value $u_a(\mathrm{x})$ is a given function. ${}_{Ce}\mathrm{D}_{a,t}^{\alpha}$ denotes Caputo-type derivative with exponential kernel as before and the fractional Laplacian $(-\Delta)^s$ is the standard fractional Laplacian defined by an integral [2]. In the next section, we show the above Cauchy problem (17) has an exponentially asymptotical solution.

2 Exponential Asymptotics

This section shows the asymptotic behavior of the solution to Eq. (17). We first deduce the fundamental and analytical solutions to Eq. (17).

Applying amended Laplace transform (7) and standard Fourier transform to Eq. (17), and noticing equalities (12) and (15) in [2], which yields

$$\overline{\widehat{u}}(\omega,\lambda) = \frac{(e^{\lambda}-e^{a})^{\alpha-1}}{(e^{\lambda}-e^{a})^{\alpha}+|\omega|^{2s}}\widehat{u}_a(\omega) := \overline{\widehat{G}}_a(\omega,\lambda)\widehat{u}_a(\omega),$$

where $\widehat{u}_a(\omega)$ is Fourier transform of $u_a(\mathrm{x})$.

In view of the inverse amended Laplace transform and inverse Fourier transform, we obtain

$$u(\mathrm{x},t) = G_a(\mathrm{x},t) * u_a(\mathrm{x}) := \int_{\mathbb{R}^d} G_a(\mathrm{x}-\mathrm{y},t)u_a(\mathrm{y})\mathrm{dy}. \tag{18}$$

Employing the relation (16) and using the almost same lines as the fundamental solution (18) in [2], it follows that

$$G_a(\mathrm{x},t) = \frac{1}{|\mathrm{x}|^d \pi^{\frac{d}{2}}} H_{23}^{21}\left(\frac{|\mathrm{x}|^{2s}}{2^{2s}(e^t-e^a)^{\alpha}} \middle| \begin{matrix} (1,1);(1,\alpha) \\ (1,1),(\frac{d}{2},s);(1,s) \end{matrix} \right), \tag{19}$$

where $H_{23}^{21}(z)$ denotes the Fox H-function [2,3].

The following two lemmas present estimates of the fundamental solution $G_a(\mathrm{x},t)$. For convenience, we denote $||\cdot||_p = ||\cdot||_{L^p(\mathbb{R}^d)}$ and $||\cdot||_{p,\infty} = ||\cdot||_{L^{p,\infty}(\mathbb{R}^d)}$, where $L^{p,\infty}(\mathbb{R}^d)$ represents a weak $L^p(\mathbb{R}^d)$ space [2].

Lemma 3. *Let $d \in \mathbb{N}$, $\alpha \in (0,1)$, $s \in (0,1)$, $t > a$, and $R = (e^t-e^a)^{-\alpha}|\mathrm{x}|^{2s}$. Then the fundamental solution $G_a(\mathrm{x},t)$ given by (19) satisfies the following asymptotic behavior:*

$$|G_a(\mathrm{x},t)| \leq C(e^t-e^a)^{\alpha}|\mathrm{x}|^{-d-2s},\ d \geq 1,\ 0 < s < 1, \tag{20}$$

if $R > 1$, *and*

$$|G_a(\mathrm{x},t)| \leq \begin{cases} C(e^t - e^a)^{-\alpha}|\mathrm{x}|^{-d+2s}, \ d > 2s, \\ C(e^t - e^a)^{-\alpha}\left(1 + \left|\log\left(\frac{1}{2}(e^t - e^a)^{-\alpha}|x|\right)\right|\right), \ d = 2s, \\ C(e^t - e^a)^{-\frac{\alpha}{2s}}, \ d < 2s, \end{cases} \tag{21}$$

if $R \leq 1$.

Proof. The proof can be referred to that of Theorem 3.1 in [2].

Lemma 4. *Let* $d \in \mathbb{N}$, $\alpha \in (0,1)$, *and* $s \in (0,1)$, *and*

$$\kappa(d,s) = \begin{cases} \frac{d}{d-2s}, & d > 2s, \\ \infty, & d \leq 2s. \end{cases}$$

(i) Assume that $1 \leq p < \kappa(d,s)$. *Then* $G_a(\mathrm{x},t) \in L^p(\mathbb{R}^d)$ *for each* $t > a$, *and there holds*

$$||G_a(\mathrm{x},t)||_p \leq C(e^t - e^a)^{-\frac{\alpha d}{2s}(1-\frac{1}{p})}, \ t > a. \tag{22}$$

(ii) If $p = \frac{d}{d-2s}$ *and* $d > 2s$, *then* $G_a(\mathrm{x},t) \in L^{\frac{d}{d-2s},\infty}(\mathbb{R}^d)$ *for each* $t > a$, *and there holds*

$$||G_a(\mathrm{x},t)||_{\frac{d}{d-2s},\infty} \leq C(e^t - e^a)^{-\alpha}, \ t > a. \tag{23}$$

Proof. The proof is almost the same as those of Theorems 3.3 and 3.5 in [2].

Remark 1. If $d < 2s$, the inequality (21) implies that $G_a(x,t) \in L^\infty(\mathbb{R})$ for any $t > a$, and it holds that $||G_a(x,t)||_\infty \leq C(e^t - e^a)^{-\frac{\alpha}{2s}}$.

Theorem 1. *Let* $d \in \mathbb{N}$, $\alpha \in (0,1)$, *and* $s \in (0,1)$. *Suppose that* $u_a \in L^q(\mathbb{R}^d)$ *with* $1 \leq q \leq \infty$. *Then* $u(\mathrm{x},t) = G_a(\mathrm{x},t) * u_a(\mathrm{x})$ *satisfies the following estimates:*

(i) If $q = \infty$, *then*

$$||u(\mathrm{x},t)||_\infty \leq C||u_a(\mathrm{x})||_\infty, \ t > a.$$

(ii) When $1 \leq q < \infty$, *for each* $r \in \left[q, \frac{dq}{d-2sq}\right)$ *if* $d > 2sq$, *or* $r \in [q,\infty)$ *if* $d = 2sq$, *or* $r \in [q,\infty]$ *if* $d < 2sq$, *then*

$$||u(\mathrm{x},t)||_r \leq C(e^t - e^a)^{-\frac{\alpha d}{2s}(\frac{1}{q}-\frac{1}{r})}||u_a(\mathrm{x})||_q, \ t > a. \tag{24}$$

Moreover, if $d > 2sq$, *then*

$$||u(\mathrm{x},t)||_{\frac{dq}{d-2sq},\infty} \leq C(e^t - e^a)^{-\alpha}||u_a(\mathrm{x})||_q, \ t > a. \tag{25}$$

Proof. Let $1 \leq p, q, r \leq \infty$ and

$$1 + \frac{1}{r} = \frac{1}{p} + \frac{1}{q}. \tag{26}$$

(i) In the case $q = \infty$, from the equality (26) we choose $r = \infty$ and $p = 1$, then Lemma 4 and Young's inequality (57) for convolution [3] show that

$$||u(\mathrm{x},t)||_\infty \le ||G_a(\mathrm{x},t)||_1 ||u_a(\mathrm{x})||_\infty \le C||u_a(\mathrm{x})||_\infty, \; t > a.$$

(ii) We now prove the case $1 \le q < \infty$. If $d > 2sq$, by the relation (26) and $1 \le p < \frac{d}{d-2s}$ with $d > 2s$, one deduces $q \le r < \frac{dq}{d-2sq}$. It follows from Lemma 4 and Young's inequality (57) for convolution [3] that

$$||u(\mathrm{x},t)||_r \le C(e^t - e^a)^{-\frac{\alpha d}{2s}(\frac{1}{q}-\frac{1}{r})} ||u_a(\mathrm{x})||_q, \; t > a.$$

If $d = 2sq$ and $1 \le p < \infty$ with $d = 2s$, then the relation (26) implies $r \in [q, \infty)$, so Lemma 4 and Young's inequality (57) for convolution [3] immediately yield (24). When $d < 2sq$ and $1 \le p < \infty$ with $d < 2s$, by (26) and Remark 1, we find that the result (24) also holds for any $r \in [q, \infty]$.

Finally, if $d > 2sq$ and $r = \frac{dq}{d-2sq}$, by (26), one gets $p = \frac{d}{d-2s}$ with $d > 2s$. Therefore, Lemma 4 and Young's inequality (58) for weak L^p norm [3] give

$$||u(\mathrm{x},t)||_{\frac{dq}{d-2sq},\infty} \le C||G_a(\mathrm{x},t)||_{\frac{d}{d-2s},\infty} ||u_a(\mathrm{x})||_q \le C(e^t - e^a)^{-\alpha} ||u_a(\mathrm{x})||_q, \; t > a,$$

which completes the proof.

Theorem 1 shows that the solution to Eq. (17) is exponentially asymptotic.

3 Conclusion

In this article, we construct a kind of fractional derivatives such that the solution of the corresponding Cauchy problem with such a kind of derivatives is exponentially asymptotic, which subverts the common view that the solutions to the usual fractional ordinary/partial differential equations only behave algebraically asymptotic.

References

1. Kemppainen, J., Siljander, J., Zacher, R.: Representation of solutions and large-time behavior for fully nonlocal diffusion equations. J. Diff. Equ. **263**, 149–201 (2017). https://doi.org/10.1016/j.jde.2017.02.030
2. Li, C.P., Li, Z.Q.: Asymptotic behaviors of solution to Caputo-Hadamard fractional partial differential equation with fractional Laplacian. Int. J. Comput. Math. **98**(2), 305–339 (2021). https://doi.org/10.1080/00207160.2020.1744574
3. Li, C.P., Li, Z.Q.: Asymptotic behaviors of solution to partial differential equation with Caputo-Hadamard derivative and fractional Laplacian: hyperbolic case. Discrete Cont. Dyn-S **14**(10), 3659–3683 (2021). https://doi.org/10.3934/dcdss.2021023
4. Li, C.P., Li, Z.Q., Wang, Z.: Mathematical analysis and the local discontinuous Galerkin method for Caputo-Hadamard fractional partial differential equation. J. Sci. Comput. **85**(2), 41 (2020). https://doi.org/10.1007/s10915-020-01353-3
5. Ma, Y.T., Zhang, F.R., Li, C.P.: The asymptotics of the solutions to the anomalous diffusion equations. Comput. Math. Appl. **66**, 682–692 (2013). https://doi.org/10.1016/j.camwa.2013.01.032

Detection of a Fractal Element with Complex-Conjugated Power-Law Exponents in Living Systems: Analysis of the Temporal Evolution of Impedance Measurements in the Unblown Bud Plumeria Flower (Frangipani Plumeria)

Raoul R. Nigmatullin[1(✉)], Samyadip Sarkar[2], and Karabi Biswas[1,2]

[1] Radioelectronics and Informative-Measurements Techniques Department, Kazan National Research Technical University (KNITU-KAI) named by A.N. Tupolev, Karl Marx Street 10, Kazan 420111, Tatarstan, Russia
renigmat@gmail.com

[2] Department of Electrical Engineering, Indian Institute of Technology Kharagpur, Kharagpur, West Bengal 721302, India
karabi@ee.iitkgp.ac.in

Abstract. In this paper the proposed theoretical model describing 3D-branching systems in the intermediate range of frequencies was applied in the first time for description and further fitting of the complex impedance in a living system, as unblown bud plumeria flower (Frangipani Plumeria). The proposed fitting function that follows from the applied model allows to describe completely with high accuracy (the fitting error is less than 0.25%) the whole stages of the temporal evolution of the bud plumeria flower in room conditions (T = 30 ± 1 °C, with relative humidity located in the interval [50% ≤ H ≤ 70%] during 19 days (25.03.21–12.04.21) of the impedance evolution measurements. The reason of selection of this plumeria flower taken in the form of the unblown bud is related to the fact that all possible conducting channels form approximately 3D-*heterogeneous* branching structure. The fitting function contains the power-law exponent with complex-conjugated parts and serves as the first evidence of detection of the fractal element in the living systems. Two stages as normal and quasi-chaotic were detected during 123 measurements. The basic fitting parameters that can detect these stages have been found. The qualitative explanation of the processes that take place in this living system and does not contradict with measured data is given.

Keywords: 3D-branshing/fractal systems · Impedance measurements in the unblown bud flower · New fractal element with complex-conjugated power law exponents

A. Dzielinski et al. (Eds.): ICFDA 2021, LNNS 452, pp. 118–124, 2022.
https://doi.org/10.1007/978-3-031-04383-3_13

1 The Theory of the Branching System in the Frequency Domain

In this section we want to show in brief how to generalize the theory of the branching systems developed previously in time domain [1]. The basic expression is expressed as

$$\begin{aligned} P_N(z) &= \prod_{n=-N}^{N} \left(f\left(z\xi^n\right)^{n_0(b(z))^n} \right) \\ &= \exp\left[n_0 \sum_{n=-N}^{N} (b(z))^n \ln\left[f\left(z\xi^n\right)\right] \right] \cong \exp S(z), \\ S(z) &= \lim_{N>>1} S_N(z). \end{aligned} \tag{1}$$

Here independent variable $z = i\omega$ associated with complex frequency, relaxation/transfer microscopic function $f(z)$ describes microscopic impedance of some independent block of a complex circuit, the log-periodic function $b(z\xi) = b(z)$ describes a possible branching structure that exists in the complex circuit considered. We take into account the appearance of the branching factor $n_0 b(z)^n$ in the product (1). The bio-models considered in details in the review [2] are based on presumably in consideration of 2D-plain RC-models. Highly likely, the electric circuits corresponding to large number of bio-models should have 3D conducting channels distribution. Expression (1) can serve as a specific "reflection" of 3D electric branching structure in 1D-frequency domain. We accept it as the working *hypothesis* and discuss its significance in the last section. Based on asymptotic behavior of the microscopic function $f(z)$ it becomes possible to evaluate expression (1) and obtain expression that can be suitable for the description of the complex impedances in the intermediate range of frequencies having a branching structure. We *suppose* that the microscopic/transfer function has the following asymptotic behavior

$$\begin{aligned} &\mathrm{Re}|z| << 1 \\ &f(z) \cong c_1 z^{\alpha} + c_2 z^{1+\alpha} + \ldots \\ &f(z) \cong c_0 + c_1 z + \ldots \end{aligned} \tag{2a}$$

$$\begin{aligned} &\mathrm{Re}|z| >> 1 \\ &f(z) \cong A_1 z^{-\gamma} \exp(-rz) + A_2 z^{-\gamma-1} \exp(-2rz) + \ldots \\ &f(z) \cong A_0 + A_1 z^{-1} + \ldots \end{aligned} \tag{2b}$$

In these expressions the power-law exponents $0 < \alpha < 1$ and $0 < \gamma \leq 1$ provide the desired asymptotic behavior for small and large values for the complex frequency $z = i\omega$.

In contrast to the asymptotic requirements considered in paper [1], we chose more general behavior at small and large values of variable z that are given by expressions (2a) and (2b). In the asymptotic behavior of the function $f(z)$ (the first line of (2b)) we consider two possibilities (exponential and power-law) scenario, simultaneously. Neglecting by the second terms in decompositions (2a) and (2b), one can obtain the evaluations of the limiting frequencies that define the mesoscale frequency region where (1) is valid

$$\omega_{\min} \cong \left(\min\left(\left|\frac{A_1}{A_0}\right|, \left|\frac{A_2}{A_1}\right|\right)\right) < \omega < \left(\max\left(\left|\frac{c_0}{c_1}\right|, \left|\frac{c_1}{c_2}\right|\right)\right) \cong \omega_{\max} \tag{3}$$

One can show that the sum $S(z)$ in (1) satisfies to the following functional equation (details are given in the Mathematical Appendix below in Sect. 4).

$$S(z\xi) = \frac{1}{b_R} S(z) + \mu \ln(z) + gz + X_0 \tag{4}$$

Here the parameters μ, g and X_0 are defined by the following expressions that are tightly related with asymptotic behavior (2a) and (2b) of the microscopic function $f(z)$.

$$\mu = -2n_0\left(\frac{\alpha}{b_{\min}} - \frac{\gamma}{b_{\max}}\right), g = -\frac{n_0 r}{b_{\max}}, X_0 = -\ln\left(c_1^{w_1} A_1^{w_2}\right),$$
$$w_1 = \frac{2n_0}{b_{\min}}, \ w_2 = \frac{2n_0}{b_{\max}}. \tag{5}$$

The solution of Eq. (4) obtained with the help of the uncertain coefficients method can be written as

$$\begin{aligned} &P(z) = \exp(S(z)), \\ &S(z) = z^{\nu} \Pr(\ln z) + C_1 \ln z + C_2 z + C_3, \\ &\nu = -\frac{\ln(b_R)}{\ln \xi}, C_1 \cong \frac{\mu}{1-(b_R)^{-1}}, C_2 \cong \frac{g}{\xi - (b_R)^{-1}}, \\ &C_3 = \frac{X_0 - C_1 \ln \xi}{1-(b_R)^{-1}}. \end{aligned} \tag{6}$$

Here $\Pr(\ln z \pm \ln \xi) = \Pr(\ln z)$ is log-periodic function. Extracting the real and imaginary parts from expression (6) we obtain finally

$$\begin{aligned} &\mathrm{Re}[S(x)] = Ac_0^{(r)} e^{\nu x} + \sum_{k=1}^{K} \left[Ac_k^{(r)} Fc_k(x) + As_k^{(r)} Fs_k(x)\right] + C_1 x + C_3, \\ &\mathrm{Im}[S(x)] = C_1 \frac{\pi}{2} + Ac_0^{(i)} e^{\nu x} + \sum_{k=1}^{K} \left[Ac_k^{(i)} Fc_k(x) + As_k^{(i)} Fs_k(x)\right] + C_2 e^{x}, \\ &x = \ln(\omega), \ Fc_k(x) = \exp(\nu x)\cos(\Omega_k x), Fs_k(x) = \exp(\nu x)\sin(\Omega_k x), \ \Omega_k = \frac{2\pi k}{\ln \xi}. \end{aligned} \tag{7}$$

Here the following expressions in (7) can be used as the fitting functions. After the fitting of expressions (7) the final fit for the measured complex impedance is obtained from the formulae

$$\mathrm{Re}(Z(\omega)) - i\,\mathrm{Im}(Z(\omega))$$

$$= \exp(\text{Re}\, S(i\omega)) + i \exp(\text{Im}\, S(i\omega)) \tag{8}$$

The very limited frame of this paper does not allow to show all details of the fitting of the impedance measurements $m = 1, 2, \ldots, (M = 123)$ realized during 19 days. These time- consuming measurements show the evolution of the unblown bud flower from the fresh state to the almost drying state when the unblown bud flower loses a large amount of nutrient solvent/free water that is distributed over open/closed channels. The functions (7) realized the fit with the high accuracy (the maximal value of the relative error does not exceed 0.25%).

2 The Basic Plots Demonstrating the Fit of the Found Functions Eq. (7, 8) to the Measured Data

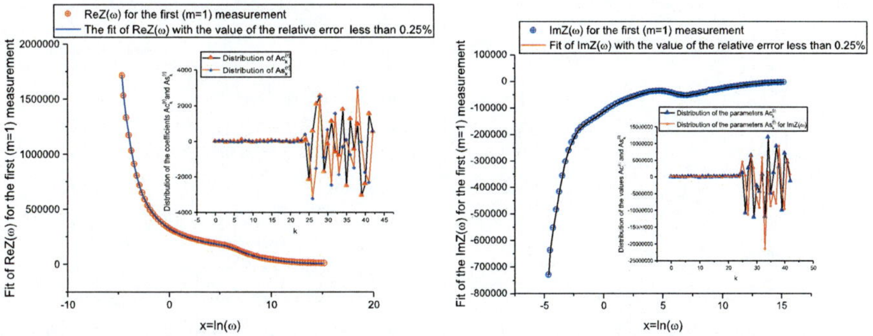

Fig. 1. On the left we demonstrate the fit of the real part of the impedance for $m = 1$. The fit of the imaginary part is shown on the right. In the small figures inside we show the distribution of the decomposition coefficients $Ac_k{}^{(r,i)}$ and $As_k{}^{(r,i)}$ figuring in (7). Starting from the 50-th measurement the character of the impedance behavior is abruptly changed

We cannot show the evolution of all parameters figuring in expression (5). However, the evolution of the ranges measured in Ohms is important for understanding the basic impedance evolution (Fig. 1).

3 The Results Obtained

In this short paper we cannot give the extended analysis of the results obtained. However, one can formulate the basic points that can help understand deeper the behavior of the 3D-branching heterogeneous system as unblown bud flower.

We *suppose* that the internal structure of a living unblown bud has specific onion-like layers divided by air/nutrient solvent. Each layer, in turns, contains a set of conducting channels full of specific nutrient solution/free water with the dissolved ions (H^+, K^+, Na^+, Ca^{++} and others) inside together with necessary organic substances that are necessary for supporting the vital activity of a bud flower in the process of its growth. In the beginning

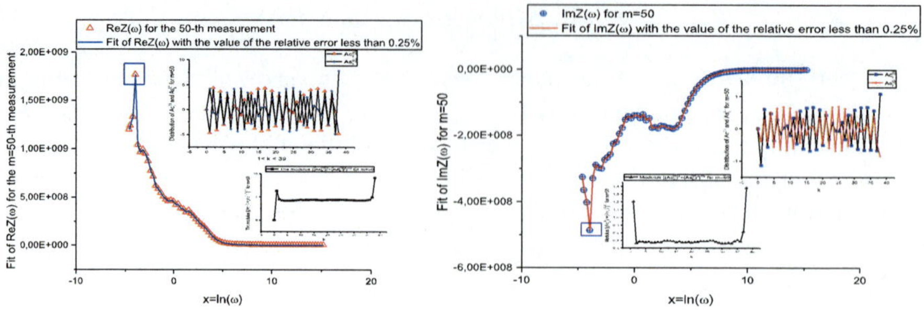

Fig. 2. On the left we show the fit of the real part for m = 50. On the right we show the fit for the imaginary part. On the small figures we show again the distribution of the coefficients $Ac_k{}^{(r,i)}$ and $As_k{}^{(r,i)}$ figuring in (7). The small figures below demonstrate the modulus of these decomposition coefficients.

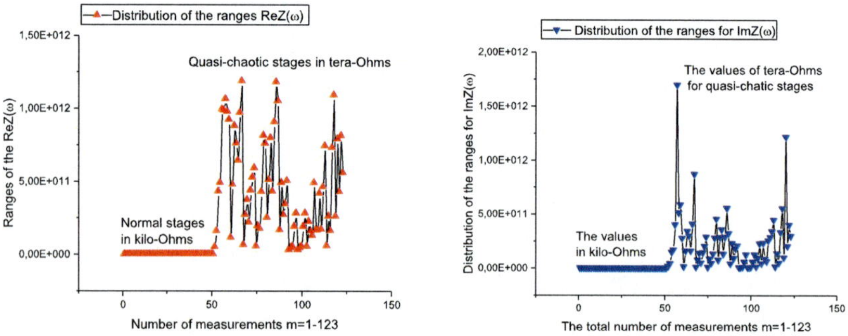

Fig. 3. These plots divide clearly the normal state (with resistance having kilo-Ohms) from the quasi-chaotic states when the resistance increases up-to the values of tera-Ohms. On the left we show the distribution of the ranges for the real part of ReZ(ω), while the distribution of the ranges for the imaginary part ImZ(ω) is shown on the right-hand side.

of the measurements, we have excess of the nutrient solution and, therefore, the value of the resistance between applied electrodes is low (~kilo-Ohms). During the five days ($m = 1, 2, \ldots, 49$), the behavior of the complex impedance can be characterized as a "normal" one. Starting from the 50-th measurement the behavior of the impedance is abruptly changed (Fig. 2). We associate these changes with the fact that part of the conducting channels is closed; some portions of light carriers are decreased and, therefore, the total resistance is essentially increased. It is clearly seen from the Fig. 3, when in the normal state we have resistance located in the limits of kilo-Ohms, while in quasi-chaotic state the resistance achieves the range of giga/tera-Ohms(!) and the almost dry flower bud becomes a 'perfect' isolator. Highly likely, in these quasi-chaotic stages ($m = 50, 51, \ldots, 123$) only heavy ions can survive as potential carriers and the most conducting channels are closed. Concluding this section, we want to highlight the following results that were found in this research.

1. Discovery of a new fractal element with complex-conjugated power-law exponents that can exist in living systems. Earlier, we considered only theoretically a possibility of the simplified version of a fractal element with complex-conjugated power-law exponents (without verification on real data) in paper [3].
2. The proposed theory of the blow-like signals was applied successfully in the frequency domain for description of evolution of 3D-heterogeneus branching system as unblown plumeria bud, taken as an example.
3. All measured data (even including outliers) are fitted completely to the proposed fitting functions (expressions (7, 8)) with the value of the relative error less than 0.25%.
4. We give a qualitative explanation of the conductive processes that do not contradict with the measured impedance data.
5. In the first time we discovered the appearance of two power-law exponents ν (as a "fractal" exponent) and μ (that can be determined as an "asymptotic" exponent). See their definitions in (5) and (6). They govern by the impedance behavior of the fitting function (8). These exponents have clear physical meaning and can be important in construction of new types of the fractional integrals with complex-conjugated terms. These new types of integrals and derivatives can enrich the exiting theory of the fractional calculus and prompt new direction in its theoretical research.

4 Mathematical Appendix

4.1 How to Obtain the Functional Eq. (4)?

Inserting the multiplier ξ we obtain the following expression

$$\begin{aligned}
S(z\xi) &= n_0 \sum_{n=0}^{N} [b(z)]^n \ln\left[f\left(z\xi^{N+1}\right)\right] + n_0 \sum_{n=-N}^{-1} [b(z)]^n \ln\left[f\left(z\xi^{N+1}\right)\right] = \\
&\overset{n+1\to n'}{=} n_0 \sum_{n=1}^{N+1} [b(z)]^{n-1} \ln\left[f\left(z\xi^{n}\right)\right] + n_0 \sum_{n=0}^{-N+1} [b(z)]^{n-1} \ln\left[f\left(z\xi^{n}\right)\right] \\
&= \frac{1}{b(z)} S(z) - \frac{2n_0}{b(z)} \ln[f(z)] + n_0 [b(z)]^N \ln\left[f\left(z\xi^{N+1}\right)\right] \\
&+ n_0 [b(z)]^{-N} \ln\left[f\left(z\xi^{-N+1}\right)\right] \cong \frac{1}{b(z)} S(z) - \frac{2n_0}{b(z)} \ln[f(z)]
\end{aligned} \tag{A1}$$

We took into account the contribution of "asymmetric" term at $n = 0$ and small contributions of the last two terms because of asymptotic decompositions (2a) and (2b) for small (2a) and large (2b) values of the modulus z. The term at $\xi = 0$ gives the desired values of μ, g and X_0. We took in account also an assumption that the function $b(z) = b(z\xi^{\pm 1})$ is the limited and identified with log-periodic function.

4.2 How to Obtain the Solution (6) of the Functional Eq. (4)?

We are looking for the solution of the Eq. (4) in the form

$$S(z) = z^{\nu} \Pr(\ln z) + C_1 \ln(z) + C_2 z + C_3 \tag{A2}$$

Substitution of solution (A2) gives the relations for unknown independent coefficients

$$\xi^{\nu} = \frac{1}{b_R}, \; C_1\left(1 - b_R^{-1}\right) = \mu,$$
$$C_2\left(\xi - b_R^{-1}\right) = g, \; C_3\left(1 - b_R^{-1}\right) = X_0 - C_1 \ln \xi \tag{A3}$$

From these relations one can obtained easily the desired solutions (6) for unknown coefficients $C_{1,2,3}$. In the last expression (A3) we replace approximately the log-periodic finite function $b(z)$ by its range $b(z) \cong b_R = \max(b(z)) - \min(b(z))$. We should stress again that the power-law exponent ν has a direct *fractal* nature, while the second power-law exponent μ reflects the asymptotic behavior of the microscopic function $f(z)$ at small and large values of the complex frequency $z = i\omega$.

References

1. Nigmatullin, R.R., Toboev, V.A., Lino, P., Maione, G.: Reduced fractal model for quantitative analysis of averaged micromotions in mesoscale: characterization of blow-like signals. Chaos, Solitons Fractals **76**, 166–181 (2015)
2. Mohsen, M., Said, L.A., Madian, A.H., Radwan, A.G., Elwakil, A.S.: Fractional-order bio-impedance modeling for interdisciplinary applications: a review. IEEE Access **9**, 33158-33168 (2021)
3. Nigmatullin, R.R., Le Mehaute, A.: Is there a geometrical/physical meaning of the fractional integral with complex exponent? J. Non-Cryst. Solids **351**, 2888–2899 (2005)

Prabhakar Discrete-Time Generalization of the Time-Fractional Poisson Process and Related Random Walks

Thomas M. Michelitsch[1], Federico Polito[2](✉), and Alejandro P. Riascos[3]

[1] Sorbonne Université, Institut Jean le Rond d'Alembert, CNRS UMR 7190, 4 Place Jussieu, 75252 Paris Cedex 05, France
michel@lmm.jussieu.fr

[2] Department of Mathematics "Giuseppe Peano", University of Torino, Via Carlo Alberto 10, 10123 Torino, Italy
federico.polito@unito.it

[3] Instituto de Física, Universidad Nacional Autónoma de México, Apartado Postal 20-364, 01000 Ciudad de México, Mexico
aperezr@fisica.unam.mx

Abstract. In recent years a huge interdisciplinary field has emerged which is devoted to the 'complex dynamics' of anomalous transport with long-time memory and non-markovian features. It was found that the framework of fractional calculus and its generalizations are able to capture these phenomena. Many of the classical models are based on continuous-time renewal processes and use the Montroll–Weiss continuous-time random walk (CTRW) approach. On the other hand their discrete-time counterparts are rarely considered in the literature despite their importance in various applications. The goal of the present paper is to give a brief sketch of our recently introduced discrete-time Prabhakar generalization of the fractional Poisson process and the related discrete-time random walk (DTRW) model. We show that this counting process is connected with the continuous-time Prabhakar renewal process by a ('well-scaled') continuous-time limit. We deduce the state probabilities and discrete-time generalized fractional Kolmogorov-Feller equations governing the Prabhakar DTRW and discuss effects such as long-time memory (non-markovianity) as a hallmark of the 'complexity' of the process.

Keywords: Discrete-time counting process · Prabhakar general fractional calculus · Long-time memory effect · Non-markovian random walks

1 Introduction

It is well-known that phenomena of anomalous transport and diffusion are not compatible with the classical picture of exponentially distributed waiting time patterns. Indeed it was found that random walks subordinated to an independent

A. Dzielinski et al. (Eds.): ICFDA 2021, LNNS 452, pp. 125–131, 2022.
https://doi.org/10.1007/978-3-031-04383-3_14

renewal process with fat-tailed waiting time densities (in other words Montroll–Weiss continuous-time random walk (CTRW) models formulated in the framework of fractional calculus) offer an appropriate description (see e.g. [1–4]). Further, these models have profound connections with semi-Markov processes [5]. In order to obtain a greater flexibility to describe real-world situations, some generalized models have emerged [9], among them the so-called Prabhakar generalization [10,12–14]. In most cases such as the Montroll–Weiss CTRW [6] the time variable is continuous. However, there are many systems in nature having intrinsically discrete time scales; such cases are rarely considered in the literature. Essential elements of this theory were developed recently [7] (see also the references therein).

In the present letter we give a brief sketch of our recent model of a Prabhakar discrete-time generalization of the fractional Poisson process and its application to stochastic motions on undirected graphs. We refer the reader to our recent paper [8] for details.

2 The Prabahakar Discrete-Time Counting Process

We consider a counting process with integer arrival times characterized by the variable

$$\mathcal{J}_n = \sum_{j=1}^{n} Z_j, \quad Z_j \in \mathbf{N} = \{1, 2, \dots\}, \quad \mathcal{J}_0 = 0, \tag{1}$$

where Z_j are IID non-zero integer inter-arrival times following the distribution $\theta_\alpha^{(\nu)}(t, \xi)$, $t \in \mathbf{N}$, $\nu > 0$, $\alpha \in (0, 1]$, $\xi > 0$, and having generating function

$$\bar{\theta}_\alpha^{(\nu)}(u) = \bar{f}(u)\bar{\varphi}_\alpha^{(\nu)}(u), \quad \bar{\varphi}_\alpha^{(\nu)}(u) = \frac{\xi^\nu}{(\xi + (1-u)^\alpha)^\nu}. \tag{2}$$

In this expression we have $\bar{f}(u) = u \sum_{k=1}^{\infty} f(k) u^{k-1}$ with $\bar{f}(u)|_{u=1} = 1$ fulfilling the desired initial condition $\bar{f}(u)|_{u=0} = f(t)\big|_{t=0} = 0$, ensuring almost sure nonzero inter-arrival times. In (2) the parameter ξ defines a characteristic time-scale and crucial here is that $\bar{f}(u)$ is free of the parameter ξ.

Consider the simplest case: $\bar{f}(u) = u$ with $\bar{\theta}_\alpha^{(\nu)}(u) = u\bar{\varphi}_\alpha^{(\nu)}(u)$. In this case, we call the counting process '*Prabhakar discrete-time process (PDTP)*'. The PDTP stands out since it contains several classical cases: for $\nu = 1$ and $0 < \alpha < 1$, it recovers the '*fractional Bernoulli process*' with the 'discrete-time Mittag–Leffler distribution' (of so-called 'type A') DML_A [7], and for $\nu = 1$ and $\alpha = 1$ the standard Bernoulli counting process. The PDTP waiting time density then has the form $\theta_\alpha^{(\nu)}(t, \xi) = \varphi_\alpha^{(\nu)}(t-1, \xi)$ with

$$\varphi_\alpha^{(\nu)}(k,\xi) = \begin{cases} \dfrac{\xi^\nu}{k!}\displaystyle\sum_{m=0}^{\infty}\frac{(-1)^m(\nu)_m\xi^m}{m!}\frac{\Gamma[\alpha(m+\nu)+k]}{\Gamma[\alpha(m+\nu)]}, & 0<\xi<1, \\ \dfrac{(-1)^k}{k!}\displaystyle\sum_{m=0}^{\infty}\frac{(-1)^m(\nu)_m\xi^{-m}}{m!}\frac{\Gamma[\alpha m+1]}{\Gamma[\alpha m-k+1]}, & \xi>1, \end{cases} \tag{3}$$

$k \in \mathbf{N}_0 = \{0,1,2,\ldots\}$ and where we employed the Pochhammer symbol $(c)_m = \Gamma(c+m)/\Gamma(c)$. The key point is that (3) and the entire class defined by (2) are discrete approximations of the so called Prabhakar or generalized Mittag–Leffler density, which is recovered by the well-scaled continuous-time limit [8]

$$\chi_{\alpha,\nu}(t) = \lim_{h\to 0}\frac{1}{h}\varphi_\alpha^{(\nu)}\left(\frac{t}{h},\xi_0 h^\alpha\right) = \lim_{h\to 0}\frac{1}{h}\theta_\alpha^{(\nu)}\left(\frac{t}{h},\xi_0 h^\alpha\right) = \xi_0^\nu t^{\nu\alpha-1}E_{\alpha,\nu\alpha}^{\nu}(-\xi_0 t^\alpha)$$

($\alpha \in (0,1],\ \nu > 0,\ \xi > 0,\ t \in h\mathbf{N} \to \mathbf{R}^+$). The multiplier h^{-1} reflects the fact that this is a density (i.e. of physical dimension sec^{-1}). Note that the above expression contains the Prabhakar function $E_{a,b}^c(z) = \sum_{m=0}^\infty \frac{(c)_m}{m!}\frac{z^m}{\Gamma(am+b)}$ ($\Re\{a\} > 0, \Re\{b\} > 0, c,\ z \in \mathbf{C}$) [15]. The continuous-time Prabhakar counting process with inter-arrival time density $\chi_{\alpha,\nu}(t)$ was first introduced by Cahoy and Polito as a generalization of the fractional Poisson process [10]. Here we call this process '*Generalized Fractional Poisson Process (GFPP)*'. The GFPP was applied to random walks on undirected graphs by Michelitsch and Riascos [13,14]. We point out that the time-scale parameter $\xi(h) = \xi_0 h^\alpha$ is re-scaled in such a way that a well-scaled continuous-time limit to the Prabhakar density exists and a new time-scale parameter $\xi_0 > 0$ of physical dimension $sec^{-\alpha}$ and independent of h comes into play. In the same way as above, we can establish the connection with the so called Prabhakar general fractional calculus by introducing the discrete-time Prabhakar kernel by $p_{a,b}^c(\mu,t) = \frac{1}{t!}\frac{d^t}{du^t}\left[\bar{f}(u)(1-u)^{-b}(1-\mu(1-u)^{-a})^{-c}\right]\big|_{u=0}$ where a well-scaled continuous-time limit to the so called Prabhakar kernel exists (see [12] for an outline of Prabhakar general fractional calculus and the references therein). By simple conditioning arguments and using the connection of generating functions with discrete-time convolutions one obtains the PDTP state probability generating function $\bar{\Phi}_{\alpha,\nu}^{(n)}(u,\xi) = \frac{(1-u\bar{\varphi}_\alpha^{(\nu)}(u))}{(1-u)}u^n\bar{\varphi}_\alpha^{(n\nu)}(u)$, which in turn leads to the PDTP state probabilities (probabilities of having n arrivals within $[0,t]$) as follows. If $t < n$, $\Phi_{\alpha,\nu}^{(n)}(t,\xi) = 0$, otherwise

$$\Phi_{\alpha,\nu}^{(n)}(t,\xi) = \frac{\Theta(t-n)\xi^{n\nu}}{(t-n)!} \sum_{m=0}^{\infty} \frac{(-1)^m (n\nu)_m \xi^m}{m!} \frac{\Gamma[\alpha(m+n\nu)+1+t-n]}{\Gamma[\alpha(m+n\nu)+1]}$$

$$-\frac{\Theta(t-n-1)\xi^{(n+1)\nu}}{(t-n-1)!} \sum_{m=0}^{\infty} \frac{(-1)^m ([n+1]\nu)_m \xi^m}{m!} \frac{\Gamma[\alpha[m+(n+1)\nu]+t-n]}{\Gamma(\alpha[m+(n+1)\nu]+1)},$$

for $0 < \xi < 1$, while for $\xi > 1$

$$\Phi_{\alpha,\nu}^{(n)}(t,\xi) = \frac{\Theta(t-n)(-1)^{t-n}}{(t-n)!} \sum_{m=0}^{\infty} \frac{(-1)^m (n\nu)_m \xi^{-m}}{m!} \frac{\Gamma[\alpha m]}{\Gamma[\alpha m-(t-n)]}$$

$$-\frac{\Theta(t-n-1)(-1)^{t-n-1}}{(t-n-1)!} \sum_{m=0}^{\infty} \frac{(-1)^m ([n+1]\nu)_m \xi^{-m}}{m!} \frac{\Gamma[\alpha m]}{\Gamma[\alpha m-(t-n-1)]}, \tag{4}$$

where $n, t \in \mathbf{N}_0$ and we introduced the discrete-time version of the Heaviside step function $\Theta(r)$ with $\Theta(r) = 1$ for $r \geq 0$ and $\Theta(r) = 0$ otherwise. Note that $\Theta(0) = 1$. We observe that the state probabilities are non-zero only for $0 \leq n \leq t$ reflecting the general feature that there cannot be more arrivals than time increments in a discrete-time counting process. Crucial are the initial condition $\Phi_{\alpha,\nu}^{(n)}(t,\xi)\big|_{t=0} = \delta_{n0}$ (with the Kronecker symbol δ_{ij}) and the normalization condition $\sum_{n=0}^{t} \Phi_{\alpha,\nu}^{(n)}(t,\xi) = 1$. The PDTP state probabilities (4) are an approximation of the GFPP state probabilities (the latter first obtained by Cahoy and Polito [10] and consult also [13,14]). The probabilities (4) are connected to the GFPP state probabilities by the well-scaled continuous-time limit

$$\Phi_{\alpha,\nu}^{(n)}(t,\xi_0)_{ct} = \lim_{h\to 0} \Phi_{\alpha,\nu}^{(n)}\left(\tfrac{t}{h}, \xi_0 h^{\alpha}\right), \quad n \in \mathbf{N}_0,\ t \in h\mathbf{N}_0 \to \mathbf{R}^+$$

$$= (\xi_0 t^{\alpha})^{n\nu} \left\{ E_{\alpha,\alpha n\nu+1}^{n\nu}(-\xi_0 t^{\alpha}) - (\xi_0 t^{\alpha})^{\nu} E_{\alpha,\alpha(n+1)\nu+1}^{(n+1)\nu}(-\xi_0 t^{\alpha}) \right\}. \tag{5}$$

The continuous-time GFPP state probabilities are plotted in Fig. 1 for different parameters. One can see that higher states n are 'less occupied' at time t.

Worthy of mention is also the universal power-law long-time asymptotic feature

$$\Phi_{\alpha,\nu}^{(n)}(t,\xi) \sim \frac{\nu}{\xi} \frac{t^{-\alpha}}{\Gamma(1-\alpha)}, \qquad \alpha \in (0,1), \tag{6}$$

which is independent of state n, i.e. at large times all states have the same power-law tail behavior. This is caused by the long-time memory and non-markovianity of the PDTP and induces a fat-tail in the inter-arrival time density $\theta_{\alpha}^{(\nu)}(t,\xi) \sim \frac{\alpha\nu}{\xi_0} \frac{t^{-\alpha-1}}{\Gamma(1-\alpha)}$, $\alpha \in (0,1)$, which is of the same type as for the Mittag–Leffler density.

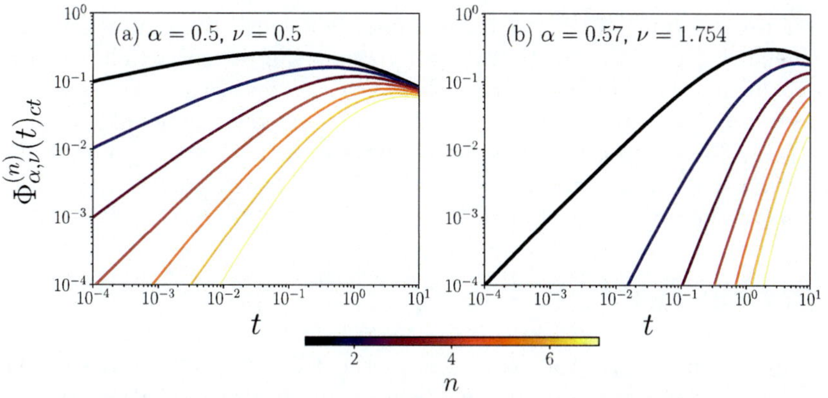

Fig. 1. GFPP state probabilities $\Phi^{(n)}_{\alpha,\nu}(t)_{ct}$ of (5) versus t for different values of $n = 1, 2, \ldots, 7$ and $\xi_0 = 1$. (a) $\alpha = 0.5$ and $\nu = 0.5$, (b) $\alpha = 0.57$ and $\nu = 1.754$.

3 Prabhakar Discrete Time Random Walk Model

Here we consider a random walk on undirected connected graphs of N nodes subordinated to an independent PDTP. The resulting walk which we call "Prabhakar Discrete-time Random Walk" (Prabhakar DTRW) is a discrete-time variant of the classical Montroll–Weiss CTRW [6]. The jumps of the walker from node $i \to j$ are governed by the $N \times N$ one-step transition matrix $\mathbf{H}$ with the elements $H_{ij} = \frac{1}{K_i} A_{ij}$, where K_i and A_{ij} indicate the degree and (in undirected graphs symmetric) adjacency matrix, respectively (see [16–18] for details). The $N \times N$ Prabhakar DTRW transition matrix then writes

$$\mathbf{P}^{\alpha,\nu}(t,\xi) = \sum_{n=0}^{t} \mathbf{H}^n \Phi^{(n)}_{\alpha,\nu}(t,\xi), \qquad P_{ij}(t)\big|_{t=0} = \delta_{ij}, \qquad t \in \mathbf{N}_0, \tag{7}$$

where this series stops at $n = t$ due to the above discussed properties of the state probabilities. Thus, (7) is a matrix polynomial in $\mathbf{H}$ of degree t. The matrix element $P^{\alpha,\nu}_{ij}(t,\xi)$ indicates the probability that the walker is present on node j at time t when having started the walk at $t = 0$ on node i. The following generalized fractional difference equation holds:

$$\hat{\mathcal{D}}^{\nu}_{\alpha} \cdot \mathbf{P}^{\alpha,\nu}(t,\xi) - \mathcal{M}_{\alpha,\nu}(t,\xi)\mathbf{1} = \mathbf{H} \cdot \mathbf{P}^{\alpha,\nu}(t-1,\xi), \qquad P^{\alpha,\nu}_{ij}(0) = \delta_{ij}. \tag{8}$$

In the above equation, $\mathcal{M}_{\alpha,\nu}(t,\xi)$ is a memory kernel and $\hat{\mathcal{D}}^{\nu}_{\alpha}$ denotes a Prabhakar generalized fractional difference operator, see [8] for their explicit forms and further details. The well-scaled continuous-time limit of the transition matrix (7), defined by $\mathbf{P}^{\alpha,\nu}(t,\xi_0)_{ct} = \lim_{h\to 0} \mathbf{P}^{\alpha,\nu}(\frac{t}{h}, \xi_0 h^{\alpha})$, yields the transition matrix of the walk subordinated to the GFPP which is a Montroll–Weiss CTRW governed by a generalized fractional Kolmogorov-Feller (KF) equation involving Prabhakar kernels. This generalized fractional KF equation is the well-scaled continuous-time limit of (8) and has the form of a generalized diffusion equation with memory (see [11], Eq. (5.2)) (and consult the review article [12] as well as [8,13,14] for the CTRW model). Well-known classical cases are contained as special cases. The choice $\nu = 1$, $\alpha \in (0,1)$ recovers the fractional KF equation (see [1], Eq. (40)) and $\nu = 1$, $\alpha = 1$ the Markovian (memoryless) limit with the classical KF equation of normal diffusion (Fick's second law). The reader may also consult the upcoming paper [21].

4 Conclusions

Discrete-time counting processes such as the PDTP and the related DTRW models open a large interdisciplinary field, including generalizations of existing models such as in biased random walks [19], pandemic spread [20], anomalous relaxation and transport, just to name a few, and generally in the modelling of 'complex systems'.

References

1. Metzler, R., Klafter, J.: The random walk's guide to anomalous diffusion: a fractional dynamics approach. Phys. Rep. **339**, 1–77 (2000)
2. Gorenflo, R., Mainardi, F.: On the fractional Poisson process and the discretized stable subordinator. Axioms **4**(3), 321–344 (2015)
3. Beghin, L., Orsingher, E.: Fractional Poisson processes and related planar random motions. Electron. J. Probab. **14**(61), 1790–1826 (2009)
4. Laskin, N.: Fractional Poisson process. Commun. Nonlinear Sci. Numer. Simul. **8**(3–4), 201–213 (2003)
5. Lévy, P.: Processus semi-Markovien. Proc. Int. Congr. Math. **3**, 416–426 (1956)
6. Montroll, E.W., Weiss, G.H.: Random walks on lattices II. J. Math. Phys **6**(2), 167–181 (1965)
7. Pachon, A., Polito, F., Ricciuti, C.: On discrete-time semi-Markov processes. Discret. Contin. Dyn. Syst. B **26**(3), 1499–1529 (2021)
8. Michelitsch, T.M., Polito, F., Riascos, A.P.: On discrete time Prabhakar-generalized fractional Poisson processes and related stochastic dynamics. Phys. A. **565**, 125541 (2021). (arXiv:2005.06925 [math.PR])
9. Michelitsch, T.M., Polito, F., Riascos, A.P.: Biased continuous-time random walks with Mittag-Leffler jumps. Fractal Fract. **4**, 51 (2020). https://doi.org/10.3390/fractalfract4040051
10. Cahoy, D.O., Polito, F.: Renewal processes based on generalized Mittag-Leffler waiting times. Commun. Nonlinear Sci. Numer. Simul. **18**(3), 639–650 (2013)
11. Sandev, T., Metzler, R., Chechkin, A.: From continuous time random walks to the generalized diffusion equation. Fract. Calc. Appl. Anal. **21**(1), 10–28 (2018)
12. Giusti, A., Colombaro, I., Garra, R., Garrappa, R., Polito, F., Popolizio, M., Mainardi, F.: A practical guide to Prabhakar fractional calculus. Fract. Calc. Appl. Anal. **23**(1), 9–54 (2020)
13. Michelitsch, T.M., Riascos, A.P.: Continuous time random walk and diffusion with generalized fractional Poisson process. Phys. A. Statist. Mech. App. **545**, 123294 (2020). (arXiv:1907.03830)

14. Michelitsch, T.M., Riascos, A.P.: Generalized fractional Poisson process and related stochastic dynamics. Fract. Calc. Appl. Anal. **23**(3), 656-693 (2020). (arXiv:1906.09704 [cond-mat.stat-mech])
15. Prabhakar, T.R.: A singular integral equation with a generalized Mittag-Leffler function in the kernel. Yokohama Math. J. **19**, 7–15 (1971)
16. Noh, J.D., Rieger, H.: Random walks on complex networks. Phys. Rev. Lett. **92**, 118701 (2004)
17. Michelitsch, T., Riascos, A.P., Collet, B., Nowakowski, A., Nicolleau, F.: Fractional Dynamics on Networks and Lattices. ISTE-Wiley, Hoboken (2019). https://doi.org/10.1002/9781119608165. (ISBN: 9781786301581)
18. Riascos, A.P., Mateos, J.L.: Fractional dynamics on networks: emergence of anomalous diffusion and Lévy flights. Phys. Rev. E **90**, 032809 (2014)
19. Riascos, A.P., Michelitsch, T.M., Pizarro-Medina, A.: Nonlocal biased random walks and fractional transport on directed networks. Phys. Rev. E. **102**, 022142 (2020). (arXiv:2004.00575 [cond-mat.stat-mech])
20. Bestehorn, M., Riascos, A.P., Michelitsch, T.M., Collet, B.A.: A Markovian random walk model of epidemic spreading. Cont. Mech. Thermodyn. **33**(4), 1207–1221 (2021). https://doi.org/10.1007/s00161-021-00970-z, (arXiv:2010.07731 [physics.soc-ph])
21. Gajda, J., Beghin, L.: Prabhakar-Lévy processes. Statist. Probab. Lett. **178**, 109162 (2021)

Check for updates

A Temporal Second-Order Scheme for Time Fractional Mixed Diffusion and Wave Equation with an Initial Singularity

Rui-lian Du[1] and Zhi-zhong Sun[2(✉)]

[1] Changzhou University, Changzhou 213164, China
[2] Southeast University, Nanjing 211189, China
zzsun@seu.edu.cn

Abstract. A Crank-Nicolson numerical approximation on graded meshes for time fractional mixed diffusion and wave equation with two Caputo derivatives of fractional order: one belonging to $(0, 1)$, the other belonging to $(1, 2)$, is established and analyzed. We derive Crank-Nicolson formulae to approximate fractional operators, which can obtain second-order accuracy. Error estimates are analyzed strictly. The main contribution of this paper is to show that the Crank-Nicolson numerical approximation for the model problem under consideration is temporal second-order accurate over graded meshes. Numerical tests are presented to support our theoretical results.

Keywords: Time fractional mixed diffusion and wave equation with an initial singularity · Crank-Nicolson formula · Error estimate

1 Model Problem

In this paper, a temporal second-order numerical scheme on graded meshes is established for solving the following time fractional mixed diffusion and wave problem [1,2]:

$$\begin{cases} {}_0^C D_t^{\alpha+1} u(x,t) + {}_0^C D_t^{\beta} u(x,t) = u_{xx}(x,t) + f(x,t), \quad x \in \Omega,\ 0 < t \le T, & (1) \\ u(x,0) = \phi(x), \quad u_t(x,0) = \varphi(x), \quad x \in \bar{\Omega} = \Omega \cup \partial\Omega, & (2) \\ u(x,t) = 0, \quad x \in \partial\Omega,\ 0 < t \le T, & (3) \end{cases}$$

where $\alpha, \beta \in (0, 1)$, $\Omega = (x_l, x_r)$ is an interval with its boundary $\partial\Omega$, $f(x,t)$ denotes the source term, $\phi(x)$ and $\varphi(x)$ are given functions. Caputo fractional derivative ${}_0^C D_t^{\gamma} u(x,t)$ with respect to t is defined

$$ {}_0^C D_t^{\gamma} u(x,t) = \int_0^t \omega_{1-\gamma}(t-s) \partial_s u(x,s) \mathrm{d}s, \quad 0 < \gamma < 1 \tag{4}$$

A. Dzielinski et al. (Eds.): ICFDA 2021, LNNS 452, pp. 132–140, 2022.
https://doi.org/10.1007/978-3-031-04383-3_15

$$ {}_0^C D_t^\gamma u(x,t) = \int_0^t \omega_{2-\gamma}(t-s)\partial_s^2 u(x,s)\mathrm{d}s, \quad 1<\gamma<2, \tag{5}$$

with the kernel

$$\omega_\nu(t) = t^{\nu-1}/\Gamma(\nu), \quad t>0.$$

In any numerical methods for solving the problem (1)–(3), a key consideration is that the solution is typically weak regular than the case for classical PDEs, see details in [3]. To the best of our knowledge, various types of L1 schemes for the problem (1)–(3) were developed and analyzed [1,2,4,5]. In most studies, a convergence rate of order $2-\alpha$ $(0<\alpha<1)$ was proved. Recently, some temporal second-order schemes were obtained [6,7]. However, the numerical methods for the problem (1)–(3) with an initial weak regularity have not appeared. The motivation of this paper is to propose and analyze a temporal second-order accurate Crank-Nicolson (C-N) numerical approximation on graded meshes for solving the model problem (1)–(3) with an initial weak regularity.

Under proper regularity and compatibility assumptions, the time fractional mixed diffusion and wave problem (1)–(3) has a unique classical solution, and there exists a constant C such that

$$|\partial_t^l u(x,t)| \le Ct^{\sigma-l}, \quad l=0,1,2,3, \tag{6}$$

where $\sigma \in (1,2)$ is a regularity parameter. Throughout the paper, $C_l\,(l=1,2,\cdots)$ are positive constants independent of the time and space step sizes.

2 Crank-Nicolson Formulae and Error Analyses

Firstly, we transform problem (1) into an equivalent problem. Introduce Riemann-Liouville fractional integral of order α

$$ {}_0I_t^\alpha f(t) = \int_0^t \omega_\alpha(t-s)f(s)ds, \quad t>0.$$

Similarly to [8,9], acting ${}_0I_t^\alpha$ on both hand sides of (1), we obtain

$$u_t(x,t) + {}_0^C D_t^{\beta-\alpha}u(x,t) = \varphi(x) + {}_0I_t^\alpha u_{xx}(x,t) + F(x,t), \tag{7}$$

where $F(x,t) = \int_0^t \omega_\alpha(t-s)f(x,s)\mathrm{d}s$.

2.1 C-N Formulae to Approximate Fractional Operators

We focus on presenting the numerical formulae for fractional operators in (7). For simplicity, we omit the spatial variable x for a while. Apply time-graded meshes with the following nodes:

$$t_n = (n\tau)^r, \quad 0\le n\le N,\ r\ge 1, \quad \tau = T^{1/r}/N,$$

where N is the number of subintervals. Denote by $\tau_n = t_n - t_{n-1}$ the length of the n-th subinterval $\mathrm{I}_n = (t_{n-1}, t_n)$ for $1 \le n \le N$. It is not difficult to show that such a time-graded mesh has the following properties:

$$t_n \le 2^r t_{n-1},\ r\tau t_{n-1}^{1-1/r} \le \tau_n \le r\tau t_n^{1-1/r},\ \tau_n - \tau_{n-1} \le C_r \tau^2 \min\{1, t_n^{1-2/r}\},\ n \ge 2, \tag{8}$$

where C_r is a positive constant that may depend on r.

For a given function v defined on the time level $[0,T]$, let $v^n = v(t_n)$ for $0 \le n \le N$. Denote

$$\nabla_\tau v^n = v^n - v^{n-1}, \quad \partial_\tau v^{n-\frac{1}{2}} = \nabla_\tau v^n / \tau_n, \quad v^{n-\frac{1}{2}} = (v^n + v^{n-1})/2.$$

Let $(\Pi_{1,j} v)(t)$ be the linear interpolant of a function $v(t)$ at two nodes t_{j-1} and t_j, i.e.,

$$\Pi_{1,j} v = v^j \frac{t - t_{j-1}}{\tau_j} + v^{j-1} \frac{t_j - t}{\tau_j}.$$

Define

$$\Pi_1 v = \Pi_{1,j} v, \quad t_{j-1} < t \le t_j,\ j \ge 1.$$

It is easy to know that

$$(\Pi_1 v)'(t) = \partial_\tau v^{j-\frac{1}{2}}, \quad t_{j-1} < t \le t_j, \quad j \ge 1.$$

We define a discrete fractional derivative

$$(\partial_\tau^{\beta-\alpha} v)^{n-\frac{1}{2}} = \frac{1}{\tau_n} \int_{t_{n-1}}^{t_n} \int_0^t \omega_{1+\alpha-\beta}(t-s)(\Pi_1 v)'(s)\mathrm{d}s\mathrm{d}t = \sum_{j=1}^{n} \bar{a}_{n-j}^{(n,\beta)} \nabla_\tau v^j, \tag{9}$$

where

$$\bar{a}_{n-j}^{(n,\beta)} = \frac{1}{\tau_n \tau_j} \int_{t_{n-1}}^{t_n} \int_{t_{j-1}}^{\min\{t,t_j\}} \omega_{1+\alpha-\beta}(t-s)\mathrm{d}s\mathrm{d}t > 0, \quad 1 \le j \le n$$

and $(\partial_\tau^{\beta-\alpha} v)^{n-\frac{1}{2}}$ is designed so that

$$\sum_{j=1}^{n} (\nabla_\tau v^j)(\partial_\tau^{\beta-\alpha} v)^{j-\frac{1}{2}} = \int_0^{t_n} (\Pi_1 v)'(t) \int_0^t \omega_\alpha(t-s)(\Pi_1 v)'(s)\mathrm{d}s\mathrm{d}t \ge 0. \tag{10}$$

This identity ensures that the discrete quadratic form on the left hand side of (10) is positive semi-definite.

Define a piecewise-constant approximation

$$\bar{v}(t) = \begin{cases} v^1, & 0 < t < t_1, \\ v^{n-\frac{1}{2}}, & t_{n-1} < t < t_n,\ n \ge 2. \end{cases}$$

Using $\bar{v}$, we define a discrete fractional integral

$$({}_0\mathrm{I}_t^\alpha v)^{n-\frac{1}{2}} = \frac{1}{\tau_n} \int_{t_{n-1}}^{t_n} \int_0^t \omega_\alpha(t-s)\bar{v}(s)\mathrm{d}s\mathrm{d}t = a_{n-1}^{(n,\alpha)} v^1 \tau_1 + \sum_{j=2}^{n} a_{n-j}^{(n,\alpha)} v^{j-\frac{1}{2}} \tau_j, \tag{11}$$

where

$$a_{n-j}^{(n,\alpha)} = \frac{1}{\tau_n \tau_j} \int_{t_{n-1}}^{t_n} \int_{t_{j-1}}^{\min\{t,t_j\}} \omega_\alpha(t-s)\mathrm{d}s\mathrm{d}t > 0, \quad 1 \le j \le n.$$

2.2 Error Analyses

Firstly, we study the error bound for the formula (9). For convenience, we put

$$\begin{aligned}\eta^n =& \frac{1}{\tau_n} \int_{t_{n-1}}^{t_n} \int_0^t \omega_{1+\alpha-\beta}(t-s)\big[v'(s) - (\Pi_1 v)'(s)\big]\mathrm{d}s\mathrm{d}t \\ =& \frac{1}{\tau_n} \int_{t_{n-1}}^{t_n} \sum_{j=1}^{n} \int_{t_{j-1}}^{\min\{t,t_j\}} \omega_{1+\alpha-\beta}(t-s)\big[v'(s) - \partial_\tau v^{j-\frac{1}{2}}\big]\mathrm{d}s\mathrm{d}t. \end{aligned} \tag{12}$$

After some manipulations, one can show that

$$v'(s) - \partial_\tau v^{j-\frac{1}{2}} = e_1(s) + e_2(s), \quad s \in \mathrm{I}_j,$$

where

$$e_1(s) = \int_s^{t_j} (\mu - s)v'''(\mu)\mathrm{d}\mu - \frac{1}{2\tau_j}\int_{t_{j-1}}^{t_j} (\mu - t_{j-1})^2 v'''(\mu)\mathrm{d}\mu, \quad e_2(s) = (s - t_{j-\frac{1}{2}})v''(t_j).$$

Consequently, we split η^n as $\eta^n = \tau_n^{-1}(\eta_1^n + \eta_2^n)$, where

$$\eta_1^n = \sum_{j=1}^{n} \int_{t_{n-1}}^{t_n} \int_{t_{j-1}}^{\min\{t,t_j\}} \omega_{1+\alpha-\beta}(t-s)e_1(s)\mathrm{d}s\mathrm{d}t, \tag{13}$$

$$\eta_2^n = \sum_{j=1}^{n} \int_{t_{n-1}}^{t_n} \int_{t_{j-1}}^{\min\{t,t_j\}} \omega_{1+\alpha-\beta}(t-s)e_2(s)\mathrm{d}s\mathrm{d}t. \tag{14}$$

Next two lemmas present the estimations of $|\eta_1^n|$ and $|\eta_2^n|$. They can be obtained following the ideas of Lemma 3.2 and 3.4 in [10], which replace α by $1+\alpha-\beta$.

Lemma 1. *For η_1^n defined as in (13) with $n \ge 1$, we have*

$$|\eta_1^n| \le C_1 \tau_n \tau^2 t_n^{\sigma+\alpha-\beta-2/r}, \quad r \ge 2/\sigma.$$

Lemma 2. *For η_2^n defined as in (14) with $n \ge 1$, we have*

$$|\eta_2^n| \le C_2 \tau^2 \tau_n t_n^{\sigma+\alpha-\beta-2/r}, \quad r \ge \max\{3/(\sigma+1), 2/\sigma\}.$$

Secondly, we give the error bound for (11).
Denote

$$\theta^n = \frac{1}{\tau_n} \int_{t_{n-1}}^{t_n} \int_0^t \omega_\alpha(t-s)\big[v(s) - \bar{v}(s)\big]\mathrm{d}s\mathrm{d}t. \tag{15}$$

Make the splitting $\theta^n = \tau_n^{-1}(\theta_1^n + \theta_2^n)$, in which

$$\theta_1^n = \int_{t_{n-1}}^{t_n} \sum_{j=1}^{n} \int_{t_{j-1}}^{\min\{t,t_j\}} \omega_\alpha(t-s)\big[v(s) - \Pi_1 v\big]\mathrm{d}s\mathrm{d}t, \tag{16}$$

$$\theta_2^n = \int_{t_{n-1}}^{t_n} \sum_{j=1}^{n} \int_{t_{j-1}}^{\min\{t,t_j\}} \omega_\alpha(t-s)\big[\Pi_1 v - \bar{v}(s)\big]\mathrm{d}s\mathrm{d}t. \tag{17}$$

The estimations of θ_1^n and θ_2^n are given by following lemmas, which followed the ideas of Lemma 3.2 and Theorem 3.3 in [9].

Lemma 3. *For θ_1^n defined as in (16) with $n \geq 1$, we have*

$$|\theta_1^n| \leq C_3 \tau_n \tau^2 t_n^{\sigma+\alpha-2/r}, \quad r \geq 2/(\sigma+\alpha-1).$$

Lemma 4. *For θ_2^n defined as in (17) with $n \geq 1$, we have*

$$\sum_{n=2}^{m} |\theta_2^n| \leq C_{\alpha,T} C_{\sigma,r} \begin{cases} \tau^{r(\sigma+1)}, & r < 2/(\sigma+1), \\ \tau^2 \ln(t_n/t_1), & r = 2/(\sigma+1), \\ \tau^2, & r > 2/(\sigma+1), \end{cases}$$

where $C_{\alpha,T}$, $C_{\sigma,r}$ are positive constants independent of the time step size.

3 A Numerical Scheme on Graded Meshes

To define our time-stepping numerical scheme, integrating equality (7) over the time interval I_n leads to

$$\begin{aligned} &\frac{1}{\tau_n}\int_{t_{n-1}}^{t_n} \partial_s u(x,s)\mathrm{d}s + \frac{1}{\tau_n}\int_{t_{n-1}}^{t_n}\int_0^t \omega_{1+\alpha-\beta}(t-s)\partial_s u(x,s)\mathrm{d}s\mathrm{d}t \\ =&\varphi(x) + \frac{1}{\tau_n}\int_{t_{n-1}}^{t_n}\int_0^t \omega_\alpha(t-s)u_{xx}(x,s)\mathrm{d}s\mathrm{d}t + \frac{1}{\tau_n}\int_{t_{n-1}}^{t_n} F(x,s)\mathrm{d}s. \end{aligned} \tag{18}$$

Take a positive integer M, let $h = (x_r - x_l)/M$ be the spatial step size, and the mesh $\bar{\Omega}_h = \{x_i \,|\, x_i = x_l + ih,\ 0 \leq i \leq M\}$, $\Omega_h = \bar{\Omega}_h \cap \Omega$ and $\partial\Omega_h = \bar{\Omega}_h \cap \partial\Omega$. Define mesh function spaces $\mathcal{V}_h = \{v \,|\, v = (v_0, v_1, \cdots, v_{M-1}, v_M),\ v_i = v(x_i),\ x_i \in \bar{\Omega}_h\}$, $\mathring{\mathcal{V}}_h = \{v \,|\, v \in \mathcal{V}_h,\ v_0 = v_M = 0\}$. For any $v \in \mathcal{V}_h$, introduce

$$\delta_x v_{i-\frac{1}{2}} = (v_i - v_{i-1})/h, \quad \delta_x^2 v_i = (v_{i+1} - 2v_i + v_{i-1})/h^2.$$

Consider (18) at the point $x = x_i$. Applying (9), (11) and the trapezoidal formula, we have the following C-N time-stepping scheme for the problem (1)–(3):

$$\begin{cases} \partial_\tau u_i^{n-\frac{1}{2}} + \sum_{j=1}^{n} \bar{a}_{n-j}^{(n,\beta)} \nabla_\tau u_i^j = \varphi_i + a_{n-1}^{(n,\alpha)} \delta_x^2 u_i^1 \tau_1 \\ + \sum_{j=2}^{n} a_{n-j}^{(n,\alpha)} \delta_x^2 u_i^{j-\frac{1}{2}} \tau_j + F_i^{n-\frac{1}{2}}, \quad 1 \le i \le M-1, \ 1 \le n \le N, & (19) \\ u_i^0 = \phi(x_i), \quad 1 \le i \le M-1, & (20) \\ u_0^n = 0, \quad u_M^n = 0, \quad 0 \le n \le N. & (21) \end{cases}$$

From lemmas 1–4, we obtain that the truncation error R_i^n of the formula (19) meets the following inequality:

$$|R_i^n| \le C_4(\tau^2 + h^2), \quad 1 \le i \le M-1, \quad 1 \le n \le N.$$

4 Numerical Examples

This section is devoted to performing numerical experiments to support the theoretical results.

To measure the error of the numerical solution, we computed

$$E_{N,M} = \max_{1 \le n \le N} \|U^n - u(t_n)\|,$$

The convergence rate r_t (in time) is calculated from the relation

$$r_t = \log_2(E_{N,M}/E_{2N,M}).$$

Example 1. Take $T = 1$ and $\Omega = (0, \pi)$. The exact solution is

$$u(x,t) = \frac{1}{4}\omega_{1+\sigma}(t)\sin(2x),$$

where $\sigma = 3/2$. The source term is given by

$$f(x,t) = \left[\frac{t^\sigma}{\Gamma(1+\sigma)} + \frac{t^{\sigma-1-\alpha}}{4\Gamma(\sigma-\alpha)} + \frac{t^{\sigma-\beta}}{4\Gamma(1+\sigma-\beta)}\right]\sin(2x).$$

We fixed sufficiently small spatial (uniform) step size $h = 1/5000$, so that the time errors are dominant. Therefore, we expect to observe $O(\tau^2)$-rates of convergence for different values of α, β. As predicted, $O(\tau^2)$-rates were observed for $r \geq 2$ in Table 1, which is consistent with theoretical results.

Table 1. Errors and convergence rates with different choices of α, β, r in Example 1

(α, β)	N	$r = 1$		$r = 2$		$r = 3$		$r = 4$	
		$E_{N,M}$	r_t	$E_{N,M}$	r_t	$E_{N,M}$	r_t	$E_{N,M}$	r_t
$(0.1, 0.3)$	80	8.95e−6		7.01e−6		1.58e−5		2.75e−5	
	160	4.41e−6	1.02	1.89e−6	1.89	4.35e−6	1.86	7.66e−6	1.85
	320	1.78e−6	1.31	4.97e−7	1.93	1.17e−6	1.89	2.09e−6	1.88
	640	6.67e−7	1.41	1.22e−7	2.03	3.04e−7	1.95	5.53e−7	1.91
$(0.8, 0.4)$	80	1.56e−5		6.30e−6		1.40e−5		2.41e−5	
	160	5.60e−6	1.48	1.57e−6	2.00	3.53e−6	1.99	6.09e−6	1.98
	320	1.99e−6	1.49	3.85e−7	2.03	8.79e−7	2.00	1.53e−6	2.00
	640	7.04e−7	1.50	8.85e−8	2.12	2.13e−7	2.05	3.76e−7	2.02
$(0.7, 0.9)$	80	1.43e−5		9.17e−6		2.04e−5		3.56e−5	
	160	5.33e−6	1.42	2.30e−6	1.99	5.16e−6	1.98	9.05e−6	1.98
	320	1.93e−6	1.46	5.70e−7	2.01	1.29e−6	2.00	2.28e−6	1.99
	640	6.93e−7	1.48	1.36e−7	2.07	3.17e−7	2.03	5.66e−7	2.01

Example 2. Take $T = 1$, $\Omega = (0, 2)$ and the regularity parameter $\sigma = \alpha + 1$. The exact solution is

$$u(x, t) = t^{\alpha+1} \sin(\frac{\pi}{2}x).$$

The source term is given by

$$f(x, t) = \Big[\Gamma(2 + \alpha) + \frac{\Gamma(2 + \alpha)}{\Gamma(2 + \alpha - \beta)} t^{1+\alpha-\beta} + \frac{\pi^2}{4} t^{\alpha+1}\Big] \sin(\frac{\pi}{2}x).$$

The results for the scheme (19)–(21) at fixed spatial step size $h = 1/5000$ with several sets of values of (α, β) are showed in Table 2. It is obvious that the numerical convergence orders match well with the theoretical results.

Table 2. Errors and convergence rates with different choices of α, β, r in Example 2

(α, β)	N	$r=1$		$r=2$		$r=3$		$r=4$	
		$E_{N,M}$	r_t	$E_{N,M}$	r_t	$E_{N,M}$	r_t	$E_{N,M}$	r_t
$(0.1, 0.3)$	80	1.18e−4		2.07e−5		4.61e−5		8.00e−5	
	160	7.78e−5	0.60	5.59e−6	1.89	1.26e−5	1.87	2.21e−5	1.86
	320	4.15e−5	0.91	1.48e−6	1.91	3.39e−6	1.89	6.01e−6	1.88
	640	2.05e−5	1.02	3.82e−7	1.96	8.95e−7	1.92	1.60e−6	1.90
$(0.8, 0.4)$	80	2.49e−5		3.68e−5		6.93e−5		1.13e−4	
	160	5.87e−6	2.08	9.24e−6	1.99	1.75e−5	1.99	2.87e−5	1.98
	320	1.47e−6	2.00	2.31e−6	2.00	4.39e−6	1.99	7.22e−6	1.99
	640	4.03e−7	1.87	5.72e−7	2.01	1.09e−6	2.00	1.81e−6	2.00
$(0.7, 0.9)$	80	2.37e−5		4.95e−5		1.03e−4		1.76e−4	
	160	7.62e−6	1.63	1.25e−5	1.99	2.61e−5	1.98	4.47e−5	1.98
	320	2.44e−6	1.64	3.12e−6	2.00	6.55e−6	1.99	1.13e−5	1.99
	640	7.67e−7	1.67	7.77e−7	2.01	1.64e−6	2.00	2.83e−6	1.99

5 Conclusion

A temporal second-order difference scheme on graded meshes for time fractional mixed diffusion and wave equation with an initial singularity is developed. The sharp error estimates are provided. The presented numerical tests confirmed the theoretical results. Unfortunately, we have not completed the rigorous theoretical analyses on stability and convergence of the numerical scheme. We plan to deal these issues in the near future.

References

1. Bazhlekova, E., Bazhlekov, I.: Viscoelastic flows with fractional derivative models: computational approach by convolutional calculus of Dimovski. Fract. Calc. Appl. Anal. **17**(4), 954–976 (2014)
2. Feng, L., Liu, F., Turner, I., et al.: Novel numerical analysis of multi-term time fractional viscoelastic non-Newtonian fluid models for simulating unsteady MHD Couette flow of a generalized Oldroyd-B fluid. Fract. Calc. Appl. Anal. **21**(4), 1073–1103 (2018)
3. Sakamoto, K., Yamamoto, M.: Initial value/boundary value problems for fractional diffusion wave equations and applications to some inverse problems. J. Math. Anal. Appl. **382**(1), 426–447 (2011)
4. Feng, L., Liu, F., Turner, I.: Finite difference/finite element method for a novel 2D multi-term time-fractional mixed sub-diffusion and diffusion-wave equation on convex domains. Commun. Nonlinear Sci. Numer. Simul. **70**, 354–371 (2019)
5. Wang, Y., Mei, L.: Generalized finite difference/spectral Galerkin approximations for the time-fractional telegraph equation. Adv. Differ. Equ. **281**, 1–6 (2017). https://doi.org/10.1186/s13662-017-1348-2

6. Du, R.L., Sun, Z.Z.: Temporal second-order difference method for solving multi-term time fractional mixed diffusion and wave equations. Numer. Algorithm **88**(1), 191–226 (2021). https://doi.org/10.1007/s11075-020-01037-x
7. Du, R.L., Sun, Z.Z.: A fast temporal second-order compact Adi scheme for time fractional mixed diffusion and wave equation. East Asian J. Appl. Math. **11**, 647–673 (2021). https://doi.org/10.4208/eajam.271220.090121. (Accepted in Jan)
8. Kopteva, N.: Error analysis of the L1 method on graded and uniform meshes for a fractional derivative problem in two and three dimensions. Math. Comput. **88**(319), 2135–2155 (2019)
9. McLean, W., Mustapha, K.: A second-order accurate numerical method for a fractional wave equation. Numer. Math. **105**(3), 481–510 (2007)
10. Mustapha, K.: An L1 approximation for a fractional reaction-diffusion equation, a second-order error analysis over time-graded meshes. SIAM J. Numer. Anal. **58**(2), 1319–1338 (2020)

Optimized Current and Speed Fractional-Order PID Control in Electrical Drives

Boris Jakovljević[1], Paolo Lino[2], and Guido Maione[2](✉)

[1] University of Novi Sad, Faculty of Technical Sciences, Computing and Control Department, Trg D. Obradovića 6, Novi Sad, Serbia
bjakov@uns.ac.rs

[2] Politecnico di Bari, Department of Electrical and Information Engineering, Via E. Orabona 4, Bari 70125, Italy
{paolo.lino,guido.maione}@poliba.it

Abstract. Fractional-order PID controllers are suitably designed for cascade control of permanent magnet synchronous motor drives. The stator current and the speed are controlled in two nested loops. Six parameters characterizing the two controllers (the proportional, integral, and derivative gains, the non-integer orders of integration and differentiation, and a filter time constant) are determined by optimizing different cost criteria for the two loops. To this aim, a particle swarm optimization is applied in a single step, instead of performing two successive steps, the first for the inner loop, the second for the external loop. Simulation by an accurate model confirms effectiveness of the approach and shows improvements with respect to PID optimized controllers.

Keywords: Fractional-order PID controllers · Permanent magnet synchronous motors · Sensitivity functions · Particle swarm optimization

1 Introduction

Permanent magnet synchronous motors (PMSM) are used in many applications because of fast response, high power/weight ratio, reduced size and high efficiency. They must achieve performance in reference tracking and in disturbance rejection, because the control loops in which they are applied—for example in industry—are often subject to disturbances. The PMSM control system is represented by two nested loops, the inner is devoted to controlling the stator current (i_{sq}) in the reference frame fixed to the rotor, while the external loop is for controlling the rated angular speed (ω_r) (see Fig. 1). The nested configuration allows to reduce sensitivity to uncertain and varying parameters. However, the common practice is to employ PI or PID controllers that are tuned by well-established rules. For example, the inner PI controller (C_1) parameters are usually tuned by the *optimum modulus criterion* [4], whereas the external PI controller (C_2)

A. Dzielinski et al. (Eds.): ICFDA 2021, LNNS 452, pp. 141–152, 2022.
https://doi.org/10.1007/978-3-031-04383-3_16

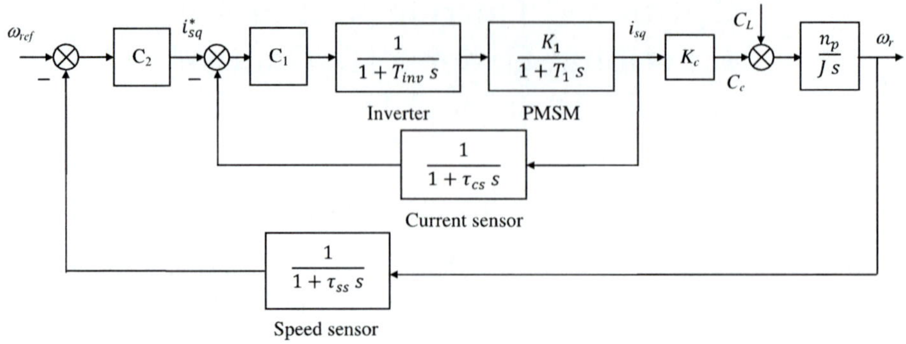

Fig. 1. PMSM control scheme: C_1 and C_2 are PID or FOPID controllers

parameters are determined by the *symmetrical optimum criterion* [4]. Moreover, a set-point pre-filter (not shown in Fig. 1) is applied to decrease the overshoot in the reference step response. Despite the results guaranteed by these controllers, research is proposing new techniques to improve performance and robustness in control of PMSM drives [10]. Some attempts were based on controllers with non-integer orders of differentiation and integration [2,3,5,11–15].

The basic idea was to profitably employ fractional-order PI (or FOPI) controllers $G_c(s) = K_p + \frac{K_i}{s^\lambda}$ or, more generally, fractional-order PID (or FOPID) controllers $G_c(s) = K_p + \frac{K_i}{s^\lambda} + K_d\, s^\mu$. Moreover, the controller can be generalized to a distributed-order PID structure by using more than two differentiators or integrators of non-integer order. In [5], the authors proposed an optimization strategy to achieve a better tradeoff between dynamic performance and robust stability, for both loops of the considered control system. Obviously, the optimized controllers were realized by approximating irrational transfer functions through suitable rational transfer functions. They are intrinsically more complex than standard PI/PID controllers, but the increased flexibility due to the non-integer orders allows to improve transient response, disturbance rejection, and other characteristics. More specifically, in [5], the optimization approach is in two steps. In the first one, the controller C_1 is optimized. Hence, the inner closed-loop transfer function depending on C_1 is computed, because it is part of the external loop controlled system. In the second step, the parameters of C_2 are optimized based on the inner closed-loop system and the other elements in the external loop. In this case, the transfer function of the controlled system is of non-integer order, then it is natural to think about using a fractional-order controller in the external loop (the best controller for a fractional-order system should be a fractional-order controller). To synthesize, the two-steps optimization problem was divided into two sub-problems: the first is the inner controller optimization for tracking a current reference, the second is the external controller optimization for rejecting step disturbances. The aim of this paper is to propose a slightly modified approach to perform the optimization process in one step only. In this way, all the parameters of the two FOPID controllers will be determined at once, which should improve the results.

The rest of the paper is organized as follows. Section 2 provides some preliminaries on the PMSM mathematical model. Section 3 introduces the controllers and describes the optimization problem with the employed cost function and optimization constraints. Section 4 describes the peculiar optimization technique based on a particle swarm optimization algorithm, which allows to set the controllers parameters. Section 5 shows some simulation results to evaluate the effectiveness and performance of the FOPID controllers. Simulation runs on an accurate model of the PMSM drive. Section 6 gives some final remarks.

2 Mathematical Preliminaries

The controlled variable in the inner loop is the PMSM stator current along the quadrature axis (q-axis) of the reference frame, indicated as i_{sq}. The reference i^*_{sq} is given by the external controller C_2. Moreover, the inner loop includes an inverter in the forward path and a current sensor in the feedback path. A similar loop is considered for the stator current along the direct axis (d-axis) of the reference frame, indicated as i_{sd}, which is not shown here, for sake of clarity. Putting all elements together, the PMSM mathematical model is

$$G(s) = \frac{K_1}{1 + T_1 s} \, , \tag{1}$$

where $K_1 = \frac{1}{R_s}$, R_s is the stator resistance, $T_1 = \frac{L_s}{R_s}$, L_s is the stator inductance. The inverter and the current sensor operations determine small time delays, then can be represented by first-order systems with small time constants (T_{inv} and τ_{cs} respectively). $T_e = T_{inv} + \tau_{cs}$ is the equivalent time constant of such operations. Then, the open-loop transfer function in the inner loop is $G_{c1}(s)\, G_1(s)$, being $G_{c1}(s)$ the transfer function of controller C_1 and with

$$G_1(s) = \frac{K_1}{(1 + T_1\, s)\,(1 + T_e\, s)} \, , \tag{2}$$

where $T_e < T_1$. With these premises, $G_{f1}(s)$ is the closed-loop transfer function of the inner loop determined by $G_{c1}(s)\, G_1(s)$:

$$G_{f1}(s) = \frac{1}{G_{cs}(s)} \frac{G_{c1}(s)\, G_1(s)}{1 + G_{c1}(s)\, G_1(s)} \, , \tag{3}$$

where $G_{cs}(s) = \frac{1}{1+\tau_{cs} s}$ is the current sensor transfer function. Then, since $G_{cs}^{-1}(s) = 1 + \tau_{cs} s \approx \frac{1}{1-\tau_{cs} s}$, it holds

$$G_{f1}(s) = \frac{1}{1 - \tau_{cs} s} G_{FOS}(s) \, , \tag{4}$$

where $G_{FOS}(s) = \frac{G_{c1}(s)\, G_1(s)}{1+G_{c1}(s)\, G_1(s)}$ is the closed-loop fractional-order system resulting from application of a fractional-order controller to the PMSM motor.

In the external loop, a constant K_c determines the conversion of the stator current to the electromagnetic torque C_e, which is counteracted by the load disturbance torque C_L. The difference determines the PMSM angular speed ω_r if the number of pole pairs n_p and the moment of inertia J are properly taken into account. Finally, a speed sensor is characterized by a time constant τ_{ss} and provides the feedback useful for comparison with the speed reference and generation of the error input to the external controller C$_2$. Then, the open-loop transfer function in the external loop is $G_{c2}(s)\, G_2(s)$ with

$$G_2(s) = G_{f1}(s)\,\frac{K_c\, n_p}{J\, s\,(1+\tau_{ss}\, s)}\,, \tag{5}$$

and can be used to design the external controller transfer function, say $G_{c2}(s)$.

3 Controllers Optimization

The controllers in the two nested loops are fractional-order PID controllers, introduced by Podlubny [7]. Their transfer functions are respectively given by:

$$G_{c1}(s) = \frac{K_{p_1} + \frac{K_{i_1}}{s^{\lambda_1}} + K_{d_1} s^{\mu_1}}{(1+T_{f_1} s)^{\mu_1}} \tag{6}$$

and

$$G_{c2}(s) = \frac{K_{p_2} + \frac{K_{i_2}}{s^{\lambda_2}} + K_{d_2} s^{\mu_2}}{(1+T_{f_2} s)^{\mu_2}}\,, \tag{7}$$

where λ_1 and λ_2 are non-integer orders of integration, μ_1 and μ_2 are non-integer orders of differentiation, K_{p_1}, K_{p_2}, K_{i_1}, K_{i_2}, K_{d_1}, K_{d_2} are positive controller gains, T_{f1} and T_{f2} are filter time constants and the orders of the filters are the same as the orders of differentiation. In this way, high-frequency noise amplification is finite and constant. Moreover, it is lower than the amplification when filtering the derivative action only.

The controllers are optimized by considering both performance and robustness indexes [9]. Firstly, the maximum noise sensitivity is considered:

$$M_n = \max_{\omega} \left| \frac{G_c(j\omega)}{1+G_c(j\omega)G_p(j\omega)} \right|\,, \tag{8}$$

where the authors in [6] showed that, with FOPID controllers, $M_n = \frac{K_d}{T_f^{\mu}}$ is obtained as $\omega \to \infty$, as for most processes. On the other hand, with PID controllers, $M_n = \frac{K_d}{T_f}$, namely the order of differentiation is $\mu = 1$ and the order of the filter is the same. Secondly, as robustness index, the maximum sensitivity is considered:

$$M_s = \max_{\omega} \left| \frac{1}{1+G_c(j\omega)G_p(j\omega)} \right|\,, \tag{9}$$

and ω_s is the specific angular frequency in which the maximum is achieved. Finally, the performance index

$$Q = \max_{\omega} \left| \frac{K_i \frac{G_p(j\omega)}{j\omega}}{1 + G_c(j\omega)G_p(j\omega)} \right| \tag{10}$$

is used [6] and corresponds to the maximum of the resonant peak. In this case, ω_q is the angular frequency in which the maximum is achieved.

The controller C_1 in the inner loop is optimized such that an optimal current reference tracking is achieved. To this aim, the following cost function is minimized:

$$\mathcal{J}_1 = \sigma\, K_i + (1 - \sigma)\, \omega_B \,, \tag{11}$$

where K_i is the controller integral gain, ω_B is the system bandwidth, and σ is a parameter that is changed (typically between 0.4 and 1) to experiment different weights of K_i and ω_B. Therefore, the optimization problem for the FOPID controller in the inner loop is defined as:

$$\max_{K_{p_1}, K_{i_1}, \lambda_1, K_{d_1}, \mu_1, \omega_s, \omega_q} \mathcal{J}_1 \,, \tag{12}$$

subject to the following constraints:

$$\text{stable closed-loop system} \,, \tag{13}$$

$$M_s = \left| \frac{1}{1 + G_c(j\omega)\, G_p(j\omega)} \right|_{\omega=\omega_s} \leq M_s^{max} \,, \tag{14}$$

$$M_n = \frac{K_{d_1}}{T_{f_1}^{\mu_1}} \leq M_n^{max} \,, \tag{15}$$

$$Q = \left| \frac{K_i \frac{G_p(j\omega)}{j\omega}}{1 + G_c(j\omega)\, G_p(j\omega)} \right|_{\omega=\omega_q} \leq Q^{max} \,, \tag{16}$$

where $G_c(j\omega) = G_{c_1}(j\omega)$ as in (6), $G_p(j\omega) = G_1(j\omega)$, and M_s^{max}, M_n^{max} and Q^{max} are limits for the maximum values of the considered indexes. In case a PID controller is used in the inner loop, optimization is determined by:

$$\max_{K_{p_1}, K_{i_1}, K_{d_1}, \omega_s, \omega_q} \mathcal{J}_1 \,, \tag{17}$$

subject to the same constraints (13)–(16) but with $\lambda_1 = \mu_1 = 1$ and $M_n = \frac{K_{d_1}}{T_{f_1}}$.

For the external loop, it is important to achieve reference speed tracking and load disturbance rejection. Then, the controller should have the largest possible amplitude on the entire frequency range to minimize disturbance effect, while guaranteeing closed-loop stability. Moreover, if the controller achieves a minimal suppression of disturbance at some frequency ω_m, in which the controller amplitude is minimum, the objective is to maximize this minimal suppression, that is to maximize the minimum of controller amplitude. For PID controllers, it is proven that such minimum is equal to K_p and $\omega_m = \sqrt{K_i/K_p}$ [5], such that the optimization problem for the external PID controller C_2 becomes

$$\max_{K_{p_2}, K_{i_2}, K_{d_2}, \omega_s, \omega_q} K_{p_2} , \tag{18}$$

subject to the constraints (13)–(16), with $\lambda_2 = \mu_2 = 1$ and $M_n = \frac{K_{d_1}}{T_{f_1}}$, where $G_c(j\omega) = G_{c_2}(j\omega)$ is the classical PID transfer function from (7) with $\lambda_2 = \mu_2 = 1$, and $G_p(j\omega) = G_2(j\omega)$.

In a similar way, FOPID controllers should have the largest possible amplitude at all frequencies to reject disturbance, while maintaining closed-loop stability, and the maximization of the minimum amplitude controller is important to reduce disturbance effect. However, the cost function depends on the frequency

$$\max_{K_{p_2}, K_{i_2}, K_{d_2}, \lambda_2, \mu_2, \omega_s, \omega_q} \min |G_{c_2}(j\omega)| , \tag{19}$$

subject to the same constraints (13)–(16), where $G_c(j\omega) = G_{c_2}(j\omega)$ as in (7) and $G_p(j\omega) = G_2(j\omega)$.

Note that, given M_n^{max}, if we put $M_n = M_n^{max}$, we get $T_f = \sqrt[\mu]{K_d/M_n^{max}}$ for FOPID controllers and $T_f = K_d/M_n^{max}$ for PID controllers.

4 Solving the Optimization Problem

Based on the previous formalization of the controllers optimization problem, the parameters to be optimized in the inner and external loops are: $[K_{p_1}, K_{i_1}, \lambda_1, K_{d_1}, \mu_1, T_{f_1}, K_{p_2}, K_{i_2}, \lambda_2, K_{d_2}, \mu_2, T_{f_2}]$. Since the optimization problem is non-linear and can not be easily solved by classical gradient-based methods, a generalized Particle Swarm Optimization (PSO) algorithm is employed. Namely, PSO is successful with multiple optimum points—as it is the case—and multi-dimensional search spaces [1]. Other important characteristics are the reduced computational cost, low sensitivity to the population size, higher efficiency with respect to evolutionary algorithms, suitability to PID tuning. The specific generalized version of the algorithm was proposed in [8].

Here is how the algorithm works. It looks for the optimal solution in a search space, which is defined a priori. A set of candidate solutions constitutes the swarm of particles that are updated at each iteration of the algorithm. Each particle has a position and a velocity. The position is a possible solution to the optimization problem. The algorithm uses a criterion (the previously defined cost function) to evaluate the "fitness" of each particle. Each particle has memory of its best position—that is the "personal best" position, indicated by $p_b(t)$—and shares this information with all other particles in the swarm. Moreover, the algorithm determines and updates the "global best" position among all particles in the swarm, indicated by $g_b(t)$. The particles exchange information and explore the search space such that the swarm converges to the optimal solution. To summarize, the position of every particle is updated as follows:

$$\chi(t) = 2\,\rho\,\zeta\,\chi(t-1) - \rho^2\,\chi(t-2) + (1 - 2\,\rho\,\zeta + \rho^2)\,[c\,p_b(t) + (1-c)\,g_b(t)] , \tag{20}$$

where $0 < \rho < 1$ is the damping factor, $-1 < \zeta < 1$ is the oscillation factor, and $0 < c < 1$ is the relative cognitive factor, which determines how the personal and global best positions affect the particle dynamics. Specific choices are as follows: ρ is linearly varied from 0.95 to 0.6, c from 0.8 to 0.2, ζ is sampled from a uniform distribution over $[-0.9, 0.2]$ at each iteration.

Each algorithm iteration determines an optimal solution, which is then improved in the following iteration. The best optimal and final solution is determined by the last iteration. However, this is not absolutely the best solution one can find. Namely, it depends on the number of optimized parameters, the search space, the number of particles, the number of iterations, and the stochastic nature of the PSO algorithm. The number of iterations (I) and the number of particles (P) can be varied until acceptable results are obtained. For example, $50 \leq I \leq 1000$, $50 \leq P \leq 200$. Then, the best results coming out from the combinations of I and P are selected. In particular, the results reported here are with $I = 600$ and $P = 100$. Moreover, the constrained optimization problem is solved by enforcing the following conditions: positive controllers' gains; stable poles in the controllers; positive phase margin.

All the PSO computations were performed by using the simplified control system model previously described and by using the following values for the maximum performance and robustness indexes: $M_s^{max} = 2\,(6\,\text{dB})$, $M_n^{max} = 20\,(26\,\text{dB})$, $Q^{max} = 1.01$.

The obtained values of the FOPID controllers parameters are shown in the first two lines of Table 1.

The performance indexes obtained after optimizing the inner FOPID controller are: $M_s = 0\,\text{dB}$, $M_n = 26\,\text{dB}$. Optimization of the external FOPID controller gives: $M_s = 6.5\,\text{dB}$, $M_n = 24.66\,\text{dB}$. The magnitudes of the sensitivity and noise sensitivity functions are shown in Figs. 2, 3, 4 and 5.

PID controllers were also optimized and their parameters are shown in the last two lines of Table 1.

Table 1. Optimized parameters of PID and FOPID controllers

Controller	Parameters					
	K_p	K_i	K_d	λ	μ	T_f
$G_{c1}(s)$: FOPID	1.1454	2.9178	6.0797	0.6070	1.2321	0.3804
$G_{c2}(s)$: FOPID	19.2452	5.4396	0.5399	0.2908	1.9955	0.1636
$G_{c1}(s)$: PID	0.4814	428	0.0019	1	1	9.5×10^{-5}
$G_{c2}(s)$: PID	0.6	1.0337	0.0478	1	1	0.0024

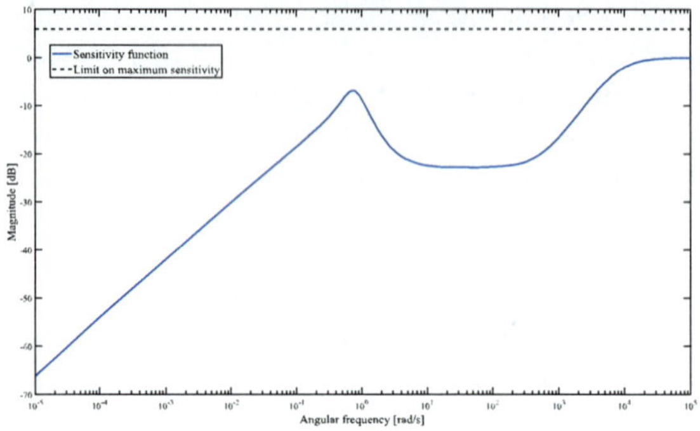

Fig. 2. Sensitivity function in the inner loop

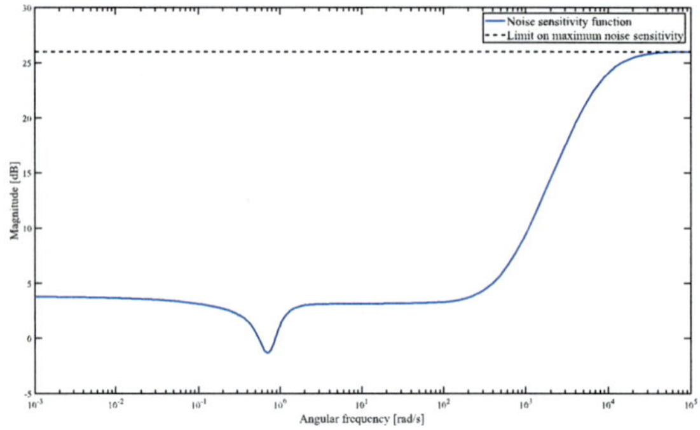

Fig. 3. Noise sensitivity function in the inner loop

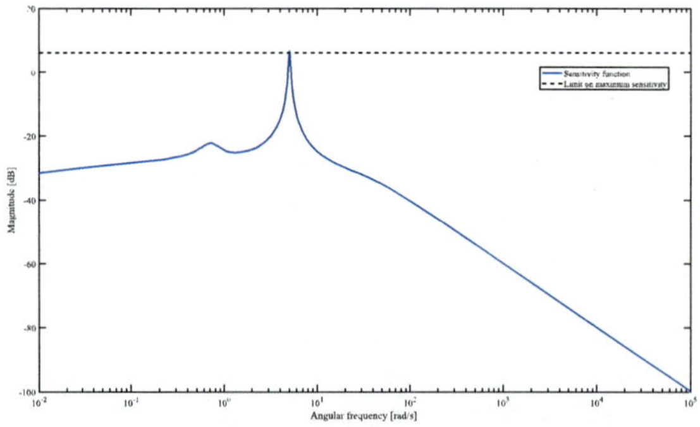

Fig. 4. Sensitivity function in the external loop

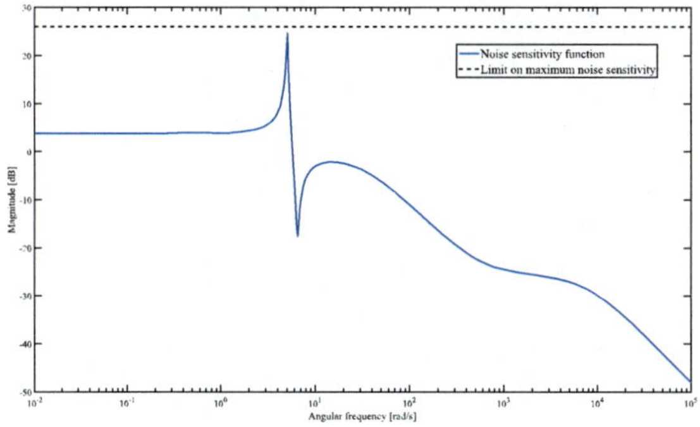

Fig. 5. Noise sensitivity function in the external loop

5 Simulation Results

A nonlinear accurate simulation model of the double-loop control system with the PMSM was used to evaluate the optimized FOPID controllers that were designed by the simplified model described in Sect. 2. The simulation model represents a laboratory equipment [2]. The main PMSM parameters are $R_s = 1.55\,\Omega$, $L_s = 2.5\,\text{mH}$, $K_1 = 0.6452\,\Omega^{-1}$, $T_1 = 1.6\,\text{ms}$, $n_p = 2$, $J = 2.51 \times 10^{-5}\,\text{kg}\,\text{m}^2$, $K_c = 0.27\,\text{N}\,\text{m/A}$, $\tau_{cs} = 0.16\,\text{ms}$, $\tau_{ss} = 0.796\,\text{ms}$.

The results obtained by using optimized FOPID controllers were compared to those with optimized PID controllers, which are frequently used for controlling PMSM drives.

A time-domain test was performed to evaluate performance of the proposed controllers. A reference step step input of 1256.6 rad/s, corresponding to 200 Hz, is applied at $t = 0\,\text{s}$ and a nominal load disturbance $C_L = 0.6\,\text{N}\,\text{m}$ is superimposed starting at $t = 0.4\,\text{s}$. The behaviors obtained for the current i_{sq} by applying the optimized FOPID controllers are shown in Fig. 6, in which zooms of the initial no-load transient and of the transient immediately after load application are also displayed. As one can note from the first zoom, the current does not exactly follow the reference produced by the external controller C_2 since there are saturation effects, because the current in the real system is limited to 2 times the nominal value (4.123 A) of the stator current while the reference goes a bit beyond this value. However, the current output reaches steady-state very quickly and by lower and less oscillations than the provided reference. Moreover, something similar occurs after load application (see the second zoom). The disturbance is very quickly rejected and the current settles to a new value which is associated with the torque necessary to balance the torque depending on the applied load. Moreover, oscillations are much reduced such that possible damages due to high current values to the motor are prevented.

A comparison with the current obtained by applying the optimized PID controllers is shown in Fig. 7. The response by FOPID controllers is much faster and better rejects the disturbance. Note that the reference signal i_{sq}^* for the inner current loop is the trajectory given by the output of the external loop controller

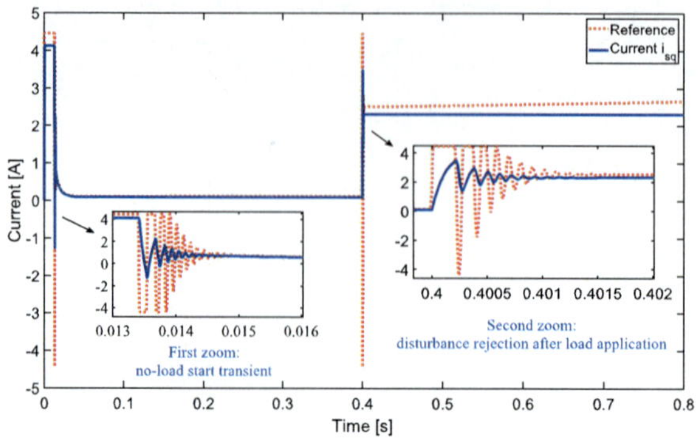

Fig. 6. Current controlled by optimized FOPID

C_2. Then the PID and FOPID controllers in the outer loop generate different references and the obtained output current i_{sq} is also different in the two cases.

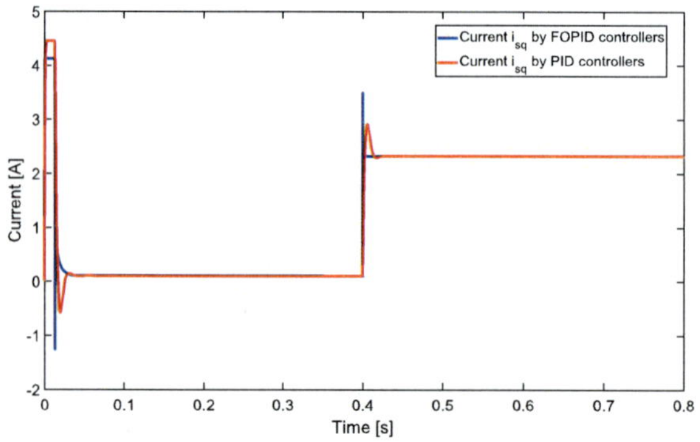

Fig. 7. Current controlled by optimized PID and FOPID

The speed obtained by FOPID controllers is shown in Fig. 8. A first zoom clearly shows that the response has small oscillations and a very limited maximum overshoot. The second zoom shows how the disturbance effect is rejected. The comparison with the optimized PID controllers is shown in Fig. 9. It is clear that PID yield a higher maximum overshoot $S\%$, higher undershoot after load application, and also a slower response. In details, $S\% = 6.29\%$ for PID, while $S\% = 0.22\%$ for FOPID. The settling time to 1% of the steady-state value is 23.3 ms by FOPID controllers and 22.9 ms by PID controllers. Finally, the load disturbance effect is more evident with PID controllers.

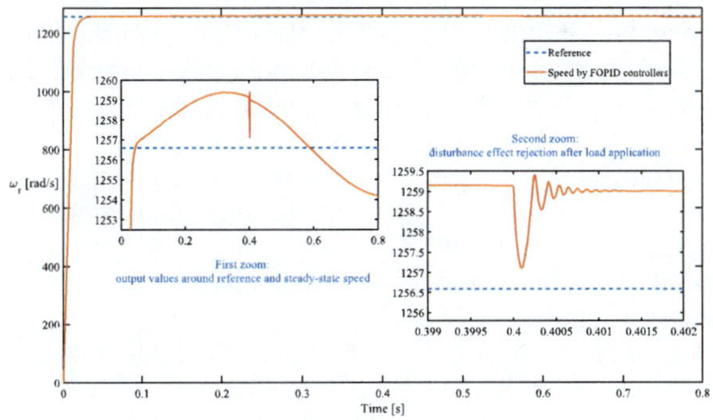

Fig. 8. Speed controlled by optimized FOPID

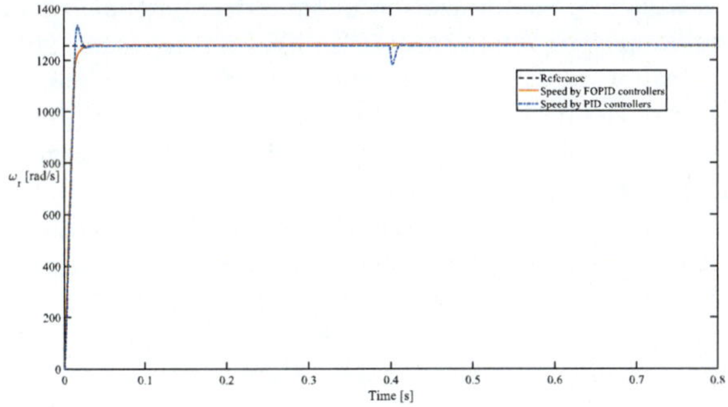

Fig. 9. Speed controlled by optimized PID and FOPID

6 Conclusion

This paper analyzes the control problem of a permanent magnet synchronous motor drive. Fractional-order PID controllers are employed to control the stator current in the reference frame fixed to the rotor, in the inner loop, and the angular rated speed, in the external loop. The controllers are designed by an optimization strategy that takes into account performance and robustness measures based on sensitivity functions and maximum resonant peak. Optimization is performed by a suitable particle swarm optimization algorithm. The same strategy is applied to integer-order PID controllers for comparison purpose. The controllers are tested by a Simulink accurate model of a real system. Results indicate effectiveness of the strategy and improvements provided by fractional-order PID controllers, especially in disturbance rejection and speed of response.

References

1. Clerc, M., Kennedy, J.: The particle swarm - explosion, stability, and convergence in a multidimensional complex space. IEEE Trans. Evol. Comput. **6**(1), 58–73 (2002). https://doi.org/10.1109/4235.985692

2. Lino, P., Maione, G., Padula, F., Stasi, S., Visioli, A.: Synthesis of fractional-order PI controllers and fractional-order filters for industrial electrical drives. IEEE/CAA J. Autom. Sin. **4**(1), 58–69 (2017). https://doi.org/10.1109/JAS.2017.7510325
3. Luo, Y., Chen, Y.Q., Ahn, H.-S., Pi, Y.G.: Fractional order robust control for cogging effect compensation in PMSM position servo systems: stability analysis and experiments. Control Eng. Pract. **18**(9), 1022–1036 (2010). https://doi.org/10.1016/j.conengprac.2010.05.005
4. Lutz, H., Wendt, W.: Taschenbuch der Regelungstechnik, 4th edn. Verlag Harri Deutsch, Frankfurt am main (2002)
5. Jakovljević, B., Lino, P., Maione, G.: Control of double-loop permanent magnet synchronous motor drives by optimized fractional D fractional and distributed-order PID controllers. Eur. J. Control **58**, 232–244 (2021). https://doi.org/10.1016/j.ejcon.2020.06.005
6. Jakovljević, B.B., Rapaić, M.R., Jeličić, Z.D., Šekara, T.B.: Optimization of fractional PID controller by maximization of the criterion that combines the integral gain and closed-loop system bandwidth. In: Proceedings of the 2014 18th Int. Conf. on System Theory, Control and Computing (ICSTCC), Sinaia, Romania, 17–19 October 2014, pp. 64–69 (2014). https://doi.org/10.1109/ICSTCC.2014.6982392
7. Podlubny, I.: Fractional order systems and $PI^{\lambda}D^{\mu}$-controllers. IEEE Trans. Autom. Control **44**(1), 208–214 (1999). https://doi.org/10.1109/9.739144
8. Rapaić, M.R., Kanović, Z., Jeličić, Z.D., Petrovački, D.: Generalized PSO algorithm – An application to Lorenz system identification by means of neural networks. In: Proceedings of the Ninth Symposium on Neural Network Applications in Electrical Engineering NEUREL 2008, pp. 31–35, Belgrade, Serbia, 25–27 September 2008. https://doi.org/10.1109/NEUREL.2008.4685554
9. Šekara, T.B., Mataušek, M.R.: Optimization of PID controller based on maximization of the proportional gain under constraints on robustness and sensitivity to measurement noise. IEEE Trans. Autom. Control **54**(1), 184–189 (2009). https://doi.org/10.1109/TAC.2008.2008359
10. Xia, X., Zhang, B., Li, X.: High precision low-speed control for permanent magnet synchronous motors. Sensors **20**(5), 1526 (2020). https://doi.org/10.3390/s20051526
11. Xie, Y., Tang, X., Song, B., Zhou, X., Guo, Y.: Data-driven adaptive fractional order PI control for PMSM servo system with measurement noise and data dropouts. ISA Trans. **75**, 172–188 (2018). https://doi.org/10.1016/j.isatra.2018.02.018
12. Yu, W., Luo, Y., Chen, Y.Q., Pi, Y.G.: Frequency domain modelling and control of fractional-order system for permanent magnet synchronous motor velocity servo system. IET Control Theory Appl. **10**(2), 136–143 (2016). https://doi.org/10.1049/iet-cta.2014.1296
13. Yu, W., Luo, Y., Pi, Y.G.: Fractional order modeling and control for permanent magnet synchronous motor velocity servo system. Mechatronics **23**(7), 813–820 (2013). https://doi.org/10.1016/j.mechatronics.2013.03.012
14. Zhang, B.T., Pi, Y.G., Luo, Y.: Fractional order sliding-mode control based on parameters auto-tuning for velocity control of permanent magnet synchronous motor. ISA Trans. **51**(5), 649–656 (2012). https://doi.org/10.1016/j.isatra.2012.04.006
15. Zheng, W.J., Luo, Y., Chen, Y.Q., Pi, Y.G.: Fractional-order modeling of permanent magnet synchronous motor speed servo system. J. Vib. Control **22**(9), 2255–2280 (2016). https://doi.org/10.1177/1077546315586504

Operational Calculus with Applications to Generalized Two-Sided Fractional Derivative

Gabriel Bengochea[1(✉)] and Manuel Ortigueira[2]

[1] Academia de Matemáticas, Universidad Autónoma de la Ciudad de México, México D.F., Mexico
gabriel.bengochea@uacm.edu.mx

[2] CTS-UNINOVA and DEE of Faculdade de Ciências e Tecnologia da UNL, Campus da FCT da UNL, Quinta da Torre, 2829-516 Caparica, Portugal
mdo@fct.unl.pt

Abstract. The paper contains the construction of an operational calculus used to solve fractional differential equations based on a generalized two-sided fractional derivative. The Riesz and Feller potential are particular cases of this derivative. It is proved that generalized two-sided fractional derivative is a *right shift operator*. Our resulting solutions are expressed in terms of series. In some cases the close form is obtained.

1 Introduction

The Riesz potential [1] is a two-sided anti-derivative [2] and it was introduced by M. Riesz to solve normal hyperbolic differential equations. In [3], it is applied to the theories of the electromagnetic and the meson fields. Although in a different scientific area and with different motivations, Feller proposed a similar operator, the Feller potential, which is a continuation of Riesz work [4]. The Riesz-Feller operator is a linear combination of both operators that is applied in diffusion [5–8] and quantum mechanics [9,10]. This new operator [6,11] can be considered as a general two-sided derivative (TSFD). It is defined by two parameters: the derivative order and the asymmetry. The introduction of an asymmetry parameter allows us to recover also the one-sided derivatives. Therefore, the Riesz and Feller potencials, Liouville (causal/anti-causal) antiderivative, and Liouville-Caputo (causal/anti-causal) are particular cases of TSFD. A recent application can be found in [12].

Let $\phi_n(t), n \in \mathbb{Z}$ be an sequence of functions. When an operator L satisfies $L\phi_n(t) = \phi_{n-1}(t), n \in \mathbb{Z}$, it is called *left shift operator*. In the case $L\phi_n(t) = \phi_{n+1}(t), n \in \mathbb{Z}$, it is called *right shift operator*. In [13], an operational calculus for some left shift operators is developed, see [14,15]. In this paper, we prove that generalized two-sided fractional derivative is a *right shift operator* for a suitable sequence of functions $\psi_{n\theta}^{n\beta}(t), n \in \mathbb{Z}$. With them, we construct an operational calculus to be used with *right shift operators* (Riesz potential, Feller potential and TSFD). This method permits us to solve fractional differential equations

A. Dzielinski et al. (Eds.): ICFDA 2021, LNNS 452, pp. 153–160, 2022.
https://doi.org/10.1007/978-3-031-04383-3_17

focussed in the TSFD. Our operational approach does not require neither the concept of integral nor of integral transform. We use algebraic operations only.

The paper is organized as follows. In Sect. 2 a background of the TSFD is presented and shown that it can be considered as a *right shift operator*. Section 3 presents the modifications realized to [13] for be applied to differential equations based on the TSFD. In Sect. 4, we present two illustrative examples of how the method is applied to differential equations. Finally, Sect. 5 includes our conclusions and future works.

2 Generalized Two-Sided Fractional Derivative

In [2] the definition of generalized two-sided fractional derivative was introduced by means of the Fourier transform.

Definition 1. *The generalized two-sided fractional derivative is defined as*

$$\mathcal{F}\left[D_\theta^\beta f(x)\right] = |\kappa|^\beta e^{i\frac{\pi}{2}\theta sgn(\kappa)} F(\kappa)$$

where β and θ are any real numbers called derivative order and asymmetry parameter, respectively.

Define $\Psi_\theta^\beta(x) = |\kappa|^\beta e^{i\frac{\pi}{2}\theta sgn(\kappa)}$. This function is called *frequency response* of the TSFD system. Some properties of $D_\theta^\beta f(x)$ and $\Psi_\theta^\beta(x)$ are listed below.

–
$$D_\theta^\beta D_{(n-1)\theta}^{(n-1)\beta} f(x) = D_{n\theta}^{n\beta} f(x)$$

–
$$D_\theta^\beta e^{i\kappa x} = |\kappa|^\beta e^{i\frac{\pi}{2}\theta sgn(\kappa)} e^{i\kappa x}, \qquad \kappa, x \in \mathbb{R}.$$

–
$$\Psi_\theta^\beta(x) = \cos\left(\theta\frac{\pi}{2}\right)\Psi_0^\beta(x) + \cos\left(\theta\frac{\pi}{2}\right)\Psi_1^\beta(x).$$

For computing the inverse Fourier transform of frequency response we consider two cases. For $\beta \in \mathbb{R} - \mathbb{Z}$, we have that

$$\psi_\theta^\beta(x) = \mathcal{F}^{-1}\left[\Psi_\theta^\beta\right] = \frac{\sin\left(\left(\beta + \theta sgn(x)\right)\frac{\pi}{2}\right)}{2\sin(\beta\pi)\Gamma(-\beta)}|x|^{-\beta-1}.$$

For $\beta \in \mathbb{Z}$, see the Table 1.

Table 1. Inverse Fourier transform of frequency response when $\beta \in \mathbb{Z}$

β	$\mathcal{F}^{-1}\left[\Psi_\theta^\beta\right] = \psi_\theta^\beta(x)$
$\beta = 2N$	$\cos\left(\theta\frac{\pi}{2}\right)(-1)^N\delta^{(2N)}(x) - \sin\left(\theta\frac{\pi}{2}\right)\left(\frac{(-1)^N(2N)!}{\pi}\|x\|^{-2N-1}sgn(x)\right)$
$\beta = 2N+1$	$\cos\left(\theta\frac{\pi}{2}\right)\left(\frac{(-1)^N(2N+1)!}{\pi}\|x\|^{-2N-2}\right) - i\sin\left(\theta\frac{\pi}{2}\right)(-1)^N\delta^{(2N+1)}(x)$
$\beta = -2N$	$\cos\left(\theta\frac{\pi}{2}\right)\frac{(-1)^N\|x\|^{2N-1}}{2(2N-1)!} - \sin\left(\theta\frac{\pi}{2}\right)\frac{(-1)^N x^{2N-1}}{\pi(2N)!}\left[\ln\|x\| - \Gamma'(1)\right]$
$\beta = -2N-1$	$\cos\left(\theta\frac{\pi}{2}\right)\frac{(-1)^N x^{2N}}{\pi(2N)!}\left[\ln\|x\| - \Gamma'(1)\right] - \sin\left(\theta\frac{\pi}{2}\right)\frac{(-1)^N}{2(2N)!}\|x\|^{2N}sgn(x)$

The case $\beta = 0$ will be defined by using the first row of the previous table. For $\beta \in \mathbb{R} - \mathbb{Z}$, the generalized TSFD has the integral representation

$$D_\theta^\beta f(x) = \frac{1}{2\sin(\beta\pi)\Gamma(-\beta)}\int_{\mathbb{R}} f(x-y)\sin\left[(\beta + sgn(y)\theta\pi/2)\right]|y|^{-\beta-1}dy. \quad (1)$$

We list some important cases.

- If $\beta < 0$ and $\theta = 0$, then (1) is the Riesz potencial.
- If $\beta < 0$ and $\theta = 1$, then (1) is the Feller potencial.
- If $\beta < 0$ and $\theta = \pm\beta$, then (1) is the Liouville integral formulae.
- If $\beta > 0$ and $\theta = \pm\beta$, then (1) is the Liouville integral formulae nonregularised.

It is true that $\Psi_\theta^\beta(\kappa)\Psi_{n\theta}^{n\beta}(\kappa) = \Psi_{(n+1)\theta}^{(n+1)\beta}(\kappa)$ and from Definition 1, we have that

$$\begin{aligned}\mathcal{F}\left[D_\theta^\beta\psi_\theta^\beta(x)\right] &= \Psi_\theta^\beta(\kappa)\mathcal{F}\left[\psi_\theta^\beta(x)\right]\\ &= \Psi_\theta^\beta(\kappa)\Psi_\theta^\beta(\kappa)\\ &= \Psi_{2\theta}^{2\beta}(\kappa).\end{aligned}$$

It follows that

$$D_\theta^\beta\psi_\theta^\beta(x) = \psi_{2\theta}^{2\beta}(x)$$

This result can be generalized to obtain

$$D_\theta^\beta\psi_{n\theta}^{n\beta}(x) = \psi_{(n+1)\theta}^{(n+1)\beta}(x), \quad n \in \mathbb{Z}.$$

This shows that generalized two-sided fractional derivative D_θ^β is a *right shift operator*. Because that $D_{m\theta}^{m\beta}D_{n\theta}^{n\beta} = D_{(m+n)\theta}^{(m+n)\beta}$ we get

$$D_{m\theta}^{m\beta}\psi_{n\theta}^{n\beta}(x) = \psi_{(n+m)\theta}^{(n+m)\beta}(x), \quad n \in \mathbb{Z}. \quad (2)$$

3 Operational Calculus

Based in the theory presented in [13], we will construct an operational calculus adapted to generalized TSFD.

Let $\{\psi_{n\theta}^{n\beta}(x)\}_{n\in\mathbb{Z}}$ be a sequence of base functions for $\beta, \theta \in \mathbb{R}$. These functions will conform the set $\mathscr{F}$ of all the formal Laurent series

$$\mathcal{F} = \left\{ \sum_{n=-k}^{\infty} c_n \psi_{n\theta}^{n\beta}(x) : k \in \mathbb{N}_0, c_n \in \mathbb{C} \right\}.$$

We define the algebraic generalized convolution as

$$\psi_{n\theta}^{n\beta}(x) * \psi_{m\theta}^{m\beta}(x) = \psi_{(n+m)\theta}^{(n+m)\beta}(x)(x).$$

In terms of product $*$, the formulae (2) can be rewritten as

$$D_{n\theta}^{n\beta} \psi_{m\theta}^{m\beta}(x) = \psi_{n\theta}^{n\beta}(x) * \psi_{m\theta}^{m\beta}(x) = \psi_{(n+m)\theta}^{(n+m)\beta}(x).$$

Following the Table 1, we observe that $\psi_0^0(x) = \delta(x)$ which leads to

$$D_0^0 \psi_{m\theta}^{m\beta}(x) = \delta(x) * \psi_{m\theta}^{m\beta}(x) = \psi_{m\theta}^{m\beta}(x).$$

This means that $\delta(x)$ is the multiplicative neutral element for product $*$.

By means of Cauchy product, we extended the previous convolution to $\mathcal{F}$ as follows. If $c = \sum_{n=-k_c}^{\infty} c_n \psi_{n\theta}^{n\beta}(x)$, and $d = \sum_{n=-k_d}^{\infty} d_n \psi_{n\theta}^{n\beta}(x)$, are elements of $\mathcal{F}$, then

$$c * d = \sum_{n=-k_c-k_d+1}^{\infty} f_n \psi_{n\theta}^{n\beta}(x),$$

where

$$f_n = \sum_{-k_c \le j \le n+k_d} c_j d_{n-j}.$$

This generalized convolution satisfies the mathematical properties of associativity and commutativity. Under this multiplication, $\mathcal{F}$ turns out to be a field. Define

$$E_{\gamma,m}(x) = \sum_{n=m}^{\infty} \binom{n}{m} \gamma^{n-m} \psi_{n\theta}^{n\beta}(x).$$

Let D_γ denote the usual differentiation with respect to γ. It is not difficult to verify that

$$\frac{D_\gamma^k}{k!} E_{\gamma,0}(x) = E_{\gamma,k}(x),$$

and

$$E_{\gamma,k}(x) = \psi_{k\theta}^{k\beta}(x) * \left(E_{\gamma,0}(x)\right)_*^{k+1},$$

where $(\cdot)_*^{k+1} := (\cdot) * (\cdot) * \cdots * (\cdot)$, $(k+1)-$times. Other properties are listed below.

–
$$\left(\psi_0^0(x) - \gamma\psi_\theta^\beta(x)\right) * E_{\gamma,0}(x) = \psi_0^0(x), \tag{3}$$

–
$$\psi_{-k\theta}^{-k\beta}(x) * \left(\psi_0^0(x) - \gamma\psi_\theta^\beta(x)\right)_*^{k+1} * E_{\gamma,k}(x) = \psi_0^0(x).$$

–
$$\psi_\theta^\beta(x) * E_{\gamma_1,0}(x) * E_{\gamma_2,0}(x) = \frac{E_{\gamma_1,0} - E_{\gamma_2,0}}{\gamma_1 - \gamma_2}, \qquad \gamma_1 \neq \gamma_2. \tag{4}$$

– If $\gamma_1 \neq \gamma_2$, then

$$\psi_\theta^\beta(x) * E_{\gamma_1,m}(x) * E_{\gamma_2,n}(x) = \sum_{l=0}^{m} \frac{\binom{n+l}{l}(-1)^l E_{\gamma_1,m-l}(x)}{(\gamma_1 - \gamma_2)^{1+n+l}} + \sum_{j=0}^{n} \frac{\binom{m+j}{j}(-1)^j E_{\gamma_2,n-j}(x)}{(\gamma_2 - \gamma_1)^{1+m+j}}.$$

4 Illustrative Examples

The kind of differential equations to be resolved by means of this operational calculus have the format
$$w(D_\theta^\beta)f(x) = g(x), \tag{5}$$
where w is a polynomial with complex coefficients, f is a function to be found and $g(x)$ is a known function. The way to solve (5) can be handled in two ways. The direct form is when the function $g(x)$ can be expressed as a series in $\mathfrak{F}$. The other form is using the usual convolution. We found the solution $f(x)$ of

$$w(D_\theta^\beta)f(x) = \delta(x), \tag{6}$$

and then the solution of (5) is given by the usual convolution of $f(x)$ and $g(x)$. Due to the above, in our example we only solve differential equation of the kind (6).

Example 1. Consider the differential equation

$$D_\theta^\beta f(x) - f(x) = \delta(x). \tag{7}$$

In terms of the algebraic convolution, this system can be rewritten as

$$-\left(\psi_0^0(x) - \psi_\theta^\beta(x)\right) * f(x) = \psi_0^0(x)$$

By (3), $E_{1,0}(x)$ is the multiplicative inverse of $\left(\psi_0^0(x) - \psi_\theta^\beta(x)\right)$. Then

$$f(x) = -E_{1,0}(x) * \psi_0^0(x).$$

Since $\delta(x)$ is the multiplicative neutral, it follows that the solution of (8) is the series

$$f(x) = -E_{1,0}(x) = -\sum_{n=0}^{\infty} \psi_{n\theta}^{n\beta}(x).$$

If β is an irrational number, then

$$f(x) = -\sum_{n=0}^{\infty} \frac{\sin\left((n\beta + n\theta sgn(x))\,\frac{\pi}{2}\right)}{2\sin(n\beta\pi)\Gamma(-n\beta)} |x|^{-n\beta-1}.$$

When β is a rational number, we need to use the Table 1 to compute the solution. Below we consider some interesting cases.

- For $\beta = 1$ and $\theta = 0$, we have the series

$$f(x) = -\sum_{n=0}^{\infty} \frac{(-1)^n (2n+1)!}{\pi} |x|^{-2n-2}, \qquad x > 0.$$

  This series can be rewritten as

$$g(z) = -\frac{z}{\pi} \sum_{n=0}^{\infty} (-1)^n (2n+1)! z^n,$$

  with $z = |x|^{-2}$. By the root test, the convergence ratio is $r = 0$. Then the series $g(z)$ is not convergent for any z. We conclude that also $f(x)$ is not convergent for any x different to zero.

- For $\beta = -1$ and $\theta = 0$, we obtain, after algebraic manipulations, the series

$$f(x) = \frac{1}{2}\sin(x) + \frac{1}{\pi}\left(\cos(x) - 1\right)\left(\ln|x| - \Gamma'(1)\right), \qquad x > 0.$$

- For $\beta = -1$ and $\theta = 1$ we get that

$$f(x) = -\frac{1}{2}(\cos(x) - 1) - \frac{1}{\pi}\sin(x)\left(\ln|x| - \Gamma'(1)\right), \qquad x > 0.$$

Example 2. Consider the differential equation

$$D_{2\theta}^{2\beta} f(x) + f(x) = \delta(x). \tag{8}$$

In therms of algebraic convolution

$$\left(\psi_0^0(x) - i\psi_\theta^\beta(x)\right) * \left(\psi_0^0(x) + i\psi_\theta^\beta(x)\right) * f(x) = \psi_0^0(x)$$

The multiplicative inverse of operator $\left(\psi_0^0(x) - i\psi_\theta^\beta(x)\right) * \left(\psi_0^0(x) + i\psi_\theta^\beta(x)\right)$ is given by $E_{i,0}(x) * E_{-i,0}(x)$. Then

$$f(x) = E_{i,0}(x) * E_{-i,0}(x) * \psi_0^0(x)$$

From (4), we have that $E_{i,0}(x) * E_{-i,0}(x) = \sum_{n=0}^{\infty} (-1)^n \psi_{2n\theta}^{2n\beta}(x)$. Finally

$$f(x) = \sum_{n=0}^{\infty} (-1)^n \psi_{2n\theta}^{2n\beta}(x)$$

If β is a irrational number, then

$$f(x) = \sum_{n=0}^{\infty} (-1)^n \frac{\sin\left((2n\beta + 2n\theta sgn(x))\,\frac{\pi}{2}\right)}{2\sin(2n\beta\pi)\Gamma(-2n\beta)} |x|^{-2n\beta-1}.$$

Some important cases for β rational are included below.

- For $\beta = -1$ and $\theta = 0$, we obtain, after algebraic manipulations, the series

$$f(x) = -\frac{1}{2}\sin(x), \qquad x > 0.$$

- For $\beta = -1$ and $\theta = 1$ we get that

$$f(x) = \frac{\cos(x)}{x\pi}\left(\ln|x| - \Gamma'(1)\right), \qquad x > 0.$$

5 Conclusions

Based in the fact that generalized two-sided fractional derivative is a right shift operator, we developed an operational method to solve differential equations based on the TSFD. This method do not require advanced mathematical tools neither the integral concept. We have solved some examples and from these we conclude that the operational method is easy to apply. Our solutions are expressed as series and in some cases we have obtained the closed form.

Future works:

- Extend this operational calculus to case of two variables and applying to space-time fractional diffusion equation and Maxwell's equations.
- In [16], several operators and special functions are exposed which we consider candidates to apply our operational method. We will study them and we will determine if our method can be applied to these operators and special functions.

Acknowledgments. The second author was partially funded by Portuguese National Funds through the FCT-Foundation for Science and Technology within the scope of the CTS Research Unit - Center of Technology and Systems/UNINOVA/FCT/NOVA, under the reference UIDB/00066/2020.

References

1. Riesz, M.: L'intégrale de Riemann-Liouville et le problème de Cauchy. Acta Math. **81**, 1–222 (1949). https://doi.org/10.1007/BF02395016
2. Ortigueira, M.: Two-sided and regularised Riesz-Feller derivatives. Math. Methods Appl. Sci. **44**(10), 8057–8069 (2021). https://doi.org/10.1002/mma.5720
3. Fremberg, N.: Some applications of the Riesz potential to the theory of the electromagnetic field and the meson field. Proc. R. Soc. Lond. A: Math. Phys. Eng. Sci. **188**(1012), 18–31 (1946). https://doi.org/10.1098/rspa.1946.0094

4. Feller, W.: On a generalization of Marcel Riesz potentials and the semigroups, generated by them. In: Communications du seminaire mathematique de universite de Lund, vol. 21, pp. 72–81 (1952)
5. Jianping, D., Mingyu, X.: Space-time fractional schrödinger equation with time-independent potentials. J. Math. Anal. Appl. **344**(2), 1005–1017 (2008). https://doi.org/10.1016/j.jmaa.2008.03.061
6. Gorenflo, R., Mainardi, F.: Some recent advances in theory and simulation of fractional diffusion processes. J. Comput. Appl. Math. **229**(2), 400–415 (2009). https://doi.org/10.1016/j.cam.2008.04.005
7. Jarosz, S., Vaz, J.: Fractional schrödinger equation with Riesz-Feller derivative for delta potentials. J. Math. Phys. **57**(12), 123506 (2016). https://doi.org/10.1063/1.4972291
8. Luchko, Y., Gorenflo, R.: Scale-invariant solutions of a partial differential equation of fractional order. Fract. Calc. Appl. Anal. **1**(1), 63–78 (1998)
9. Laskin, N.: Fractional quantum mechanics and Lévy path integrals. Phys. Lett. A **268**(4), 298–305 (2000). https://doi.org/10.1016/S0375-9601(00)00201-2
10. Laskin, N.: Fractional Schrödinger equation. Phys. Rev. E **66**, 056108 (2002). https://doi.org/10.1103/PhysRevE.66.056108
11. Mainardi, F.: Applications of integral transforms in fractional diffusion processes. Integral Transform. Spec. Funct. **15**(6), 477–484 (2004). https://doi.org/10.1080/10652460412331270652
12. Gulgowski, J., Kwiatkowski, D., Stefański, T.: Signal propagation in electromagnetic media modelled by the two-sided fractional derivative. Fractal Fract. **5**(1), 10 (2021). https://doi.org/10.3390/fractalfract5010010
13. Bengochea, G., Verde-Star, L.: Linear algebraic foundations of the operational calculi. Adv. Appl. Math. **47**, 330–351 (2011). https://doi.org/10.1016/j.aam.2010.08.001
14. Bengochea, G.: An operational approach with application to fractional bessel equation. Fract. Calc. Appl. Anal. **18**(5), 1201–1211 (2015). https://doi.org/10.1515/fca-2015-0069
15. Bengochea, G., Ortigueira, M.: An operational approach to solve fractional continuous-time linear systems. Int. J. Dyn. Control **5**(1), 61–71 (2017). https://doi.org/10.1007/s40435-015-0220-z
16. Kiryakova, V.: Unified approach to fractional calculus images of special functions-a survey. Mathematics **8**(12), 2260 (2020). https://doi.org/10.3390/math8122260

A Remark on the Memory Property of Fractional Difference Operators

Jagan Mohan Jonnalagadda(✉)

Department of Mathematics, Birla Institute of Technology and Science Pilani, Hyderabad 500078, Telangana, India
j.jaganmohan@hotmail.com

Abstract. Fractional difference operators possess nonlocal structure which largely affects and complicates the qualitative analysis of fractional difference equations. To understand the effect of this memory property, in this article, we establish sufficient conditions on asymptotic behaviour of the nabla fractional difference equation

$$\big(\nabla^{\nu}_{\rho(a)}u\big)(t) = c(t)u(t-1), \quad t \in \mathbb{N}_{a+1}, \tag{1}$$

and compare these properties with that of the corresponding first order nabla difference equation

$$\big(\nabla u\big)(t) = c(t)u(t-1), \quad t \in \mathbb{N}_{a+1}.$$

Here $c : \mathbb{N}_{a+1} \to \mathbb{R}$ and $0 < \nu < 1$.

Keywords: Nabla fractional difference · Nonlocal · Memory · Solution · Asymptotic behaviour

1 Introduction and Preliminaries

The notion of fractional derivative [13,15] dates back to the works of Euler but the concept of nabla fractional difference [10] is just a decade old. Nabla fractional calculus represents a natural instrument to model nonlocal phenomena either in space or time. The nabla fractional difference of any function contains information about the function at earlier points, so it possesses a long memory effect. From science to engineering, there are many processes that involve different space or time scales. In many problems of this context, the dynamics of the system can be formulated by fractional difference equations which include the nonlocal effects.

Throughout the article, we shall use the following notations, definitions, and known results of nabla fractional calculus [10]. For any real number a, denote by $\mathbb{N}_a = \{a, a+1, a+2, \ldots\}$. The backward jump operator $\rho : \mathbb{N}_{a+1} \to \mathbb{N}_a$ is defined by

$$\rho(t) = t - 1, \quad t \in \mathbb{N}_{a+1}.$$

A. Dzielinski et al. (Eds.): ICFDA 2021, LNNS 452, pp. 161–170, 2022.
https://doi.org/10.1007/978-3-031-04383-3_18

Define the μth-order nabla fractional Taylor monomial by

$$H_{\mu}(t,a) = \frac{(t-a)^{\overline{\mu}}}{\Gamma(\mu+1)} = \frac{\Gamma(t-a+\mu)}{\Gamma(t-a)\Gamma(\mu+1)}, \quad \mu \in \mathbb{R} \setminus \{\dots,-2,-1\},$$

provided the right-hand side exists. Here $\Gamma(\cdot)$ denotes the Euler gamma function. Observe that $H_{\mu}(a,a) = 0$ and $H_{\mu}(t,a) = 0$ for all $\mu \in \{\dots,-2,-1\}$ and $t \in \mathbb{N}_a$. From [12], we have that

$$\lim_{t\to\infty} H_{\mu-1}(t,a) = 0, \quad 0 < \mu < 1. \tag{2}$$

Definition 1 (See [7]). *Let $u : \mathbb{N}_a \to \mathbb{R}$ and $N \in \mathbb{N}_1$. The first order backward (nabla) difference of u is defined by*

$$\big(\nabla u\big)(t) = u(t) - u(t-1), \quad t \in \mathbb{N}_{a+1},$$

and the Nth-order nabla difference of u is defined recursively by

$$\big(\nabla^{N} u\big)(t) = \Big(\nabla\big(\nabla^{N-1}u\big)\Big)(t), \quad t \in \mathbb{N}_{a+N}.$$

Definition 2 (See [10]). *Let $u : \mathbb{N}_{a+1} \to \mathbb{R}$ and $\nu > 0$. The νth-order nabla sum of u based at a is given by*

$$\big(\nabla_a^{-\nu} u\big)(t) = \sum_{s=a+1}^{t} H_{\nu-1}(t,\rho(s))u(s), \quad t \in \mathbb{N}_a,$$

where by convention $\big(\nabla_a^{-\nu} u\big)(a) = 0$ and $\nabla_a^{-0} u = u$.

Definition 3 (See [10]). *Let $u : \mathbb{N}_{a+1} \to \mathbb{R}$, $\nu > 0$ and choose $N \in \mathbb{N}_1$ such that $N-1 < \nu \leq N$. The νth-order Riemann–Liouville nabla difference of u based at a is given by*

$$\big(\nabla_a^{\nu} u\big)(t) = \Big(\nabla^{N}\big(\nabla_a^{-(N-\nu)}u\big)\Big)(t), \quad t \in \mathbb{N}_{a+N}.$$

Theorem 1 (See [2]). *Let $u : \mathbb{N}_{a+1} \to \mathbb{R}$, $\nu > 0$ and choose $N \in \mathbb{N}_1$ such that $N-1 < \nu < N$. Then*

$$\big(\nabla_a^{\nu} u\big)(t) = \sum_{s=a+1}^{t} H_{-\nu-1}(t,\rho(s))u(s), \quad t \in \mathbb{N}_{a+1}.$$

Clearly, $\big(\nabla_a^{\nu} u\big)(t)$ depends on the values of u at the points $a+1$, $a+2$, ..., t unlike the first order nabla difference $\big(\nabla u\big)(t)$, which just depends on the values of u at the points $t-1$ and t only. Hereafter, we say that the memory of $\big(\nabla_a^{\nu} u\big)(t)$ is $(t-a)$.

Lemma 1 [10]. *Let ν, $\mu \in \mathbb{R}$ and $n \in \mathbb{N}_1$ such that $N-1 < \nu < N$. Then,*

$$\nabla_a^{\nu} H_{\mu}(t,a) = H_{\mu-\nu}(t,a), \quad t \in \mathbb{N}_{a+1}.$$

Atici and Eloe [4] initiated the study of asymptotic behaviour of the two-term linear homogeneous nabla fractional difference equation

$$\big(\nabla_{-1}^{1/2}u\big)(t) = \lambda u(t), \quad t \in \mathbb{N}_1, \quad \lambda \neq 1, \tag{3}$$

with the initial condition $u(0) > 0$. Later, Čermák et al. [8] derived a sharp condition for the asymptotic stability, and gave a precise asymptotic description of the solutions of the two-term linear homogeneous nabla fractional difference equation

$$\big(\nabla_{-1}^{\nu}u\big)(t) = \lambda u(t), \quad t \in \mathbb{N}_1, \quad \lambda \neq 1, \tag{4}$$

where $0 < \nu < 1$. In this line, Jia et al. [12] and Wu [19] obtained asymptotic results for the nabla fractional difference equation

$$\big(\nabla_{\rho(a)}^{\nu}u\big)(t) = c(t)u(t), \quad t \in \mathbb{N}_{a+1}, \tag{5}$$

with the initial condition $u(0) > 0$. Here $c : \mathbb{N}_{a+1} \to \mathbb{R}$ and $0 < \nu < 1$. Following these works, the authors in [11,16,17] established some criteria for boundedness, stability properties, and separation of solutions for various classes of nabla fractional difference equations using the variation of constants formula.

The memory property of nabla fractional difference operators play an important role in describing the behaviour of solutions of the corresponding nabla fractional difference equations. We demonstrate this fact through the following examples.

Example 1. Assume $0 < \nu, \mu < 1$. Consider the first order and the νth-order nabla difference equations

$$\big(\nabla u\big)(t) = H_{\mu-1}(t, \rho(a)), \quad t \in \mathbb{N}_{a+1}, \tag{6}$$

$$\big(\nabla_{\rho(a)}^{\nu}u\big)(t) = H_{\mu-1}(t, \rho(a)), \quad t \in \mathbb{N}_{a+1}. \tag{7}$$

For $u_0 \in \mathbb{R}$, every solution of (6) is of the form

$$u(t, a, u_0) = u_0 + H_{\mu}(t, \rho(a)), \quad t \in \mathbb{N}_a, \tag{8}$$

which tends to ∞ as $t \to \infty$ [7,9]. Every solution of (7) is of the form

$$u(t, a, u_0) = u_0 H_{\nu-1}(t, \rho(a)) + H_{\nu+\mu-1}(t, \rho(a)), \quad t \in \mathbb{N}_a. \tag{9}$$

Assume $0 < \nu+\mu < 1$. Then, from (2), every solution of (7) tends to 0 as $t \to \infty$.

Example 2. Assume $0 < \nu < 1$, $c : \mathbb{N}_{a+1} \to \mathbb{R}$ such that $1-c(t) \neq 0$ for $t \in \mathbb{N}_{a+1}$. Consider the first order and the νth-order nabla difference equations

$$\big(\nabla u\big)(t) = c(t)u(t), \quad t \in \mathbb{N}_{a+1}, \tag{10}$$

$$\big(\nabla_{\rho(a)}^{\nu}u\big)(t) = c(t)u(t), \quad t \in \mathbb{N}_{a+1}. \tag{11}$$

From [7,9], we know that every solution of (10) tends to 0 as $t \to \infty$ if

$$\lim_{t\to\infty} \left| \prod_{s=a+1}^{t} \left[\frac{1}{1-c(s)}\right] \right| = 0.$$

Also, it follows from [19] that every solution of (11) tends to 0 as $t \to \infty$ if $|1 - c(t)| \geq 1$ for $t \in \mathbb{N}_{a+1}$. For example, if we take $c(t) = 2$ for $t \in \mathbb{N}_1$, every solution of

$$(\nabla u)(t) = 2u(t), \quad t \in \mathbb{N}_1,$$

doesn't tends to zero as $t \to \infty$ whereas every solution of

$$(\nabla^{\nu}_{\rho(0)} u)(t) = 2u(t), \quad t \in \mathbb{N}_1,$$

tends to 0 as $t \to \infty$.

Motivated by these facts, in this article, we establish sufficient conditions on asymptotic behaviour of solutions of the following nabla fractional difference equation

$$(\nabla^{\nu}_{\rho(a)} u)(t) = c(t)u(t-1), \quad t \in \mathbb{N}_{a+1}, \tag{12}$$

and compare these properties with that of the corresponding first order nabla difference equation

$$(\nabla u)(t) = c(t)u(t-1), \quad t \in \mathbb{N}_{a+1}. \tag{13}$$

to study the effect of this memory property. Here $c : \mathbb{N}_{a+1} \to \mathbb{R}$ and $0 < \nu < 1$.

2 Main Results

Define a sequence recursively as follows: $\tilde{E}_{c,\nu}(a, a) = 1$ and

$$\tilde{E}_{c,\nu}(t, a) = c(t)\tilde{E}_{c,\nu}(t-1, a) - \sum_{s=a}^{t-1} H_{-\nu-1}(t, \rho(s))\tilde{E}_{c,\nu}(s, a), \quad t \in \mathbb{N}_{a+1}. \tag{14}$$

Theorem 2. *Assume $c : \mathbb{N}_{a+1} \to \mathbb{R}$ and $N - 1 < \nu < N$, $N \in \mathbb{N}_1$. Then, the unique solution of the nabla fractional initial value problem*

$$\begin{cases} (\nabla^{\nu}_{\rho(a)} u)(t) = c(t)u(t-1), & t \in \mathbb{N}_{a+1}, \\ u(a) \in \mathbb{R}, \end{cases} \tag{15}$$

is given by

$$u(t) = u(a)\tilde{E}_{c,\nu}(t, a), \quad t \in \mathbb{N}_a, \tag{16}$$

where $\tilde{E}_{c,\nu}(t, a)$ is defined by (14).

Proof. From Theorem 1, we have

$$\begin{aligned} (\nabla^{\nu}_{\rho(a)} u)(t) &= \sum_{s=a}^{t} H_{-\nu-1}(t, \rho(s))u(s) \\ &= \sum_{s=a}^{t-1} H_{-\nu-1}(t, \rho(s))u(s) + u(t), \quad t \in \mathbb{N}_{a+1}. \end{aligned} \tag{17}$$

Then, it follows from (15) and (17) that

$$u(t) = c(t)u(t-1) - \sum_{s=a}^{t-1} H_{-\nu-1}(t, \rho(s))u(s), \quad t \in \mathbb{N}_{a+1}. \tag{18}$$

Take $t = a + n$, $n \in \mathbb{N}_1$. From (16), it is enough to prove that the statement

$$u(a+n) = u(a)\tilde{E}_{c,\nu}(a+n, a), \quad n \in \mathbb{N}_1, \tag{19}$$

is true by using the principle of strong mathematical induction on n. From (14), (17) and (18), we obtain

$$\begin{aligned}
&u(a+1)\\
&= c(a+1)u(a) - \sum_{s=a}^{a} H_{-\nu-1}(a+1, \rho(s))u(s)\\
&= u(a)\left[c(a+1) - \sum_{s=a}^{a} H_{-\nu-1}(a+1, \rho(s))\tilde{E}_{c,\nu}(s, a)\right]\\
&= u(a)\left[c(a+1)\tilde{E}_{c,\nu}(a, a) - \sum_{s=a}^{a} H_{-\nu-1}(a+1, \rho(s))\tilde{E}_{c,\nu}(s, a)\right]\\
&= u(a)\left[c(a+1)\tilde{E}_{c,\nu}(a+1-1, a) - \sum_{s=a}^{a+1-1} H_{-\nu-1}(a+1, \rho(s))\tilde{E}_{c,\nu}(s, a)\right]\\
&= u(a)\tilde{E}_{c,\nu}(a+1, a).
\end{aligned}$$

Thus, the statement (19) is true for $n = 1$. Again, from (14), (17) and (18), we obtain

$$\begin{aligned}
&u(a+2)\\
&= c(a+2)u(a+1) - \sum_{s=a}^{a+1} H_{-\nu-1}(a+2, \rho(s))u(s)\\
&= u(a)\left[c(a+2)\tilde{E}_{c,\nu}(a+1, a) - \sum_{s=a}^{a+1} H_{-\nu-1}(a+2, \rho(s))\tilde{E}_{c,\nu}(s, a)\right]\\
&= u(a)\left[c(a+2)\tilde{E}_{c,\nu}(a+2-1, a) - \sum_{s=a}^{a+2-1} H_{-\nu-1}(a+2, \rho(s))\tilde{E}_{c,\nu}(s, a)\right]\\
&= u(a)\tilde{E}_{c,\nu}(a+2, a).
\end{aligned}$$

Thus, the statement (19) is true for $n = 2$. Assume the statement (19) is true for $k = 3, 4, \cdots, n-1$. That is,

$$u(a+k) = u(a)\tilde{E}_{c,\nu}(a+k, a), \quad k \in \mathbb{N}_1^{n-1}. \tag{20}$$

Now, we prove the statement (19) is true for n. It follows from (14), (17) and (18) that

$$\begin{aligned}
&u(a+n) \\
&= c(a+n)u(a+n-1) - \sum_{s=a}^{a+n-1} H_{-\nu-1}(a+n,\rho(s))u(s) \\
&= u(a)\left[c(a+n)\tilde{E}_{c,\nu}(a+n-1,a) - \sum_{s=a}^{a+n-1} H_{-\nu-1}(a+n,\rho(s))\tilde{E}_{c,\nu}(s,a)\right] \\
&= u(a)\tilde{E}_{c,\nu}(a+n,a).
\end{aligned}$$

Hence, by the principle of strong mathematical induction on n, the statement (19) is true for each $n \in \mathbb{N}_1$. Since we obtain the solution by the method of steps,

$$u(t) = u(a)\tilde{E}_{c,\nu}(t,a), \quad t \in \mathbb{N}_a,$$

is the unique solution of the initial value problem (15). The proof is complete.

Theorem 3. *Assume* $0 < \nu < 1$ *and* $|c(t)+\nu| \leq \nu$ *for* $t \in \mathbb{N}_{a+1}$. *Then, the solutions of*

$$\left(\nabla^{\nu}_{\rho(a)}u\right)(t) = c(t)u(t-1), \quad t \in \mathbb{N}_{a+1}, \tag{21}$$

satisfy

$$\lim_{t\to\infty} u(t) = 0. \tag{22}$$

Proof. We show that

$$\left|\tilde{E}_{c,\nu}(t,a)\right| \leq H_{\nu-1}(t,\rho(a)), \quad t \in \mathbb{N}_{a+1}. \tag{23}$$

Take $t = a+n$, $n \in \mathbb{N}_1$. From (23), it is enough to prove that the statement

$$\left|\tilde{E}_{c,\nu}(a+n,a)\right| \leq H_{\nu-1}(a+n,\rho(a)), \quad n \in \mathbb{N}_1, \tag{24}$$

is true by using the principle of strong mathematical induction on n. For $n = 1$, from (14), we have

$$\begin{aligned}
\left|\tilde{E}_{c,\nu}(a+1,a)\right| &= \left|c(a+1)\tilde{E}_{c,\nu}(a,a) - \sum_{s=a}^{a} H_{-\nu-1}(a+1,\rho(s))\tilde{E}_{c,\nu}(s,a)\right| \\
&= \left|(c(a+1)+\nu)\tilde{E}_{c,\nu}(a,a)\right| \\
&= |c(a+1)+\nu| \\
&\leq \nu = H_{\nu-1}(a+1,\rho(a)).
\end{aligned}$$

Thus, the statement (24) is true for $n = 1$. For $n = 2$, from (14), we have

$$\begin{aligned}\left|\tilde{E}_{c,\nu}(a+2,a)\right| &= \left|c(a+2)\tilde{E}_{c,\nu}(a+1,a) - \sum_{s=a}^{a+1} H_{-\nu-1}(a+2,\rho(s))\tilde{E}_{c,\nu}(s,a)\right| \\ &= \left|c(a+2)\tilde{E}_{c,\nu}(a+1,a) + \frac{\nu(1-\nu)}{2}\tilde{E}_{c,\nu}(a,a) + \nu\tilde{E}_{c,\nu}(a+1,a)\right| \\ &= \left|(c(a+2)+\nu)\tilde{E}_{c,\nu}(a+1,a) + \frac{\nu(1-\nu)}{2}\right| \\ &\leq |c(a+2)+\nu|\left|\tilde{E}_{c,\nu}(a+1,a)\right| + \frac{\nu(1-\nu)}{2} \\ &\leq \nu^2 + \frac{\nu(1-\nu)}{2} \\ &= \frac{\nu(\nu+1)}{2} = H_{\nu-1}(a+2,\rho(a)).\end{aligned}$$

Thus, the statement (24) is true for $n = 2$. Assume the statement (24) is true for $k = 3, 4, \cdots, n-1$. That is,

$$\left|\tilde{E}_{c,\nu}(a+k,a)\right| \leq H_{\nu-1}(a+k,\rho(a)), \quad k \in \mathbb{N}_1^{n-1}. \tag{25}$$

Now, we prove the statement (24) is true for n. Consider

$$\begin{aligned}&\left|\tilde{E}_{c,\nu}(a+n,a)\right| \\ &= \left|c(a+n)\tilde{E}_{c,\nu}(a+n-1,a) - \sum_{s=a}^{a+n-1} H_{-\nu-1}(a+n,\rho(s))\tilde{E}_{c,\nu}(s,a)\right| \\ &= \left|c(a+n)\tilde{E}_{c,\nu}(a+n-1,a) - \sum_{k=0}^{n-1} H_{-\nu-1}(a+n,\rho(a+k))\tilde{E}_{c,\nu}(a+k,a)\right|.\end{aligned} \tag{26}$$

Observe that

$$\begin{aligned}-H_{-\nu-1}(a+n,\rho(s)) &= -\frac{1}{\Gamma(-\nu)}(a+n-s+1)^{\overline{-\nu-1}} \\ &= -\frac{\Gamma(a+n-s-\nu)}{\Gamma(a+n-s+1)\Gamma(-\nu)} \\ &= \nu\frac{\Gamma(a+n-s-\nu)}{\Gamma(a+n-s+1)\Gamma(1-\nu)}\end{aligned}$$

Since $0 < \nu < 1$ and $a \leq s \leq a+n-1$, we have $1-\nu > 0$, $a+n-s+1 > 0$ and $a+n-s-\nu > 0$, implying that

$$-H_{-\nu-1}(a+n,\rho(s)) > 0.$$

Then, from (26) and (25), we have

$$\begin{aligned}
&\left|\tilde{E}_{c,\nu}(a+n,a)\right| \\
&= \Big|c(a+n)\tilde{E}_{c,\nu}(a+n-1,a) + \nu\tilde{E}_{c,\nu}(a+n-1,a) \\
&\quad - \sum_{k=0}^{n-2} H_{-\nu-1}(n,\rho(k))\tilde{E}_{c,\nu}(a+k,a)\Big| \\
&= \left|(c(a+n)+\nu)\tilde{E}_{c,\nu}(a+n-1,a) - \sum_{k=0}^{n-2} H_{-\nu-1}(n,\rho(k))\tilde{E}_{c,\nu}(a+k,a)\right| \\
&\leq |c(a+n)+\nu|\left|\tilde{E}_{c,\nu}(a+n-1,a)\right| - \sum_{k=0}^{n-2} H_{-\nu-1}(n,\rho(k))\left|\tilde{E}_{c,\nu}(a+k,a)\right| \\
&\leq \nu H_{\nu-1}(a+n-1,\rho(a)) - \sum_{k=0}^{n-2} H_{-\nu-1}(n,\rho(k))H_{\nu-1}(a+k,\rho(a)) \\
&= H_{-\nu-1}(n,\rho(n-1))H_{\nu-1}(a+n-1,\rho(a)) \\
&\quad - \sum_{k=0}^{n-2} H_{-\nu-1}(n,\rho(k))H_{\nu-1}(a+k,\rho(a)) \\
&= -\sum_{k=0}^{n-1} H_{-\nu-1}(n,\rho(k))H_{\nu-1}(a+k,\rho(a)) \\
&= -\sum_{k=0}^{n} H_{-\nu-1}(n,\rho(k))H_{\nu-1}(a+k,\rho(a)) + H_{-\nu-1}(n,\rho(n))H_{\nu-1}(a+n,\rho(a)) \\
&= -\sum_{k=0}^{n} H_{-\nu-1}(n,\rho(k))H_{\nu-1}(k,\rho(0)) + H_{\nu-1}(a+n,\rho(a)) \\
&= -\nabla^{\nu}_{\rho(0)} H_{\nu-1}(n,\rho(0)) + H_{\nu-1}(a+n,\rho(a)) \\
&= -H_{\nu-\nu-1}(n,\rho(0)) + H_{\nu-1}(a+n,\rho(a)) \quad \text{(By Lemma 1)} \\
&= -0 + H_{\nu-1}(a+n,\rho(a)) = H_{\nu-1}(a+n,\rho(a)).
\end{aligned}$$

Hence, by the principle of strong mathematical induction on n, the statement (24) is true for each $n \in \mathbb{N}_1$. Thus, we have (23). From Theorem 2 and (2), it follows that

$$0 \leq |u(t)| = |u(a)||\tilde{E}_{c,\nu}(t,a)| \leq |u(a)|\,|H_{\nu-1}(t,\rho(a))| \to 0 \text{ as } t \to \infty.$$

The proof is complete.

3 Conclusion

Assume $0 < \nu < 1$, $c : \mathbb{N}_{a+1} \to \mathbb{R}$. Consider the first order and the νth-order nabla difference equations

$$\big(\nabla u\big)(t) = c(t)u(t), \quad t \in \mathbb{N}_{a+1}, \tag{27}$$

$$\big(\nabla^{\nu}_{\rho(a)} u\big)(t) = c(t)u(t), \quad t \in \mathbb{N}_{a+1}. \tag{28}$$

From [7,9], we know that every solution of (27) tends to 0 as $t \to \infty$ if

$$\lim_{t\to\infty} \left| \prod_{s=a+1}^{t} [1 + c(s)] \right| = 0.$$

It follows from Theorem 3 that every solution of (28) tends to 0 as $t \to \infty$ if $|c(t) + \nu| \leq \nu$ for $t \in \mathbb{N}_{a+1}$. For example, if we take $c(t) = 0$ for $t \in \mathbb{N}_1$, every solution of

$$\big(\nabla u\big)(t) = 0, \quad t \in \mathbb{N}_1,$$

doesn't tends to zero as $t \to \infty$ whereas every solution of

$$\big(\nabla^{\nu}_{\rho(0)} u\big)(t) = 0, \quad t \in \mathbb{N}_1,$$

tends to 0 as $t \to \infty$.

References

1. Abdeljawad, T., Atici, F.M.: On the definitions of nabla fractional operators. In: Abstract and Applied Analysis 2012, p. 13 (2012). Art. ID 406757
2. Ahrendt, K., Castle, L., Holm, M., Yochman, K.: Laplace transforms for the nabla-difference operator and a fractional variation of parameters formula. Commun. Appl. Anal. **16**(3), 317–347 (2012)
3. Atici, F.M., Eloe, P.W.: Discrete fractional calculus with the nabla operator. Electron. J. Qual. Theory Differ. Equ. (3), 12 (2009). Special Edition I
4. Atici, F.M., Eloe, P.W.: Linear systems of fractional nabla difference equations. Rocky Mt. J. Math. **41**(2), 353–370 (2021)
5. Atici, F.M., Atici, M., Nguyen, N., Zhoroev, T., Koch, G.: A study on discrete and discrete fractional pharmacokinetics-pharmacodynamics models for tumor growth and anti-cancer effects. Comput. Math. Biophys. **7**, 10–24 (2019)
6. Baleanu, D., Wu, G.C.: Some further results of the Laplace transform for variable-order fractional difference equations. Fract. Calc. Appl. Anal. **22**(6), 1641–1654 (2019). https://doi.org/10.1515/fca-2019-0084
7. Bohner, M., Peterson, A.: Dynamic Equations on Time Scales. An Introduction with Applications. Birkhäuser Boston Inc., Boston (2001)
8. Čermák, J., Kisela, T., Nechvátal, L.: Stability and asymptotic properties of a linear fractional difference equation. Adv. Differ. Equ. **122**, 14 (2012). https://doi.org/10.1186/1687-1847-2012-122
9. Elaydi, S.: An Introduction to Difference Equations. Undergraduate Texts in Mathematics, 3rd edn. Springer, New York (2005). https://doi.org/10.1007/0-387-27602-5

10. Goodrich, C., Peterson, A.C.: Discrete Fractional Calculus. Springer, Cham (2015). https://doi.org/10.1007/978-3-319-25562-0
11. Jonnalagadda, J.M.: Asymptotic behaviour of linear fractional nabla difference equations. Int. J. Differ. Equ. **12**(2), 255–265 (2017)
12. Jia, B., Erbe, L., Peterson, A.: Comparison theorems and asymptotic behavior of solutions of discrete fractional equations. Electron. J. Qual. Theory Differ. Equ. 18 (2015). Paper No. 89
13. Kilbas, A.A., Srivastava, H.M., Trujillo, J.J.: Theory and Applications of Fractional Differential Equations. North-Holland Mathematics Studies. Elsevier Science B. V., Amsterdam (2006)
14. Huang, L.-L., Wu, G.-C., Baleanu, D., Wang, H.-Y.: Discrete fractional calculus for interval-valued systems. Fuzzy Sets Syst. **404**, 141–158 (2021)
15. Podlubny, I.: Fractional Differential Equations. Mathematics in Science and Engineering. Academic Press Inc., San Diego (1999)
16. The Anh, P., Babiarz, A., Czornik, A., Niezabitowski, M., Siegmund, S.: Some results on linear nabla Riemann-Liouville fractional difference equations. Math. Methods Appl. Sci. **43**(13), 7815–7824 (2020)
17. Shobanadevi, N., Mohan, J.J.: Stability of linear nabla fractional difference equations. In: Proceedings of the Jangjeon Mathematical Society, vol. 17, issue number 4, pp. 651–657 (2014)
18. Wu, G.C., Baleanu, D.: Discrete fractional logistic map and its chaos. Nonlinear Dyn. **75**, 283–287 (2014). https://doi.org/10.1007/s11071-013-1065-7
19. Wu, H.: Asymptotic behavior of solutions of fractional nabla difference equations. Appl. Math. Lett. **105**, 106302, 6 (2020)

Mikusiński's Operational Calculus Applied in General Classes of Fractional Calculus

Arran Fernandez(✉)

Department of Mathematics, Eastern Mediterranean University,
Northern Cyprus, via Mersin-10, 99628 Famagusta, Turkey
arran.fernandez@emu.edu.tr
https://staff.emu.edu.tr/arranfernandez/en/

Abstract. Mikusiński's operational calculus is an algebraic formalism for integral and derivative operators, constructed in the mid 20th century, which can be used for solving differential equations. It is formally similar to the method of Laplace transforms, but applicable to a bigger set of problems due to its setting in a larger function space. In the 1990s, this formalism was extended to fractional integral and derivative operators and used to solve fractional differential equations. Current research is extending the method to more general types of fractional-calculus operators, introducing new function spaces and techniques to handle such generalised settings.

Keywords: Fractional differential equations · Operational calculus · Generalised fractional calculus

1 Introduction

Mikusiński's operational calculus is a methodology which uses algebraic abstraction to understand the operators of calculus and to solve differential equations. It was posed by the Polish mathematician J. Mikusiński in the 1950s [1], but it never gained much traction compared with the much more popular Laplace transform method, even though Mikusiński's method is more widely applicable as it works within larger function spaces.

This methodology was applied to the operators of fractional calculus, and used to solve fractional differential equations, by Y. Luchko and collaborators in the 1990s [2–4]. They showed how Mikusiński's operational calculus can be used in the setting of Riemann–Liouville fractional calculus, Caputo fractional calculus, and others. More recent work in this topic has focused on generalisations of the classical fractional calculus, such as Hilfer fractional calculus [5], fractional calculus with respect to functions [6,7], and fractional calculus with different kernel functions [8]. We aim here to present the state of the art in applying Mikusiński's operational calculus in fractional calculus, from the origins of this topic to recent and ongoing research.

A. Dzielinski et al. (Eds.): ICFDA 2021, LNNS 452, pp. 171–176, 2022.
https://doi.org/10.1007/978-3-031-04383-3_19

2 The Riemann–Liouville and Caputo Settings

2.1 Informal Description

We begin this section with an executive summary of the essential ideas of Mikusiński's operational calculus. For a more detailed description, we refer to the works of Y. Luchko with various collaborators [2–4].

The starting point is a space of functions which is defined by their growth behaviour near an initial value, controlling any possible singularity by a given power function. It can be proved that this space of functions forms a commutative rng (ring without identity) under the algebraic operations of addition and Laplace-type convolution. Formally, this ring can be extended to a field by using the well-known field of fractions formalism. In this way, we obtain a symbolic inverse to the operation of convolution (treated as a sort of multiplication), in the setting of an abstract field which contains a rng of functions inside it.

The definition of the Riemann–Liouville fractional integral leads to a natural representation of this operator by a fractional power function within the aforementioned function space. By considering composition properties of the Riemann–Liouville fractional integral operator with the fractional derivative operators of both Riemann–Liouville and Caputo type, it is possible to describe both the Riemann–Liouville fractional derivative and the Caputo fractional derivative using elements of the abstract field constructed above.

Fractional differential equations, with either Riemann–Liouville or Caputo derivatives, can now be posed and solved purely algebraically, performing multiplication and division within the field of fractions. In this way, a solution is obtained which is expressed as an element of this abstract field, and can therefore be described as a weak solution. It is then possible, in many important cases, to show that this abstract field element is actually within the rng that was the original function space, and therefore it provides a direct solution of the original fractional differential equation.

2.2 Formal Analysis

Formally, we define the function space C_{-1} with a rng structure as follows:

$$C_{-1} = \left\{ f : [0, \infty) \to \mathbb{R} \;\middle|\; f(x) = x^p f_1(x),\ p > -1,\ f_1 \text{ continuous on } [0, \infty) \right\}, \tag{1}$$

$$(f + g)(x) = f(x) + g(x),$$

$$(f * g)(x) = \int_0^x f(y)g(x - y)\,\mathrm{d}y.$$

It is straightforward to prove that the space C_{-1} is closed under both of these two algebraic operations, and that they satisfy all the remaining axioms (commutativity, associativity, etc.) to make this function space a commutative rng. i.e. a ring without a multiplicative identity element.

The field of fractions for C_{-1} is then defined as the set of pairs $(f, g) \in C_{-1} \times C_{-1}$, quotiented by the equivalence relation that identifies (f_1, g_1) with (f_2, g_2) iff $f_1 * g_2 = g_1 * f_2$, and with the addition and multiplication operations given in the natural way interpreting pairs (f, g) as fractions $\frac{f}{g}$:

$$\frac{f_1}{g_1} + \frac{f_2}{g_2} = \frac{f_1 * g_2 + g_1 * f_2}{g_1 * g_2},$$
$$\frac{f_1}{g_1} * \frac{f_2}{g_2} = \frac{f_1 * f_2}{g_1 * g_2}.$$

This is a true algebraic field, and it forms the setting for applying the method of Mikusiński's operational calculus.

The Riemann–Liouville fractional integral to order μ of a function f, with constant of integration 0, namely ${}^{RL}_{0}I^{\mu}_{x}f(x)$, is defined for any $f \in C_{-1}$, and it is equivalent to convolution (multiplication in C_{-1}) with the power function $h_\mu(x) = \frac{x^\mu}{\Gamma(\mu+1)}$. Then the Riemann–Liouville fractional derivative and Caputo fractional derivative can be embedded into the field M_{-1} by using their composition properties with the Riemann–Liouville fractional integral:

$${}^{RL}_{0}I^{\mu}_{x}\,{}^{RL}_{0}D^{\mu}_{x}f(x) = f(x) - \frac{x^{\mu-1}}{\Gamma(\mu)} \lim_{x\to 0} {}^{RL}_{0}I^{1-\mu}_{x}f(x), \qquad 0 < \mu \leq 1; \tag{2}$$

$${}^{RL}_{0}I^{\mu}_{x}\,{}^{C}_{0}D^{\mu}_{x}f(x) = f(x) - f(0), \qquad 0 < \mu \leq 1. \tag{3}$$

Thanks to these relations, it is possible to write all three of the basic operations of fractional calculus as elements of the field M_{-1}:

$${}^{RL}_{0}I^{\mu}_{x} = h_\mu \in C_{-1};$$
$${}^{RL}_{0}D^{\mu}_{x}(f) = \frac{f - F_\mu f}{h_\mu} \in M_{-1};$$
$${}^{C}_{0}D^{\mu}_{x}(f) = \frac{f - f_0}{h_\mu} \in M_{-1},$$

where $F_\mu f$ is the so-called projector function $\frac{x^{\mu-1}}{\Gamma(\mu)} \lim_{x\to 0} {}^{RL}_{0}I^{1-\mu}_{x}f(x)$ and f_0 is the constant function $f(0)$.

Now, many fractional differential equations posed using Riemann–Liouville or Caputo fractional derivatives can be written as algebraic equations in the field M_{-1}, which often enables their solution to be easily written by algebraic manipulation. We omit the details of such applications here, but refer the interested reader to the later sections of papers such as [2–4] to see many fractional differential equations solved using Mikusiński's operational calculus.

3 The General Setting of Fractional Calculus with Respect to Functions

The fractional version of integration by substitution or the chain rule is given by defining fractional integrals and derivatives of one function with respect to another function. For the duration of this section, let $\phi : [0,\infty) \to \mathbb{R}$ be a smooth function, with $\phi(0)=0$, whose derivative ϕ' exists and is positive almost everywhere. We shall consider the structure and operators of fractional calculus with respect to the function ϕ, as introduced by Osler [9] and developed by other later authors [10–12]. The purpose of this section is to understand how a version of Mikusiński's operational calculus may be applied in this generalised setting, a project for which a more detailed exposition can be found in [6,7].

The Riemann–Liouville fractional integral to order μ of a function f with respect to ϕ, with constant of integration 0, is defined as

$$ {}^{RL}_{\;\;0}I^{\mu}_{\phi(x)}f(x)=\int_0^x \frac{\big(\phi(x)-\phi(y)\big)^{\mu-1}}{\Gamma(\mu)}f(y)\phi'(y)\,\mathrm{d}y, \qquad \mu>0. $$

This is not directly a convolution of f with another function, but it can be written using a different kind of convolution, introduced in the last few years [13,14], which we call ϕ-convolution and which is ideally suited for dealing with fractional calculus with respect to ϕ.

Using this concept of ϕ-convolution, we can also define an algebraic rng structure on a function space $C_{-1,\phi}$ which is a generalisation of the space C_{-1} defined by (1) in the previous section:

$$ C_{-1,\phi}=\left\{f:[0,\infty)\to\mathbb{R}\;\middle|\; f(x)=\phi(x)^p f_1(x),\ p>-1,\ f_1 \text{ continuous on } [0,\infty)\right\}, $$

$$ \begin{aligned} (f+g)(x)&=f(x)+g(x),\\ (f*g)(x)&=\int_0^x f\big(\phi^{-1}\left(\phi(x)-\phi(y)\right)\big)g(y)\phi'(y)\,\mathrm{d}y. \end{aligned} $$

Again, it is possible to prove that this space with these two algebraic operations is a commutative rng. Any function $f \in C_{-1,\phi}$ can be fractionally integrated with respect to ϕ, and the function ${}^{RL}_{\;\;0}I^{\mu}_{\phi(x)}f(x)$ is exactly the ϕ-convolution of $f(x)$ with the function $h_{\mu,\phi}(x)=\frac{\phi(x)^{\mu}}{\Gamma(\mu+1)}$. Constructing the field of fractions leads to an abstract field $M_{-1,\phi}$ into which the rng $C_{-1,\phi}$ embeds naturally. Therefore, once again we have defined an algebraic structure into which the operators of fractional integration (this time with respect to a function) fit naturally.

For the fractional derivatives with respect to ϕ, we must make use of the conjugation relations for fractional calculus with respect to functions. These simple but powerful relations were discussed in [10,11,15], and their significance to the study of fractional calculus with respect to functions is immense. Using them, we can immediately extend the relations (2)–(3) to the following relations for fractional integrals and derivatives with respect to ϕ:

$$ {}^{RL}_{\;\;0}I^{\mu}_{\phi(x)}\,{}^{RL}_{\;\;0}D^{\mu}_{\phi(x)}f(x)=f(x)-\frac{\phi(x)^{\mu-1}}{\Gamma(\mu)}\lim_{x\to0}{}^{RL}_{\;\;0}I^{1-\mu}_{\phi(x)}f(x), \qquad 0<\mu\le1; \quad (4) $$

$$ {}^{RL}_{\;\;0}I^{\mu}_{\phi(x)}\,{}^{C}_{0}D^{\mu}_{\phi(x)}f(x)=f(x)-f(0), \qquad 0<\mu\le1. \quad (5) $$

These enable us to embed the fractional derivatives with respect to ϕ, of both Riemann–Liouville and Caputo type, into the field $M_{-1,\phi}$. Then it is possible to solve various fractional differential equations in the setting of these operators, with the results often expressible in terms of Mittag-Leffler functions or their generalisations. For more details, we refer the reader to [6,7].

4 Other Generalisations and Extensions

Fractional calculus is no longer restricted to the traditional operators of Riemann–Liouville and Caputo, as many other types of fractional integrals and derivatives have been proposed, their mathematical properties studied, and their applications explored. The many diverse ways of defining such operators can be gathered into several broad classes [16], of which the class of fractional integrals and derivatives with respect to functions (used in the previous section) is one. It has already been seen that the method of Mikusiński's operational calculus extends naturally to these generalised operators and the associated fractional differential equations.

Another such class can be that of the so-called Prabhakar fractional integrals and derivatives [17], and a currently ongoing research project is to apply the method of Mikusiński's operational calculus to solve fractional differential equations in this setting. The techniques discovered here may then be useful in applying Mikusiński's operational calculus to even more general operators, such as the fractional integrals with general analytic kernels [18].

As well as exploring different types of operators, there are also extensions and further work to be undertaken even for the operators where Mikusiński's operational calculus was already applied in the previous two sections. For example, all of the above work assumes the initial value of $x = 0$, and the work on fractional calculus with respect to ϕ assumes that $\phi(0) = 0$. Is it possible to extend this work to a general constant of integration $c \in \mathbb{R}$, or even to the infinite lower bound $c = -\infty$? This would be useful because then the work on fractional calculus with respect to functions would also include Hadamard fractional calculus, where the initial value behaviour is infinite ($\log(x) \to -\infty$ as $x \to 0^+$).

So much work remains to be done in this direction that the short form of the current submission does not even allow all of the emerging ideas to be described. The author is hopeful that the idea of applying Mikusiński's operational calculus to fractional differential equations will soon become a thriving research topic in the pure mathematics of fractional calculus, unifying algebraists with analysts for an interdisciplinary field of mathematical research.

Notes and Comments. The work described in this short abstract summarises several different research projects: the work of Y. Luchko and collaborators in the 1990s [2–4]; the recent work of H.M. Fahad and the author [6,7]; and currently ongoing projects of the author with N. Rani, I.A. Saleh, and J.-D. Djida.

References

1. Mikusiński, J.: Operational Calculus. Pergamon Press, Oxford (1959)
2. Hadid, S.B., Luchko, Y.F.: An operational method for solving fractional differential equations of an arbitrary real order. Panamer. Math. J. **6**, 57–73 (1996)
3. Luchko, Y.F.: Operational method in fractional calculus. Frac. Calc. Appl. Anal. **2**(4), 463–488 (1999)
4. Luchko, Y.F., Gorenflo, R.: An operational method for solving fractional differential equations. Acta Math. Vietnamica **24**, 207–234 (1999)
5. Hilfer, R., Luchko, Y., Tomovski, Ž: Operational method for the solution of fractional differential equations with generalized Riemann-Liouville fractional derivatives. Fract. Calc. Appl. Anal. **12**(3), 299–318 (2009)
6. Fahad, H.M., Fernandez, A.: Operational calculus for Riemann-Liouville fractional calculus with respect to functions and the associated fractional differential equations. Frac. Calc. Appl. Anal. **24**(2), 518–540 (2021)
7. Fahad, H.M., Fernandez, A.: Operational calculus for Caputo fractional calculus with respect to functions and the associated fractional differential equations. Appl. Math. Comput. **409**, 126400 (2021)
8. Luchko, Y.: Operational calculus for the general fractional derivative and its applications. Frac. Calc. Appl. Anal. **24**(2), 338–375 (2021)
9. Osler, T.J.: Leibniz rule for fractional derivatives generalized and an application to infinite series. SIAM J. Appl. Math. **18**(3), 658–674 (1970)
10. Samko, S.G., Kilbas, A.A., Marichev, O.I.: Fractional Integrals and Derivatives: Theory and Applications. Gordon and Breach, Yverdon (1993)
11. Kilbas, A.A., Srivastava, H.M., Trujillo, J.J.: Theory and Applications of Fractional Differential Equations. Elsevier, North-Holland (2006)
12. Almeida, R.: A Caputo fractional derivative of a function with respect to another function. Commun. Nonlinear Sci. Numer. Simul. **44**, 460–481 (2017)
13. Jarad, F., Abdeljawad, T.: Generalized fractional derivatives and Laplace transform. Discrete Contin. Dyn. Syst. Ser. S **13**(3), 709–722 (2020)
14. Fahad, H.M., Rehman, M.u., Fernandez, A.: On Laplace transforms with respect to functions and their applications to fractional differential equations. Math. Meth. Appl. Sci. (2021), https://doi.org/10.1002/mma.7772.
15. Fahad, H.M., Fernandez, A., Rehman, M.u., Siddiqi, M.: Tempered and Hadamard-type fractional calculus with respect to functions. Medit. J. Math. **18** 143 (2021)
16. Baleanu, D., Fernandez, A.: On fractional operators and their classifications. Mathematics **7**(9), 830 (2019)
17. Fernandez, A., Baleanu, D.: Classes of operators in fractional calculus: a case study. Math. Meth. Appl. Sci. **44**(11), 9143–9162 (2021)
18. Fernandez, A., Özarslan, M.A., Baleanu, D.: On fractional calculus with general analytic kernels. Appl. Math. Comput. **354**, 248–265 (2019)

Control of All Axis in 3D Crane Using FOPID Controllers Optimized with GWO Algorithm

Jakub Żegleń-Włodarczyk(✉) and Klaudia Dziedzic

Department of Automatics and Robotics, AGH University, Kraków, Poland
zeglenjakub@gmail.com

Abstract. This study provides the design and implementation of control of all axis in the 3D crane. Recently used fractional calculus is used to design the FOPID controller, which has added two parameters for fractional order. The 3D crane and FOPID controller are implemented in MATLAB env. The parameters to tune the FOPID controller are optimized using the Grey Wolf Algorithm. We analyze our method efficiency with simulations. The comparison of the proposed controller and the conventional PID controller is presented.

Keywords: 3D crane · FOPID · Optimization · GWO · ORA

1 Introduction

In recent years, the world has become a global village. More and more people are ordering items from the other side of the world. Due to the growing competition, the time of delivery of parcels is very important. Often, transporters use the same ports, so queues for loading and transshipment are made. Responsible for this process are 3D cranes. During reloading, it is important to resist any disturbance that creates oscillations - the wind, the imperfection of the crane or the load [1].

Many regulators can be used to control different systems. PID [2] has been the most popular for many years. According to many studies, fractional calculus allows to improve the operation of controllers, including PID or adaptive controller [3,4]. The regulators will not work well without the proper coefficients. We use the Gray Wolf Optimizer (GWO) to optimize them based on research [5].

The authors first describe the FOPID and the GWO algorithm. Later, we presents the mathematical model of the 3D crane. The implementation and the results of the work are analyzed at the end. In relation to the previous work, the authors extended the operation of controllers from XY axes to XYZ axes [6].

2 FOPID Controller

The classic PID controller was expanded by two additional parameters to be the FOPID controller [3]. The equation of FOPID is presented in (1).

$$G_c(s) = K_P + \frac{K_I}{s^{\lambda}} + K_D s^{\mu} \tag{1}$$

A. Dzielinski et al. (Eds.): ICFDA 2021, LNNS 452, pp. 177–182, 2022.
https://doi.org/10.1007/978-3-031-04383-3_20

K_P, K_I and K_D define the coefficients for the proportional, integral, and derivative terms. λ and μ mean the order of integral and differentiation respectively.

For application of the FOPID controller, we need to approximate non-integer order of s. Oustaloup Recursive Approximation is used in this article [7]. This method is done in the frequency range of ω:

$$s^{\lambda} \approx C_o \Pi \frac{s + \omega_k'}{s + \omega_k} \tag{2}$$

where each coefficient is described by:

$$\omega_k' = \omega_b \frac{\omega_h}{\omega_b}^{\frac{k+N+0.5(1-r)}{2N+1}}, \omega_k = \omega_b \frac{\omega_h}{\omega_b}^{\frac{k+N+0.5(1-r)}{2N+1}}, C_o = \frac{\omega_h}{\omega_b}^{\frac{-r}{2}} \Pi \frac{\omega_k}{s + \omega_k'}. \tag{3}$$

3 Grey Wolf Optimizer

Grey Wolf Optimizer is based on the behavior of the grey wolf pack [8]. The herd has two distinctive features. One of them is the strict hierarchy of animals. The herd contains: the alpha wolf, the beta and the delta. The weakest wolves are omega. Beyond the hierarchy, there is also a strictly defined hunting order.

1. Tracking and approaching a victim
2. Circling and harassing
3. Attacking the victim

$$D = |C * X_v(t) - X(t)| \quad C = 2 * r_2 \tag{4}$$

$$X(t+1) = X_v(t) - A * D \quad A = 2a * r_1 - a \tag{5}$$

$$X(t+1) = \frac{X_\alpha + X_\beta + X_\delta}{3}, \tag{6}$$

where Xp is the prey location, X is the wolf position, t defines the current iteration, $r_1, r_2 \in [0, 1]$ are random numbers, X_α, X_β and X_δ define the location vector of the alpha, beta and delta, the a decrease in each iteration from 2 to 0.

The algorithm is finished when the stop criterion is achieved. In this article, we assume the criterion as maximum iteration number - 30.

4 Mathematical Model of the 3D Crane

3D crane is a system that can move in three dimensions. The real system prepared by INTECO is shown in Fig. 1 with the schematic shown in Fig. 2.

In the 3D crane system, 5 quantities can be measured using sensors [9,10]. The parameters in the figure have the following meanings:

y_w - the distance of the rail with the cart from the center; x_w - the distance of the cart from; R - the length of the winch line; α - the angle between the x axis and the winch line; β - the angle between the negative direction on the z axis

Fig. 1. INTECO 3D crane

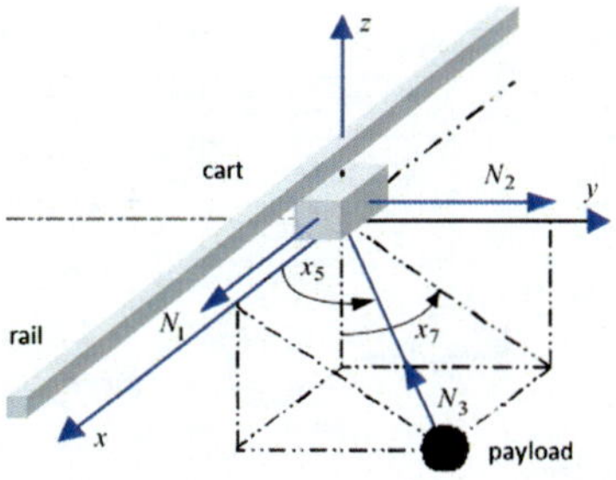

Fig. 2. 3D crane: coordinates and forces [9]

and the projection of the winch line onto the yz plane; m_c, m_w, m_s - mass of the payload, the cart and the moving rail; x_c, y_c, z_c - coordinates of the payload; S - reaction force in the lift-line acting on the cart; F_x - force driving cart; F_y - force driving the rail with cart; F_R - force controlling the length of the lift-line; T_x, T_y, T_R - friction forces; u_1 - control on the x axis; u_2 - control on the y axis; u_3 - control on the z axis.

The below nonlinear differential equations describes the 3D crane system:

$$\begin{cases} x_1 = x_w \\ x_2 = \dot{x}_1 = \dot{x}_w \\ x_3 = y_w \\ x_4 = \dot{x}_3 = \dot{y}_w \\ x_5 = \alpha x_6 = \dot{x}_5 = \dot{\alpha} \\ x_7 = \beta \\ x_8 = \dot{x}_7 = \dot{\beta} \\ x_9 = R \\ x_{10} = \dot{x}_9 = \dot{R} \end{cases} \tag{7}$$

$$\begin{cases} x_1 = x_2 \\ x_2 = N_1 + \mu_1 c_5 N_3 \\ x_3 = x_4 \\ x_4 = N_2 + \mu_2 s_5 s_7 N_3 \\ x_5 = x_6 \\ x_6 = (s_5 N_1 - c_5 s_7 N_2 + \\ (\mu_1 - \mu_2 s_7^2) c_5 s_5 N_3 \\ + V_5)/x_9 \\ x_7 = x_8 \\ x_8 = -(c_7 N_2 + \mu_2 s_5 c_7 \\ \times s_7 N_3 + V_6)/(s_5 x_9) \\ x_9 = x_{10} \\ x_{10} = -c_5 N_1 - s_5 s_7 N_2 \\ -(1 + \mu_1 c_5^2 + \mu_2 s_5^2 s_7^2) \\ \times N_3 + V_7 \end{cases} \tag{8}$$

where:

$\mu_1 = m_c/m_w$

$\mu_2 = m_c/(n_m + m_s)$

$s_n \equiv \sin x_n$

$c_n \equiv \cos x_n$

$V_5 = c_5 s_5 x_8^2 x_9 - -2 x_{10} x_6 + g c_5 c_7$

$V_6 = 2 x_8 (c_5 s_6 x_9 + s_5 x_{10}) + g s_7$

$V_7 = s_5^2 x_8^2 x_9 + g s_5 c_7 + x_6^2 x_9$

$N_1 = u_1 - k_1 \dot{x}_w = u_1 - k_1 x_2$

$N_2 = u_2 - k_2 \dot{y}_w = u_2 - k_2 x_4$

$N_3 = u_3 + k_3 \dot{R} = u_3 + k_3 x_{10}$

5 Implementation of Simulation

Main goal of this paper is to find suitable coefficients for FOPID controllers for 3D crane system. In order to determine those parameters, a suitable simulation

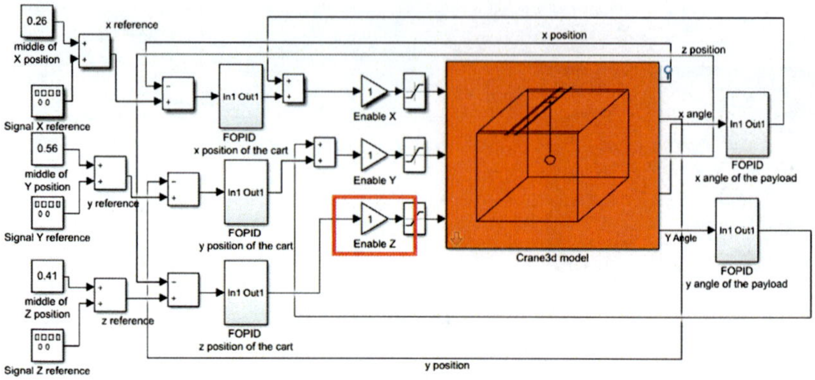

Fig. 3. Model of the 3D crane with FOPID controllers

was needed (Fig. 3). The model contains X, Y, Z position and also X and Y angle. Unlike in previous article [6], control for Z axis is enabled, which can be seen by red rectangle marking in Fig. 3.

In order to correctly implement the GWO optimization, it is necessary to define additional parameters $K_P \in [0 : 50]$, $K_I \in [0 : 50]$, $K_D \in [0 : 50]$, $\lambda \in [0 : 1]$, $\mu \in [0 : 1]$.

The cost function is shown by the below formula:

$$J = \int_{0}^{\infty} \left(|e_x(t)| + |e_y(t)| + |e_z(t)| + |\alpha(t)| + |\beta(t)|\right) dt. \tag{9}$$

where: $e_x(t)$, $e_y(t)$, $e_z(t)$ are difference between the set point and the actual value of the X, Y and Z axis, $\alpha(t)$ and $\beta(t)$ are X and Y angles of the payload.

6 Results of Simulation

Figure 4 shows a comparison of the system response on the X axis with a changing set point (blue color). Both FOPID controllers obtained an almost identical system response, which quite quickly reaches values close to the set value. Stabilization takes place immediately. In the PID controller, overshoot can be observed. The process of approaching the set point is not as smooth as in the case of FOPID controllers - minimal oscillations.

Figure 5 shows the responses of the Y axis. FOPID controllers reach the set point very quickly. The only difference is that there is an overshoot with FOPID-2. On the other hand, the PID controller behaves much worse. The time needed for its stabilization is more than twice as large and there is a relatively large offset between the final position and the set value. Only during the third change in set point value, the system is starting to respond a little better, but there is still an overshoot and the operation time of the controller is very long.

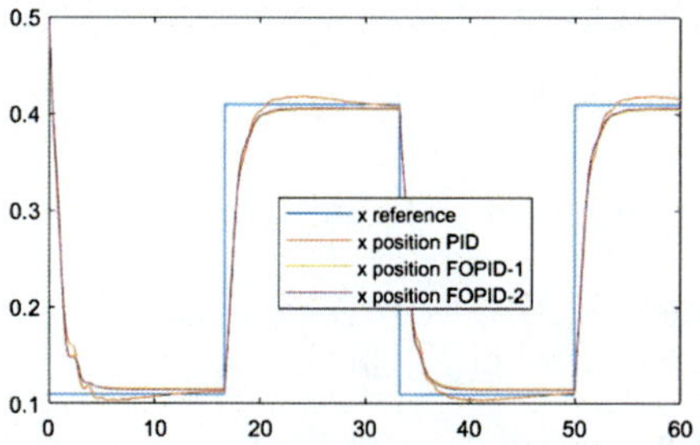

Fig. 4. Comparison of response for X axis

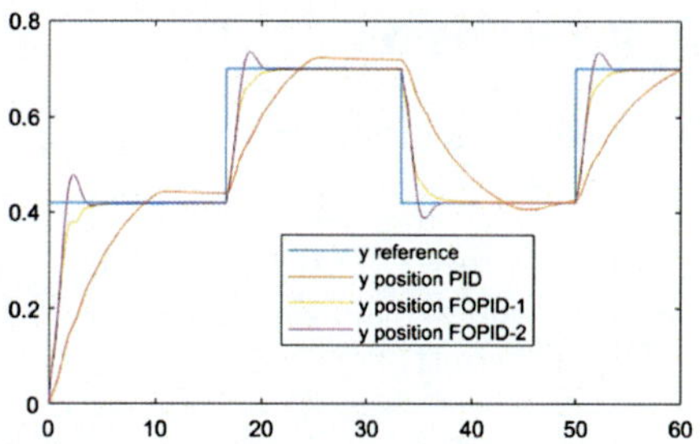

Fig. 5. Comparison of response for Y axis

Figure 6 shows the responses of the Z axis. FOPID controllers work very quickly and effectively - after the initial aggressive approach, there is a smoothing and gentle approach to the set point. In the PID controller, we observe a typical approach to the set point - the graph is rounded. Additionally, there is a small overshoot with difference in the operation time.

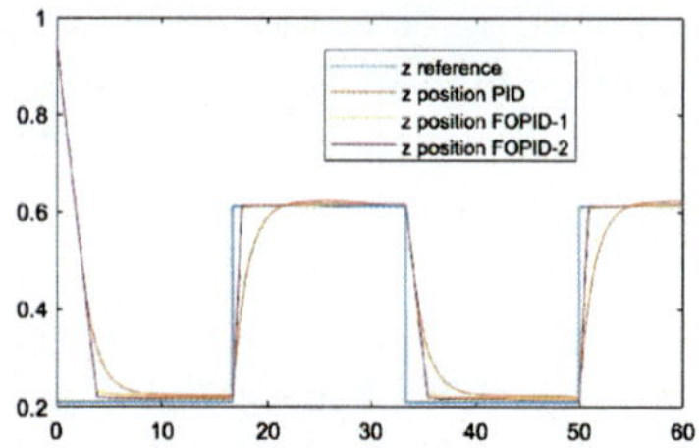

Fig. 6. Comparison of response for Z axis

The comparison of alpha and beta angles is presented in Figs. 7 and 8. In FOPID controllers, there is a large tick at each change of settings, because they approach the X, Y, Z set points more aggressively. The consequence are relatively large changes in the angles. However, the system returns to the angle of 0 after a few oscillations. In the PID controller, the later part of the plot looks worse. The alpha angle has problems with damping, as can be seen from long-term oscillations. Further, the occurring oscillations are not around the value of 0, but at a certain offset alternating between positive and negative values.

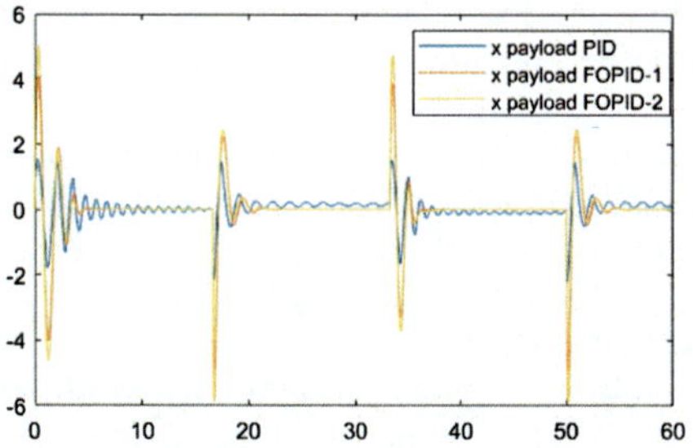

Fig. 7. Response for X payload

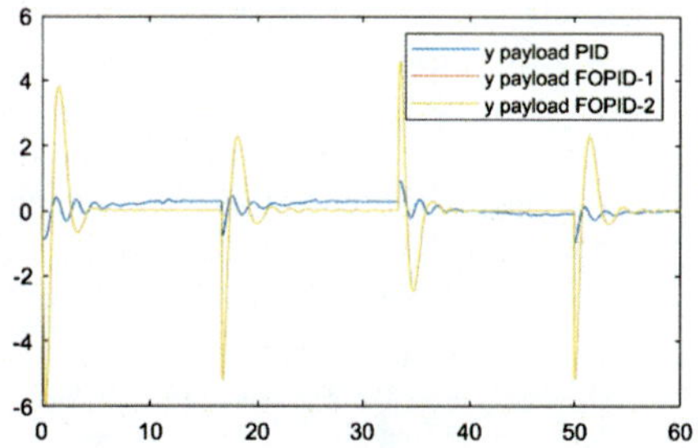

Fig. 8. Response for Y payload

7 Conclusion

The main conclusions drawn from the article:

- three control versions for 3D crane were prepared - one in the PID version and two in the FOPID version
- Grey Wolf Optimizer can be used to tune the PID/FOPID controllers
- FOPID controllers allow for more precise control of the 3D crane

Future work will focus on the preparation of the control algorithm for the real system. The plan is to reuse both types of controllers (PID and FOPID). In this way it will be possible to compare them to this work in order to confirm the simulations. In addition, the PID controller itself is very widespread, thus it works very well as a reference for the version with FOPID controllers. Minor adjustments to the controllers will be made if necessary. The 3D crane from INTECO (shown in Fig. 1) will be used as the real system.

References

1. Pauluk M.: Uproszczony regulator liniowo-kwadratowy w sterowaniu ramowa suwnicala 3D. Pomiary Automatyka Robotyka **12**(2006), 60–67 (2006)
2. Cellmer A., Banach B., Piotrowski R.: Design of modified PID controllers for 3D crane control. In: Trends in Advanced Intelligent Control, Optimization and Automation, KKA 2017. Advances in Intelligent Systems and Computing, vol. 577, pp. 77-86 (2017)
3. Vinagre, B.M., Petraš, I., Podlubny, I., Chen, Y.O.: Using fractional order adjustment rules and fractional order reference models in model-reference adaptive control. Nonlinear Dyn. **29**, 269–279 (2002)
4. Safikhani, Mohammadzadeh H., Tabatabaei, M.: Design of non-overshooting fractional order PD and PID controllers for special case of fractional order plants. J. Control Autom. Electr. Syst. **30**, 611–621 (2019)
5. Dziedzic K.: Identification of fractional order transfer function model using biologically inspired algorithms: progress in automation, robotics and measurement techniques. In: Automation, pp. 47–57 (2019)
6. Żegleń-Włodarczyk J., Dziedzic. K.: Optimization of the FOPID parameters of the 3D crane control system by using GWO. In: Paper Proposed to Conference the 25th International Conference on Methods and Models in Automation and Robotics (2021)
7. Oustaloup, A., Levron, F., Mathieu, B., Nanot, F.M.: Frequency-band complex noninteger differentiator: characterization and synthesis. IEEE Trans. Circ. Syst. I. Fundam. Theory Appl. **47**(1), 25–39 (2000)
8. Mirjalili, S., Mirjalili, S.M., Lewis, A.: Grey wolf optimizer. Adv. Eng. Softw. (2014). https://doi.org/10.1016/j.advengsoft.2013.12.007
9. Pauluk M., Korytowski A., Turnau A., Szymkat M.: Time optimal control of 3D crane. In: MMAR 2001: Proceedings of the 7th IEEE International Conference on Methods and Models in Automation and Robotics Marine Automation; Control Engineering; Identification and Signal Processing, Miedzyzdroje, 28–31 August 2001, vol. 2, pp. 927–932 (2001)
10. INTECO: 3DCrane User's Manual. http://www.inteco.com.pl/Docs/3DCrane_um.pdf. Accessed 31 Mar 2021

Axial Stabilizing CRONE Controller of an Active Magnetic Bearing with Strong Performance and Control-Effort Specifications

Patrick Lanusse[1(✉)], Salim Benbouzid[2], and Joaquim Da Silva[2]

[1] Bordeaux INP, IMS CNRS UMR 5218, University of Bordeaux, Bordeaux, France
patrick.lanusse@ims-bordeaux.fr
[2] SKF Magnetic Mechatronics, Vernon, France
salim.benbouzid@skf.com

Abstract. This paper presents the design of a fractional order based controller for an axial Active Magnetic Bearing (AMB) control loop used to maintain the rotating machinery shaft centered in its axial airgap in presence of axial disturbances. The CRONE methodology makes it possible to take into account strong performance and control effort specifications which have resulted in frequency domain sensitivity constraints. Simulations results shows that the controller provides very good dynamic performance.

Keywords: Fractional order system · Robust control · Active Magnetic Bearing · Control system design methodology

1 Introduction

Active Magnetic Bearings (AMB) are contactless force actuators used in the active vibration control of rotor dynamics [1–3]. They are used to levitate the rotor within the machine's airgap and allow high speed rotation for rotating machinery because the rotor's deviations from the airgap center are minimized thanks to controlled forces. Those forces are computed in a multivariable closed loop control where the feedback is performed using measurements provided by position sensors. They measure the rotor's deviations at several rotor locations (in minimal configuration: two sensors for radial displacement according two orthogonal axis V and W, 1 sensor for axial displacements according Z axis). The main purpose of the AMB control loop's position controller is to reject the disturbance effect of the machine environment and to minimize the deviation from the airgap center.

In this paper we will limit our development and discussion to the axial AMB control loop and the design of a robust fractional order position controller improving the control loop performance. The frequency domain CRONE methodology will be used after the axial AMB's specifications were translated into sensitivity function constraints.

A. Dzielinski et al. (Eds.): ICFDA 2021, LNNS 452, pp. 183–196, 2022.
https://doi.org/10.1007/978-3-031-04383-3_21

2 Description of the Axial Rotor Dynamic and the AMB Control Loop

2.1 System Short Description

For motorized centrifugal compressors or turboexpanders, axial magnetic bearing, cf. Fig. 1, is used as active thrust bearing to reject axial disturbances generated by the process. The axial rotor dynamic that we want control is represented in Fig. 2. It is the assembly of two heavy shafts (weighing several hundred kg) coupled with a flexible coupling. The radial levitation of each shaft is performed using two radial AMBs that we will not discuss in this paper. The Thrust disc location on the compressor part, displayed in Fig. 2(b), indicates the location where the axial control force is applied. It is also important to mention that the compressor's impellers are the locations of the axial force disturbances.

Fig. 1. Example of an active axial thrust bearing (axial AMB).

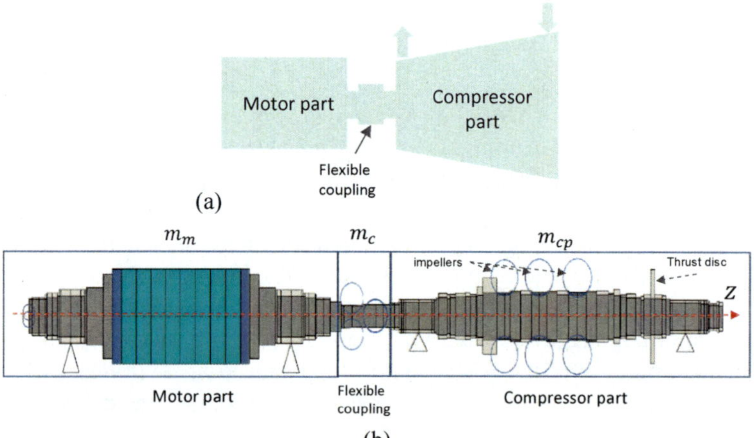

Fig. 2. Axial rotor dynamic system of a motorized centrifugal compressor. (a): Motorized centrifugal compressor with a flexible coupling. (b): Motorized centrifugal compressor with the moving parts along the Z axis.

The axial dynamic of the previous shafts assembly can be faithfully represented and modelled using the Mass-Spring-Damper system of Fig. 3.

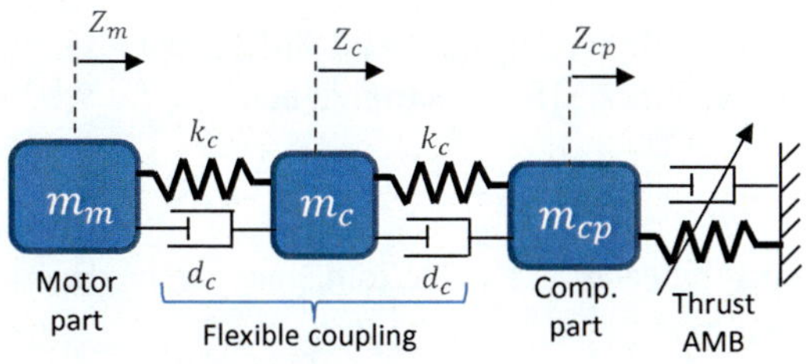

Fig. 3. Mass-Spring-Damper system representation of the axial dynamic of the shafts assembly.

The system equations directly derived from the Mass-Spring-Damper system and representing the rotor axial dynamic are:

$$
\begin{aligned}
\ddot{z}_{\mathrm{m}} &= \tfrac{k_{\mathrm{c}}}{m_{\mathrm{m}}}(z_{\mathrm{c}} - z_{\mathrm{m}}) + \tfrac{d_{\mathrm{c}}}{m_{\mathrm{m}}}(\dot{z}_{\mathrm{c}} - \dot{z}_{\mathrm{m}}) \\
\ddot{z}_{\mathrm{c}} &= -\tfrac{k_{\mathrm{c}}}{m_{\mathrm{c}}}(z_{\mathrm{c}} - z_{\mathrm{m}}) + \tfrac{k_{\mathrm{c}}}{m_{\mathrm{c}}}\left(z_{\mathrm{cp}} - z_{\mathrm{c}}\right) - \tfrac{d_{\mathrm{c}}}{m_{\mathrm{c}}}(\dot{z}_{\mathrm{c}} - \dot{z}_{\mathrm{m}}) + \tfrac{d_{\mathrm{c}}}{m_{\mathrm{c}}}\left(\dot{z}_{\mathrm{cp}} - \dot{z}_{\mathrm{c}}\right) \\
\ddot{z}_{\mathrm{cp}} &= -\tfrac{k_{\mathrm{c}}}{m_{\mathrm{cp}}}\left(z_{\mathrm{cp}} - z_{\mathrm{c}}\right) - \tfrac{d_{\mathrm{c}}}{m_{\mathrm{cp}}}\left(\dot{z}_{\mathrm{cp}} - \dot{z}_{\mathrm{c}}\right) + \tfrac{1}{m_{cp}}F_{\mathrm{AMB}}
\end{aligned}
\tag{1}
$$

where m_{m}, m_{cp} and m_{c} are respectively the masses of the motor shaft, the compressor shaft (with impellers) and the coupling. k_{c} ([N/m]) are the stiffness values of the flexible coupling. d_{c} ([N.s/m]) are the coefficient values of the viscous friction considered for this axial moving.

The measured position is the mass m_{cp} displacement.

2.2 AMB Modelling

The Axial AMB has a rotor part represented by the axial trust disc and a stator part represented by two electromagnets energized by two independent electrical amplifiers (DC choppers with a maximum voltage of 300 V and a resulted current limited to 30 A), cf. Fig. 4.

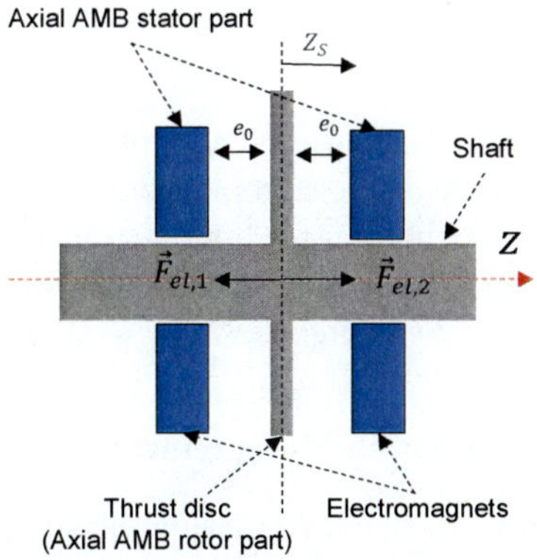

Fig. 4. Thrust bearing force (axial control force).

The axial control force provided by the Axial AMB is the resultant of two pull forces, in opposition, created by each AMB's electromagnet:

$$F_{\mathrm{AMB}}(t) = F_{\mathrm{el},2}(t) - F_{\mathrm{el},1}(t) \tag{2}$$

The force developed by each AMB electromagnet can be described as:

$$F_{\mathrm{el},k}(t) = k_{\mathrm{el},k} \frac{I_k^2(t)}{\left(e_0 \pm z_{\mathrm{cp}}(t)\right)^2} \tag{3}$$

where:

$k_{\mathrm{el},k} = \mu_0 \frac{N_{\mathrm{el},k}^2 A_{\mathrm{el},k}}{4}$
$\mu_0 = 4\pi\ 10^{-7}\left[\frac{\mathrm{Kg.m}}{\mathrm{A^2 s^2}}\right]$ or $\left[\frac{\mathrm{H}}{\mathrm{m}}\right]$, the vacuum magnetic permeability
$N_{\mathrm{el},k}$: number of turns in the electromagnet k
$A_{\mathrm{el},k}$: the projected area of the pole face of the electromagnet k
e_0, e: the nominal airgap and the airgap $e = e_0 \pm z_{\mathrm{cp}}$
z_{cp}: the variation of the nominal airgap (displacements)
I_k: the current in the electromagnet k.

As the fractional order controller design is a frequency domain based synthesis, a linear representation of this magnetic axial force is used. This linear force is derived by linearization around an operating point defined by a current I_0 and a position z_{cp} within the airgap. Equation (2) becomes:

$$F_{\mathrm{AMB}}(s) = F_{\mathrm{act}}(s) + F_{\mathrm{act,neg}}(s) = G_{\mathrm{act}}(s) F_{\mathrm{req,Z}}(s) + K_{\mathrm{neg}}(s) z_{\mathrm{cp}}(s) \tag{4}$$

where $K_{\mathrm{neg}}(s) = \left.\frac{F_{\max}}{e_0}\right|_{I_0 = \frac{I_{\max}}{2}, e = e_0}$ and where $G_{\mathrm{act}}(s)$ is the model of an idealized force generator with 2 performant inner magnetic flux control loops. It can be considered as 1st or 2nd order dynamic:

$$G_{\mathrm{act}}(s) = \frac{F_{\max}}{\frac{s}{\omega_{\mathrm{act}}} + 1} \quad \text{or} \quad G_{\mathrm{act}}(s) = \frac{F_{max}}{\frac{s^2}{\omega_{\mathrm{act}}^2} + 2\xi_{\mathrm{act}} \frac{s}{\omega_{\mathrm{act}}} + 1} \tag{5}$$

$F_{\mathrm{req,z}}$ is the normalized output of the position controller, $[-1, 1]$.

As the current limit is 30 [A], the electromagnetic maximal force delivered by the actuator is 89500 [N]. This maximum force limit is available for frequencies below 10 [Hz]. Over this frequency, the maximum force available decreases and the limitation is described by a frequency dependent template ($F_{\max}(\mathrm{j}\omega)$).

2.3 Dynamic of the Axial Deviation Measurement

The compressor mass displacements are measured using a magnetic flux based sensor technology. The airgap center is represented by an electrical zero got thanks to an established equilibrium between the magnetic fluxes of sensor's electromagnets. When a deviation from the center occurs, the equilibrium is not satisfied anymore and according to the deviation direction and the sensor sensitivity, an electrical voltage will indicate the image of the thrust disc displacement. The output sensor signal is filtered, discretized and normalized before being used by the controller. So, they must be considered in the position measurement model. The transfer function below, provides an adequate representation for this measurement dynamic:

$$\frac{Z_{\text{sens}}(s)}{Z_{\text{cp}}(s)} = \frac{a_{\text{sens}}}{s^4 + b_{\text{sens},3}s^3 + b_{\text{sens},2}s^2 + b_{\text{sens},1}s + b_{\text{sens},0}} \tag{6}$$

where Z_{sens} is the normalized input on the position controller $[-1, 1]$. The nominal air gap is around 800 [μm].

2.4 Simulator Test Bench

In very accurate modelling, often used for validation purposes with representative simulator platforms, the AMB force is derived directly from the magnetic flux and the relation between magnetic field, the currents and the axial airgap variations. Finite Elements Method Magnetics (FEMM software), which contains Maxwell equations, is used to get accurate axial actuator modelling considering magnetic saturations, Eddy current effects and leakages.

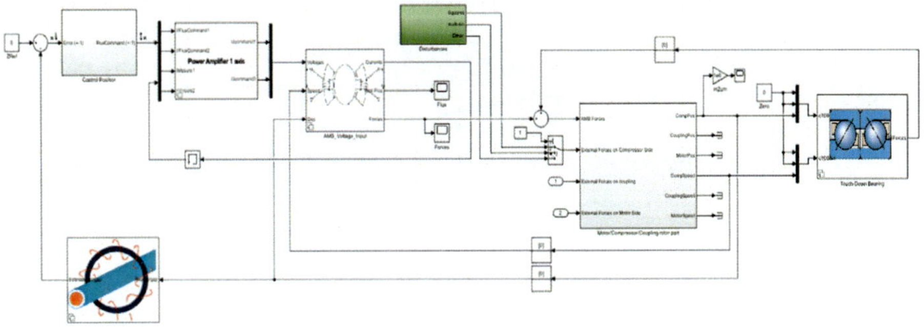

Fig. 5. Block diagram of the validation simulator.

The validation simulator is illustrated in Fig. 5. The whole axial control loop is modeled: the axial rotor dynamic, the power amplifiers, the active thrust bearing, the position sensor, the touch-down bearing and the position controller. Parameter values: $m_{\text{m}} = 3161.2$ [kg], $m_{\text{cp}} = 1250.1$ [kg], $m_{\text{c}} = 25.4$ [kg], $k_{\text{c}} = 37\ 10^5$ [N/m] and $d_{\text{c}} = 1000$ [N.s/m].

3 Design of the Control-System

3.1 Linear Model Description

Even if the system is nonlinear, the global dynamical behavior of the system to be controlled is well described by a linear model which can be decomposed as Fig. 6 shows.

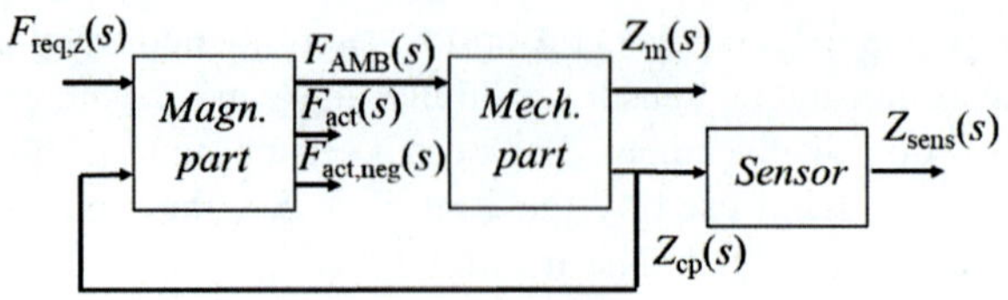

Fig. 6. Linear plant model.

The Bode diagram of this linear system is illustrated by Fig. 7.

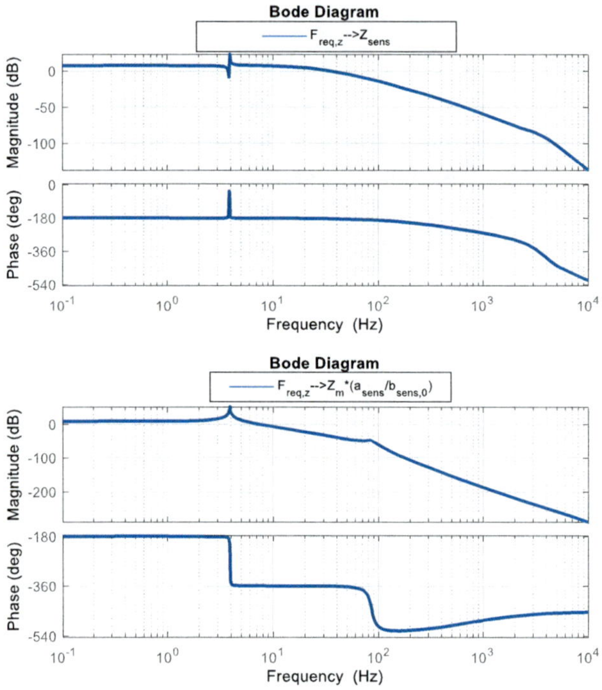

Fig. 7. Bode plot of the linear system (axial rotor dynamic with its thrust magnetic bearing and position sensor).

By approximating the frequency response of a high order state space model, the plant to be controlled is defined by a reduced order transfer function $G(s)$:

$$G(s) = \frac{Z_{\text{sens}}(s)}{F_{\text{req,z}}(s)} = G_\infty \frac{\prod_1^2 (s - z_i)}{\prod_1^9 (s - p_i)} \tag{7}$$

$G(s)$ includes (z_1, z_2) and (p_1, p_2), two lightly damped and low frequency complex conjugate zeros and pole pairs. G is unstable as (p_3, p_4) is a pair of opposite poles. The other poles are stable. The values of gain, zeros and poles of the nominal plant $G_{nom}(s)$ are:

$$G_\infty = 6.4\ 10^{26}$$
$$|z_{1,2}| = 24.167\ \text{rad/s};\ -Re\{z_{1,2}\}/|z_{1,2}| = 0.00327$$
$$|p_{1,2}| = 24.675\ \text{rad/s};\ -Re\{p_{1,2}\}/|p_{1,2}| = 0.00347$$
$$|p_{3,4}| = 185.22\ \text{rad/s}$$
$$|p_5| = 3141.6\ \text{rad/s}$$
$$|p_{6,7}| = 23562\ \text{rad/s};\ -Re\{p_{6,7}\}/|p_{6,7}| = 0.34$$
$$|p_{8,9}| = 65345\ \text{rad/s};\ -Re\{p_{8,9}\}/|p_{8,9}| = 0.69$$

3.2 CRONE Design of a Robust Feedback Digital Controller

The CRONE CSD methodology is a frequency-domain approach developed since the eighties [4–11]. It is based on the common unity-feedback structure presented by Fig. 8. In this paper, the controlled output Y_{mes} and the control input U_c represent respectively the position measure Z_{sens} and the force control $F_{req,z}$.

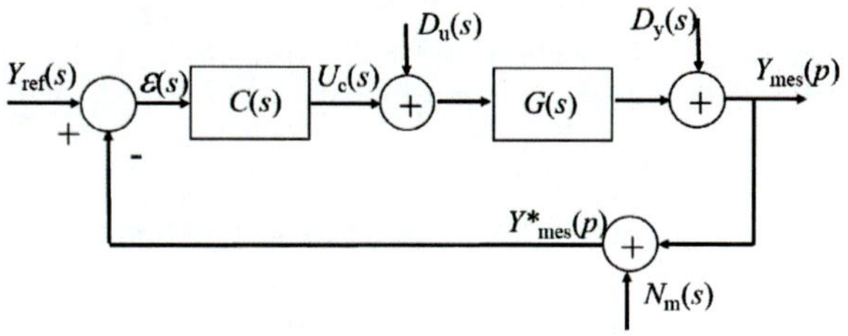

Fig. 8. CRONE feedback control structure.

Three CRONE CSD generations have been developed, successively extending the application field. The controller or open-loop transfer function is defined using fractional order integro-differentiation. It enables to design, simply and methodologically, LTI robust control-systems. Using frequency uncertainty domains (in our case provided by the perturbed black-box model), the uncertainties (or perturbation) are taken into account in a fully-structured form without overestimation, thus leading to control-systems that are as less conservative and thus as high performing as possible [11]. The third generation CRONE CSD is used for any kind of plant perturbation model: around open loop gain crossover frequency or closed loop resonant frequency, the open loop $\beta_{nom}(s)$ (for the nominal plant) is defined by a band-limited complex fractional order integrator. Integer order proportional integrator and low-pass filter define $\beta_{nom}(s)$ respectively at low and high frequencies. Right half plane or underdamped poles and zeros of the nominal model $G_{nom}(s)$ are included in the $G_{RHP\_UND}(s)$ part of $\beta_{nom}(s)$ for closed-loop internal stability and damping:

$$\beta_{\text{nom}}(s) = K\left(\frac{\omega_{N^-}}{s} + 1\right)^{n_l} \prod_{k=N^-}^{N^+} \left(\frac{1+s/\omega_{k+1}}{1+s/\omega_k}\right)^{a_k} \left(\Re e_{/i}\left\{\left(\alpha_k \frac{1+s/\omega_{k+1}}{1+s/\omega_k}\right)^{ib_k}\right\}\right)^{-q_k \operatorname{sign}(b_{qk})} \left(\frac{s}{\omega_{N^++1}} + 1\right)^{-n_h} G_{\text{RHP\_UND}}(s) \tag{8}$$

where a_k and b_{qk} are real and imaginary parts of fractional order of band-limited integrators. Integer powers q_k allows to increase the imaginary part effect without using high values of b_{qk} that could lead to an unstable closed-loop system [9].

After few high-level parameters of $\beta_{\text{nom}}(s)$ being optimized for a robust and high performing closed-loop (minimization of the variations of the resonant peak of complementary sensitivity function $T = \beta/1 + \beta$ while fulfilling performance frequency domain constraints on sensitivity functions T, $S = 1/1 + \beta$, $S.G$ and $C.S$), the feedback controller C is finally defined from the frequency response of the ratio:

$$C_R(s) = \beta_{\text{nom}}(s)/G_{\text{nom}}(s). \tag{9}$$

In our particular case, the controller should have:

1. to be implemented digitally
2. to stabilize the unstable plant
3. to damp the resonant modes of the plant
4. to reject efficiently the mechanical disturbance effect
5. to limit the level of high frequency control effort
6. to ensure the robustness of the closed-loop stability degree at the time of plant parameter dispersion.

The complex order of integro-differentiation used to define the nominal open-loop transfer function makes it possible both to set a given nominal closed-loop stability degree and to easily manage its robustness. As it is able to replace many parameters defining an integer order transfer function, it makes easier the controller optimization.

Design of a Digital Controller

The sampling frequency is 14 kHz. Thus the sampling period is $T_s = 71$ μs. Even if the Nyquist frequency $\pi f_s = 43982$ rad/s seems high compared to the required bandwidth – which needs to be greater than $|p_{3,4}|$ – the design of the digital controller can take into account the sampling effects by transforming the initial digital problem into a pseudo-continuous problem. It needs to: compute the Z-transform of the continuous-time set {plant & zero order hold}; achieve a bilinear variable change, for instance $z^{-1} = (1 - w)/(1 + w)$. Thus, $G_{\text{nom}}(s)$ becomes $G_{\text{nom}}(z) = Z_{Ts}\{\text{zoh}(s)\, G_{\text{nom}}(s)\}$, then $G_{\text{nom}}(z)$ becomes $G_{\text{nom}}(w)$. As with $w = jv$ and the normalized pseudo-frequency $v = \tan(\omega T_s/2)$), $G_{\text{nom}}(w)$ is the exact frequency response of $G_{\text{nom}}(z)$ (with $z = e^{j\omega T_s}$), controller $C(w)$ can then be designed using the CRONE frequency-domain methodology. Finally, without any approximation, the digital controller $C(z)$ could be obtained by achieving the inverse variable change $w = (1 - z^{-1})/(1 + z^{-1})$. Resulting from the Z-transform, three right-half plane zeros appear in $G_{nom}(w)$: +1, +1.05 and +2.06.

Stabilizing Controller

In order to stabilize the unstable pole of G and to ensure the internal stability of the closed-loop system presented by Fig. 8, the nominal open-loop transfer function $\beta_{nom}(w)$ -whose frequency response is shaped- has to include the nominal unstable pole (+0.00662) and the three right half plane zeros of $G_{nom}(w)$, and has to meet the Nyquist stability criterion. Thus, the $G_{RHP\_UND}(w)$ part of $\beta_{nom}(w)$ is first defined by:

$$G_{RHP\_UND}(w) = -\frac{(0.00662 + w)}{(0.00662 - w)}\left(1 - \frac{w}{1}\right)\left(1 - \frac{w}{1.05}\right)\left(1 - \frac{w}{2.06}\right) \tag{10}$$

In order to meet the common stability rules of thumb, the closed-loop bandwidth should be greater than the (greatest) unstable pole and lower than the lowest right half plane zero.

Damping of Resonant Zeros and Poles

In order to ensure that internal transfer functions of the control structure (Fig. 8) are well damped, controller $K(w)$ has not to cancel the two lightly damped zeros and poles pairs of $G_{nom}(w)$. Thus, considering relations (8) and (9), $G_{RHP\_UND}(w)$ defined by (10) needs also to include

$$\frac{1 + 7.569w + 1.343\ 10^6\ w^2}{1 + 7.883w + 1.288\ 10^6\ w^2}. \tag{11}$$

Efficient Rejection of Mechanical Disturbance Effect

In order to reject mechanical disturbance and to ensure very small variations of the position output, a stiffness constraint is introduced:

$$\min|K_{elec}(j\omega)| \geq \gamma \text{ for } \omega \in [\omega_0, \omega_{cg}] \tag{12}$$

where K_{elec} is the electrical stiffness defined by

$$K_{elec}(s) = -\frac{F_{act}(s)}{Z_{cp}(s)}. \tag{13}$$

ω_{cg} denotes the open-loop gain crossover frequency which also sets the closed-loop bandwidth. From the linear model of the system and taking into account the closed loop control structure, the electrical stiffness can be computed by:

$$K_{elec}(s) = -\frac{F_{act}(s)}{F_{req,z}(s)}\frac{F_{req,z}(s)}{Z_{cp}(s)} = G_f(s)C(s)G_s(s) \tag{14}$$

with

$$G_f(s) = \frac{F_{act}(s)}{F_{req,z}(s)} \text{ and } G_s(s) = \frac{Z_{sens}(s)}{Z_{cp}(s)}. \tag{15}$$

It has been set that constraint (12) should be met if a constraint on sensitivity function SG is taken into account such as:

$$\max|SG(j\omega)| \leq \frac{|G_f(j\omega)G_s(j\omega)|}{\gamma} \text{ for } \omega \in [\omega_0, \omega_{cg}]. \tag{16}$$

The frequency response of the bound used by this constraint can be defined by:

$$W_{SG}^{-1}(s) = \frac{10^{-1.94/20}}{1+\frac{s}{3141}} \frac{1}{\left(\frac{s}{23562}\right)^2 + 2.88s10^{-5} + 1} \frac{1}{\left(\frac{s}{65345}\right)^2 + 2.11s10^{-5} + 1} \tag{17}$$

As the plant low frequency order is null, the low frequency order $n_l = 1$ of $\beta_{nom}(w)$ will ensure an integral effect in the controller C and a small value of the SG magnitude at low frequency.

Control Effort Roll-Off Specification

As the pseudo-frequency plant model order is null for v close to 1 ($\omega \approx \pi f_s/2$) and taking into account the 3 right half plane zeros included in $G_{RHP\_UND}(w)$, the high frequency order $n_h = 4$ of $\beta_{nom}(w)$ is able to decrease the magnitude of $C(jv)$ at high frequency. It is particularly required to mitigate the high frequency control effort level by ensuring that:

- $|C(j2\pi f)|$ is as close as possible to 8 dB for $f \approx 3$ kHz
- $|C(j2\pi f_c)| = -3$ dB for f_c as low as possible

Definition of a Perturbed Plant Model

In order to design a robust controller, the nominal black-box model (I1) is derived to define a perturbed plant model. Lower and upper bounds of G_∞, $p_{3,4}$ and p_5 parameters are respectively defined by their nominal values divided or multiplied by 1.1.

3.3 Optimal Loop Shaping of the CRONE Open Loop Nominal Transfer Function

For a good stability degree: the required nominal resonant peak of complementary sensitivity T_{nom} is 3 dB and the resonant peak of T is limited to 4 dB; the S sensitivity function limitation is −6 dB at low frequency and 6 dB at high frequency; the SG plant input sensitivity function is limited to −1.94 dB, the low frequency value of W_{SG}^{-1}; the limitation of the CS control effort sensitivity function is 40dB up to $v = 0.44$ (f = 1.8 kHz) and 8 dB for $v > 0.87$ ($f \approx 3$ kHz). $N^- = 0$ and $N^+ = 1$ provide enough parameters to tune well the frequency response of $\beta_{nom}(w)$ defined by:

$$\beta_{nom}(w) = K\left(\frac{v_0}{w} + 1\right)^{n_l} \prod_{k=0}^{1} \left(\frac{1+w/v_{k+1}}{1+w/v_k}\right)^{a_k} \left(\Re e_{/i}\left\{\left(\alpha_k \frac{1+w/v_{k+1}}{1+w/v_k}\right)^{ib_k}\right\}\right)^{-q_k \operatorname{sign}(b_{qk})}$$
$$\frac{-(0.00662+w)}{(0.00662-w)}\left(1-\frac{w}{1}\right)\left(1-\frac{w}{1.05}\right)\left(1-\frac{w}{2.06}\right)\frac{1+7.569w+1.343\ 10^6\ w^2}{1+7.883w+1.288\ 10^6\ w^2}\left(\frac{w}{v_2}+1\right)^{-n_h} \tag{18}$$

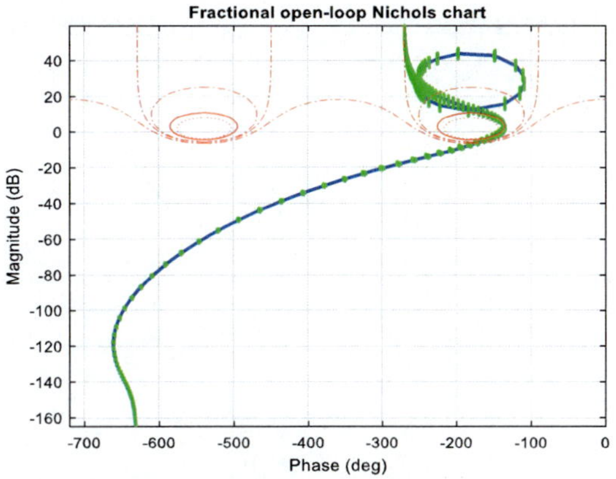

Fig. 9. Nichols plot of β_{nom} nominal frequency response (__); uncertainty domains (__); 3 dB M-contour (__).

While taking into account the frequency domain uncertainty of $G(w)$, seven independent parameters of $\beta_{\text{nom}}(w)$ are optimized in order to minimize the variations of the resonant peak of T and to meet the four sensitivity function constraints. Optimized parameter values are: $|\beta_{\text{nom}}(\text{j}\nu_\text{r})| = 2.96$ dB; $\nu_\text{r} = 0.0247$ (resonant frequency $\omega_\text{r} = 692$ rad/s); $\nu_0 = 0.00394$; $\nu_1 = 0.443$, $\nu_2 = 1.80$, $a_1 = 9.64$, $b_{\text{q}1} = 0$, $q_1 = 1$. Thus the other parameters can be computed: $a_0 = 0.85$, $b_0 = -0.21$, $q_0 = 1$ and $K = 6.23$. Figure 9 presents the Nichols plot of the optimized open loop frequency response $\beta(\text{j}\nu)$.

Figure 10 shows how the sensitivity constraints have been met: greatest values of resonant peaks of T, S, SG and CS are respectively 5.2 dB, 8 dB, −7.4 dB and 37 dB. At $f = 3$ kHz, $|C(\text{j}\nu)| = 11$ dB and $|C(\text{j}\nu)| = -3$ dB at fc = 4.3 kHz. These results confirm that the design is strongly constrained by high requirements: the stability margins are low in order to both stabilize and damp the output system and to mitigate the control effort with a very high roll-off rate of the controller gain. Figure 11 confirms that at low frequency the magnitude of sensitivity function SG met constraint W_{SG}^{-1} defined by (17) which permitted to meet the electrical stiffness constraint (12).

3.4 Rational Approximation of the Fractional Order Controller

Then, the frequency response of the fractional version $C_\text{F}(w)$ of the controller is fitted by an order seven rational and strictly proper transfer function $C_\text{R}(w)$. For a good numerical conditioning of the controller, a partial fraction decomposition of $C_\text{F}(w)$ is achieved, all the resulting first and second order transfer functions $C_{\text{R}k}(w)$ leading (as explained previously) to the digital filters $C_{\text{R}k}(z)$ which make up the digital controller $C(z)$.

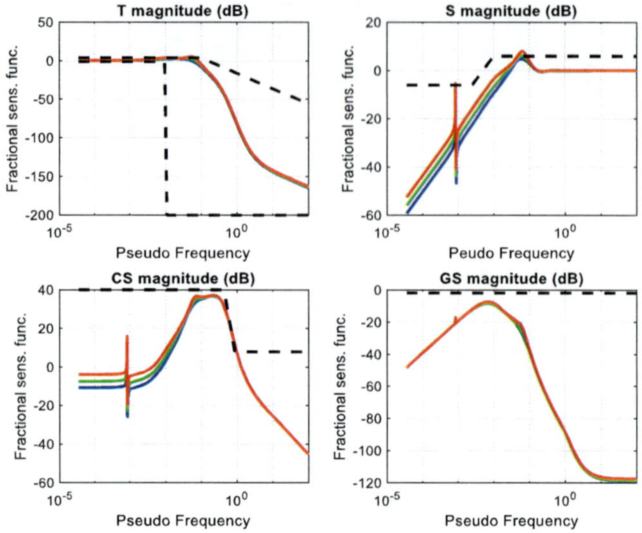

Fig. 10. Nominal and extreme sensitivity functions and constraints (- - -).

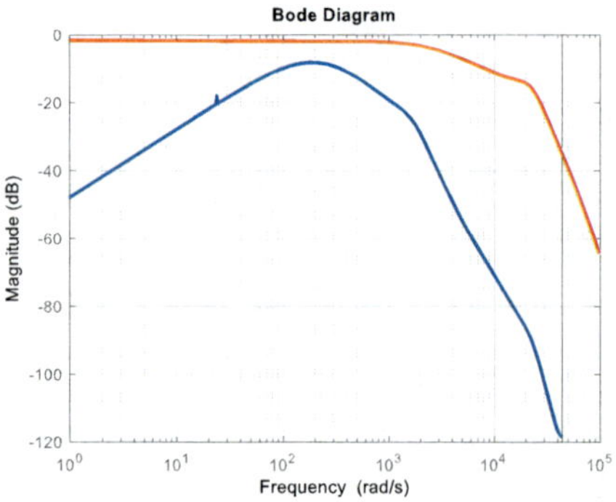

Fig. 11. *SG* sensitivity function (__) and constraint W_{SG}^{-1} (__).

4 Simulation Assessments

To assess the controller performances, some simulations have been performed using the validation simulator of Fig. 5. Figure 12 displays the closed loop responses to an axial force disturbance applied on the compressor side. This signal force is a filtered square signal with a maximal amplitude of 45000 [N]. One can observe that the deviation of the rotor compressor part in closed loop is limited to 80.62 [µm] (much lower than the 800 [µm] air gap and very far from the alarm threshold) and the disturbance force is completely well rejected. However as expected considering the resonant peak on the transfer function of Fig. 7, damped oscillations with a frequency around 3 [Hz] appear on the motor part of the rotor with a maximal amplitude of 97.56 [µm].

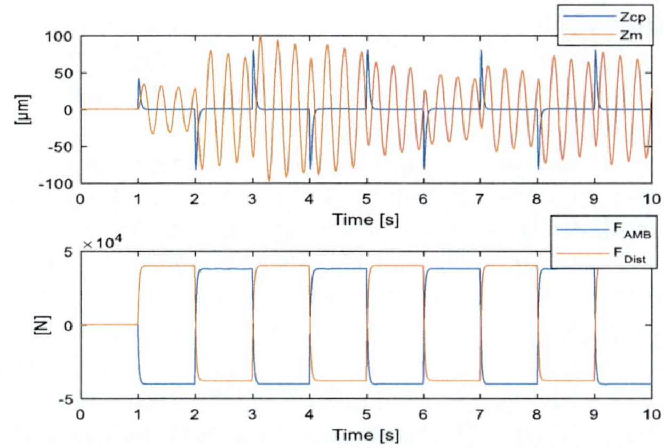

Fig. 12. Rejection of a filtered square repeated sequence as disturbance force applied on the compressor side. The maximum amplitude of the disturbance is 45000 [N] and the filter time constant is arround 20 [ms].

Figure 13 displays the closed loop responses to an axial multi-sin force disturbance applied on the compressor side. One can observe that the deviation of the rotor compressor part in closed loop is less than 61 [μm] and the disturbance force is well rejected. As in the previous simulation results, damped oscillations are still present on the motor part of the rotor with a maximal amplitude less than 97 [μm].

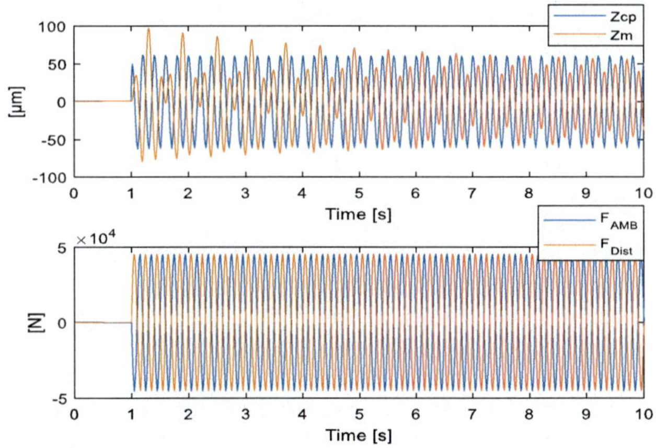

Fig. 13. Rejection of a multi-sin sequence (45000 sin(2π.5t) + 500 sin(2π.50t)) as disturbance applied on the compressor side.

These simulation results suggest as perspective, to consider the minimization of the displacement of the rotor motor part in the controller design procedure: $min\left(\left\| SG_{F_{\mathrm{Dist}}\rightarrow Z_{\mathrm{m}}}(\mathrm{j}\omega)\right\|_{\infty,\mathrm{dB}}\right)$.

5 Conclusion

The CRONE control of the axial positioning of an Active Magnetic Bearing has been studied and it confirms that even if such a system is highly challenging (natural instability, low damped modes, etc.), a fractional order based design methodology is able to take into account strong constraints while ensuring good performance. Controllers with close nominal performance have been obtained by using the H∞. They were high-order controllers even if they were designed without taking into account plant uncertainties. Fractional order controller design is really promising for AMB industrial applications.

Even if this paper focused on the axial thrust bearing control, the CRONE approach can be extended to design the radial control loop for which the rotor dynamic specificities must be considered (coupling gyroscopic effects and the eigenvalues dependency to the rotations speed) in the design. This radial control can be performed as an 8×8 MIMO controller or two 4×4 MIMO controllers. It could be presented soon.

Acknowledgment. This work took place in the framework of the collaboration combining IMS laboratory and SKF Magnetic Mechatronics.

References

1. Schweitzer, G., Maslen, E.H., Bleuler, H., Cole, M., Keogh, P., Larsonneur, R.: Magnetic Bearings, Theory, Design, and Application to Rotating Machinery. Springer, Heidelberg (2009). https://doi.org/10.1007/978-3-642-00497-1
2. Haberman, H.: Fonction de guidage en rotation – Paliers magnétiques. Techniques de l'ingénieur, Août (1984)
3. Font, S., Duc, G., Carrère, F.: Commande fréquentielle robuste-Application aux paliers magnétiques. Techniques de l'ingénieur, June 1997
4. Oustaloup, A.: Systèmes asservis linéaires d'ordre fractionnaire. Masson, Paris (1983)
5. Oustaloup, A.: La commande CRONE. Hermes Editor, Paris (1991)
6. Lanusse, P.: De la commande CRONE de première génération à la commande CRONE de troisième generation. Ph.D. thesis, Bordeaux I University, France (1994)
7. Oustaloup, A., Mathieu, B., Lanusse, P.: The CRONE control of resonant plants: application to a flexible transmission. Eur. J. Control **1**(2), 113–121 (1995)
8. Lanusse, P.: CRONE Control System Design, a CRONE toolbox for Matlab (2010). http://www.ims-bordeaux.fr/CRONE/toolbox
9. Lanusse, P., Oustaloup, A., Pommier-Budinger, V.: Stability of closed-loop fractional-order systems and definition of damping contours for the design of controllers. Int. J. Bifurcat. Chaos **22**(4), 1230013 (2012)
10. Sabatier, J., Lanusse, P., Melchior, P., Oustaloup, A.: Fractional Order Differentiation and Robust Control Design. Springer, Dordrecht (2015). https://doi.org/10.1007/978-94-017-9807-5
11. Landau, I.D., Rey, D., Karimi, A., Voda, A., Franco, A.: A flexible transmission system as a benchmark for robust digital control. Eur. J. Control **1**(2), 77–96 (1995)

Matsuda Method Adapted to Identify Fractional Order Transfer Functions

Duarte Valério[1(✉)] and Manuel D. Ortigueira[2]

[1] IDMEC, Instituto Superior Técnico, Universidade de Lisboa, Av. Rovisco Pais 1, 1049-001 Lisbon, Portugal
duarte.valerio@tecnico.ulisboa.pt

[2] Centre of Technology and Systems – UNINOVA and Department of Electrical Engineering, NOVA School of Science and Technology of NOVA University of Lisbon, Quinta da Torre, 2829-516 Caparica, Portugal
mdo@fct.unl.pt

http://web.tecnico.ulisboa.pt/duarte.valerio

Abstract. The Matsuda identification method is generalised to identify fractional transfer functions (either implicit, or explicit and commensurable) from the gain of the frequency response. Application examples are given, including filter development and the use of metaheuristics to select frequencies.

Keywords: Identification · Fractional transfer functions · Matsuda method · Continued fractions

1 Introduction

To identify a transfer function model from a frequency response, the most usual methods are the inspection of the Bode diagram and the Levy method. This is true for fractional order transfer functions as well (see [3] for the Levy method).

The Levy method requires information on both gain and phase; however, the later may not exist, or may be more affected by disturbances than the gain. Inspecting the Bode diagram can be done using only information about the gain, but requires some skill and is not a completely rigorous process.

Other identification methods from frequency responses exist, such as the Matsuda method [1,5]. It only requires the gain of the frequency response. This method is often used to obtain integer order approximations of fractional derivatives, in which case it is given the gain of s^α, or some other fractional transfer function [4].

In this paper, the Matsuda identification method is generalised so as to identify fractional transfer functions from the gain of the frequency response. These fractional transfer functions can be either implicit, or explicit and commensurable [2].

Section 2 presents the original Matsuda Method. The extension to implicit fractional order transfer functions is given in Sect. 3 and to explicit commensurable fractional order transfer functions is given in Sect. 4. Both sections give a

A. Dzielinski et al. (Eds.): ICFDA 2021, LNNS 452, pp. 197–207, 2022.
https://doi.org/10.1007/978-3-031-04383-3_22

simple example of application. Section 5 presents an additional example showing the use of metaheuristics to select frequencies, and further examples on the design of fractional order filters. Section 6 concludes the paper with a discussion of the numerical results.

2 The Original Matsuda Method

2.1 Continued Fractions

Continued fractions are mathematical entities built from two sequences, a_k and b_k, and are given by

$$a_0 + \frac{b_1}{a_1 + \frac{b_2}{a_2 + \frac{b_3}{a_3 + \frac{b_4}{a_4 + \ldots}}}} \tag{1}$$

Instead of using notation (1), continued fractions are usually notated using either the Abramowitz notation

$$a_0 + \frac{b_1}{a_1+}\,\frac{b_2}{a_2+}\,\frac{b_3}{a_3+}\,\frac{b_4}{a_4+}\cdots \tag{2}$$

or the Pringsheim notation

$$\left[a_0; \frac{b_1}{a_1}, \frac{b_2}{a_2}, \frac{b_3}{a_3}, \frac{b_4}{a_4}, \ldots\right] = \left[a_0, \frac{b_k}{a_k}\right]_{k=1}^{+\infty} \tag{3}$$

Just as a series $\sum_{k=1}^{+\infty} a_k$ built from a sequence a_k may or may not converge, so a continued fraction may converge or not.

It is not difficult to verify that, given a function $f_0(t)$, known at $t = t_0$, t_1, t_2, $\ldots t_p$, then continued fraction

$$f_0(t) \approx f_0(t_0) + \frac{t-t_0}{f_1(t_1)+}\,\frac{t-t_1}{f_2(t_2)+}\,\frac{t-t_2}{f_3(t_3)+}\cdots = \left[f_0(t_0); \frac{t-t_k}{f_{k+1}(t_{k+1})}\right]_{k=0}^{p} \tag{4}$$

$$f_{k+1}(t) = \frac{t-t_k}{f_k(t) - f_k(t_k)}, \quad k \in \mathbb{N} \tag{5}$$

is an interpolation of $f_0(t)$. This can be seen evaluating (4) at $t = t_0, t_1, t_2, \ldots t_p$.

2.2 Identification

Replacing complex numbers by their absolute values in (4), a variation of the continued fraction is obtained that approximates a given frequency behaviour $G(j\omega)$, known at frequencies $\omega = \omega_0, \omega_1, \ldots \omega_N$:

$$\hat{G}(s) = d_0(\omega_0) + \frac{s-\omega_0}{d_1(\omega_1)+}\,\frac{s-\omega_1}{d_2(\omega_2)+}\,\frac{s-\omega_2}{d_3(\omega_3)+}\cdots = \left[d_0(\omega_0); \frac{s-\omega_{k_1}}{d_k(\omega_k)}\right]_{k=1}^{N} \tag{6}$$

$$d_0(\omega) = |G(j\omega)| \tag{7}$$

$$d_k(\omega) = \frac{\omega - \omega_{k-1}}{d_{k-1}(\omega) - d_{k-1}(\omega_{k-1})}, \quad k = 1, 2, \ldots N \tag{8}$$

Approximations $\hat{G}(s)$ given by (6) are integer order transfer functions, that have $\lceil \frac{N}{2} \rceil$ zeros and $\lfloor \frac{N}{2} \rfloor$ poles. Consequently, $\hat{G}(s)$ will never be strictly proper, and will be proper only if N is even, i.e. if the number of frequencies $N + 1$ is odd.

This is the Matsuda identification method.

3 Identifying Implicit Fractional Order Transfer Functions

Let an implicit fractional order transfer function model be given by

$$\hat{G}(s) = \left(\frac{\sum_{k=0}^{m} b_k s^k}{\sum_{k=0}^{n} a_k s^k} \right)^{\alpha} \tag{9}$$

This is a fractional power of order α of an integer transfer function with n poles and m zeros, which we may denote by $G_i(s)$, i.e.

$$\hat{G}(s) = (G_i(s))^{\alpha} \tag{10}$$

Suppose that we want to approximate frequency behaviour $G(j\omega)$, known at $N + 1$ frequencies $\omega = \omega_0, \omega_1, \ldots \omega_N$, by an implicit fractional order transfer function $\hat{G}(s)$ given by (9). The gains of $\hat{G}(s)$ and $G_i(s)$ are related by

$$\left| \hat{G}(j\omega) \right| = |G_i(j\omega)|^{\alpha} \Leftrightarrow |G_i(j\omega)| = \left| \hat{G}(j\omega) \right|^{\frac{1}{\alpha}} \tag{11}$$

Consequently, we can use (6)–(8) together with gain $|G(j\omega)|^{\frac{1}{\alpha}}$ to identify $G_i(s)$, an then apply (10). That is to say, we replace (6)–(8) with

$$\hat{G}(s) = \left(d_0(\omega_0) + \frac{s - \omega_0}{d_1(\omega_1) +} \frac{s - \omega_1}{d_2(\omega_2) +} \frac{s - \omega_2}{d_3(\omega_3) +} \cdots \right)^{\alpha} = \left(\left[d_0(\omega_0); \frac{s - \omega_{k1}}{d_k(\omega_k)} \right]_{k=1}^{N} \right)^{\alpha} \tag{12}$$

$$d_0(\omega) = |G(j\omega)|^{\frac{1}{\alpha}} \tag{13}$$

$$d_k(\omega) = \frac{\omega - \omega_{k-1}}{d_{k-1}(\omega) - d_{k-1}(\omega_{k-1})}, \quad k = 1, 2, \ldots N \tag{14}$$

Example. Let $G_i(s) = s + 1$ and $\alpha = -0.2$; so, $G(s) = \left(\frac{1}{s+1}\right)^{0.2}$. We use $\omega_0 = 0.1$ rad/s and $\omega_1 = 10$ rad/s and obtain $G_i(s) = \frac{s+1}{1.095}$, and thus $\hat{G}(s) = \left(\frac{1.095}{s+1}\right)^{0.2}$. The pole is correctly identified and the error in the gain is less than 10%. The corresponding Bode diagrams are given in Fig. 1. Notice that because $\alpha < 0$ we used an even number of frequencies to arrive at a causal model.

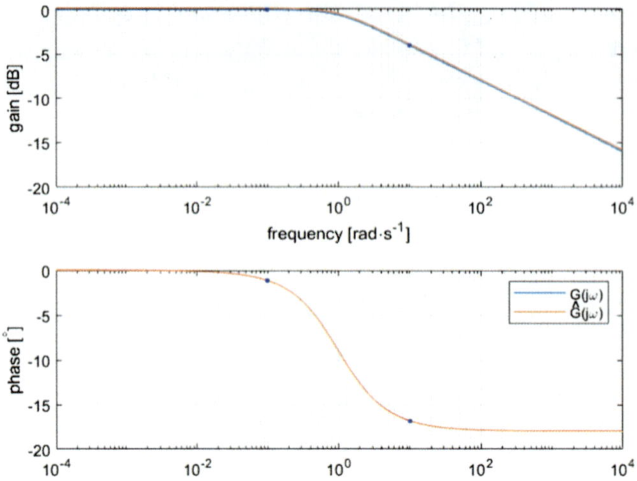

Fig. 1. Bode diagrams of the example for implicit fractional order transfer function models in Sect. 3.

4 Identifying Explicit Commensurable Fractional Order Transfer Functions

Let an explicit commensurable fractional order transfer function model be given by

$$\hat{G}(s) = \frac{\sum_{k=0}^{m} b_k s^{k\alpha}}{\sum_{k=0}^{n} a_k s^{k\alpha}} \tag{15}$$

The commensurate order is α; there are n pseudo-poles and m pseudo-zeros.

Suppose that we want to approximate frequency behaviour $G(j\omega)$, known at frequencies $\omega = \omega_0, \omega_1, \ldots \omega_N$, by an explicit commensurable fractional order transfer function $\hat{G}(s)$ given by (15). This model will now have its frequency response $\hat{G}(j\omega)$ given by a rational function of $(j\omega)^\alpha$. So (6)–(8) need to be changed to

$$\hat{G}(s) = d_0(\omega_0) + \frac{s^\alpha - \omega_0^\alpha}{d_1(\omega_1)+} \, \frac{s^\alpha - \omega_1^\alpha}{d_2(\omega_2)+} \, \frac{s^\alpha - \omega_2^\alpha}{d_3(\omega_3)+} \ldots = \left[d_0(\omega_0); \frac{s^\alpha - \omega_{k_1}^\alpha}{d_k(\omega_k)} \right]_{k=1}^{N} \tag{16}$$

$$d_0(\omega) = |G(j\omega)| \tag{17}$$

$$d_k(\omega) = \frac{\omega^\alpha - \omega_{k-1}^\alpha}{d_{k-1}(\omega) - d_{k-1}(\omega_{k-1})}, \quad k = 1, 2, \ldots N \tag{18}$$

Example. Let $G(s) = \frac{1}{s^{0.5}+1}$; so, $\alpha = 0.5$. We use $\omega_0 = 0.1$ rad/s, $\omega_1 = 10$ rad/s and $\omega_2 = 1000$ rad/s, and obtain $\hat{G}(s) = \frac{-0.001808s^{0.5}+1.066}{s^{0.5}+1.009}$. The pseudo-pole obtained at -1.009 is almost correctly identified: the error is less than 1%. The spurious pseudo-zero at 589.9094 is more than two decades away; it is clearly neglectable. As to the identified constant gain 1.0564, it is off by about 5%. The corresponding Bode diagrams are given in Fig. 2.

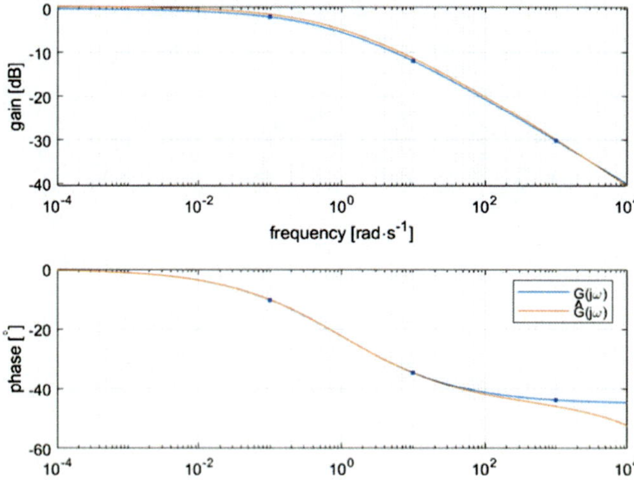

Fig. 2. Bode diagrams of the example for explicit commensurable fractional order transfer function models in Sect. 4.

5 Examples of Application

5.1 Metaheuristics

Choosing the commensurate value α, choosing the number of frequencies N, and choosing the frequencies $\omega_0, \omega_1, \ldots \omega_N$ themselves, is not always straightforward, especially if there are many experimental values. Of course it would not do to use them all, since the model would have an impossibly high order; numerical problems are likely to preclude a result for large values of N. Metaheuristics can be used to assist this process.

Example. A genetic algorithm is used to find the parameters of $G(s) = \frac{1}{s^{0.5}+1}$ from the previous example; 161 frequencies logarithmically spaced in $[10^{-4}, 10^4]$ rad/s were used, and the genetic algorithm had to select $\alpha \in [0, 2]$ and which of the 161 frequencies to use. There was a maximum number of frequencies $N_{\max}$ that could be used, but the actual value of N could be lower, i.e.

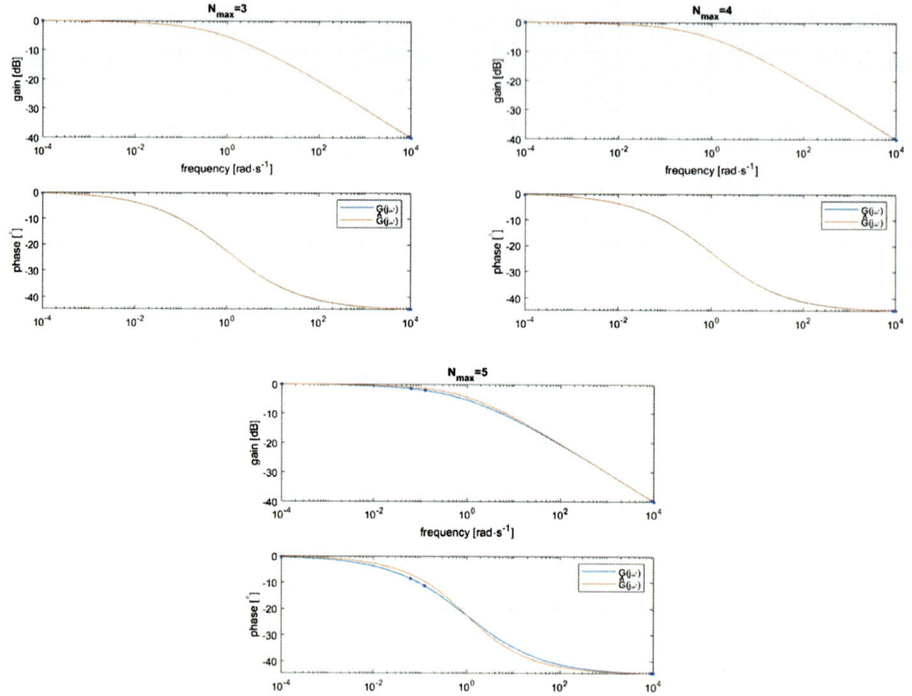

Fig. 3. Bode diagrams of the example for the application of a genetic algorithm to an explicit commensurable fractional order transfer function model in Sect. 5.1.

the algorithm could choose not to use all the frequencies. The following results were obtained:

$$\hat{G}(s) = \frac{-6.3 \times 10^{-5} s^{0.4984} + 0.9948}{s^{0.4984} + 0.9917} \text{ for } N_{\max} = 3 \tag{19}$$

$$\hat{G}(s) = \frac{-6.1 \times 10^{-5} s^{0.4985} + 0.9954}{s^{0.4985} + 0.9923} \text{ for } N_{\max} = 4 \tag{20}$$

$$\begin{aligned}\hat{G}(s) &= \frac{-7.4 \times 10^{-5} s^{0.9957} + 0.9906 s^{0.4979} - 0.3355}{s^{0.9957} + 0.6403 s^{0.4979} - 0.3343} \\ &= \frac{(-7.4 \times 10^{-5} s^{0.4979} + 0.9904)(s^{0.4979} - 0.3387)}{(s^{0.4979} + 0.9811)(s^{0.4979} - 0.3408)} \text{ for } N_{\max} = 5\end{aligned} \tag{21}$$

In both (19) and (20), commensurate order, gain and pseudo-pole are found with errors below 1%; a spurious pseudo-zero is clearly neglectable. Notice that only $N = 3$ frequencies were used when $N_{\max} = 4$. For $N_{\max} = 5$, more care is needed when interpreting the results: an unstable pseudo-pole and a pseudo-zero have to be cancelled (they are equal to less than 1%) before the same reasoning applied above leads us to the correct transfer function. The corresponding Bode diagrams, showing also the frequencies chosen by the genetic algorithm, are given in Fig. 3.

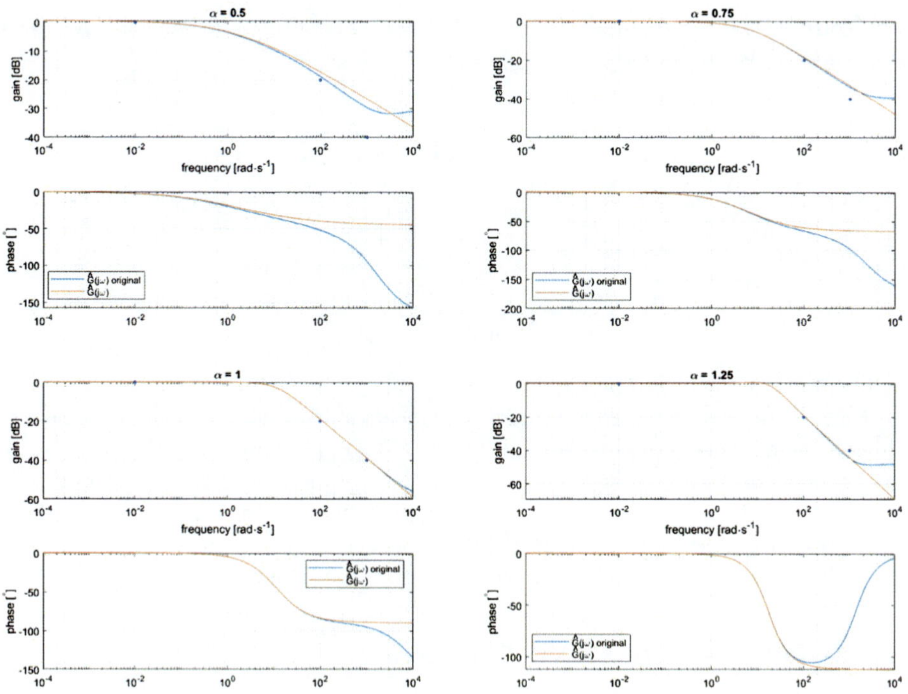

Fig. 4. Bode diagrams of the example of identification of a transfer function with one pole in Sect. 5.2.

5.2 Identifying Filters with One Pole

Transfer functions with one pole can be found from gains for only three frequencies; a spurious zero will also be obtained, which can be eliminated automatically by keeping only zeros and poles which as not clearly above the average. Accurate results are obtained when the commensurate order corresponds to the gain slope; otherwise, reasonable results are still got.

Example. The following transfer functions were obtained from three gains corresponding to a first order integer transfer function:

$$\hat{G}(s) = \frac{1.516}{s^{0.5} + 1.413} \tag{22}$$

$$\hat{G}(s) = \frac{3.927}{s^{0.75} + 3.895} \tag{23}$$

$$\hat{G}(s) = \frac{11.23}{s + 11.22} \tag{24}$$

$$\hat{G}(s) = \frac{33.71}{s^{1.25} + 33.7} \tag{25}$$

The spurious zero was eliminated keeping only poles p_k, $k = 1 \dots N$ and zeros z_k, $k = 1 \dots M$ that verify

$$\bar{x} = \sqrt[N+M]{\prod_{k=1}^{N} |p_k| \prod_{k=1}^{M} |z_k|} \tag{26}$$

$$\left|\frac{p_k}{\bar{x}}\right| \leq \sqrt{\frac{\bar{x}}{\min_k \{|p_k|, |z_k|\}}} \tag{27}$$

$$\left|\frac{z_k}{\bar{x}}\right| \leq \sqrt{\frac{\bar{x}}{\min_k \{|p_k|, |z_k|\}}} \tag{28}$$

and keeping the steady-state gain.

Figure 4 shows the data used for identification, and the frequency responses of both the original model before applying (26)–(28) as well as models (22)–(25). Notice how the steady-state gain is correctly identified; how the gain behaviour is recovered, though with larger errors for high frequencies; and how the phase behaviour, not used in the identification, is accurately retrieved for low and intermediate frequencies, though not for high frequencies.

5.3 Identifying Filters with Two Poles

Transfer functions with two poles can be found from gains for five frequencies; two spurious zeros will be obtained, and at least one can be eliminated automatically as above. Comments about accuracy for transfer functions with one pole apply here too.

Example. The following transfer functions were obtained from five gains corresponding to a transfer function with two poles and commensurate order 0.75:

$$\hat{G}(s) = \frac{0.3274}{s - 0.9048s^{0.5} + 0.4049} \tag{29}$$

$$\hat{G}(s) = \frac{0.9619}{s^{1.5} - 0.3556^{0.75} + 0.9721} \tag{30}$$

$$\hat{G}(s) = \frac{0.0796s + 2.807}{s^2 + 1.046s + 2.798} \tag{31}$$

$$\hat{G}(s) = \frac{0.3052^{1.25} + 7.366}{s^{2.5} + 4.525s^{1.25} + 7.353} \tag{32}$$

Figure 4 shows the data used for identification, and the frequency responses of both the original model before applying (26)–(28) as well as models (29)–(32).

5.4 General Fractional Order Filter Design

Transfer functions found in the previous examples can be considered as filters, but a more common approach is to set a constant gain in the desired pass band and a (clearly lower) constant gain in the desired rejection band. In practice,

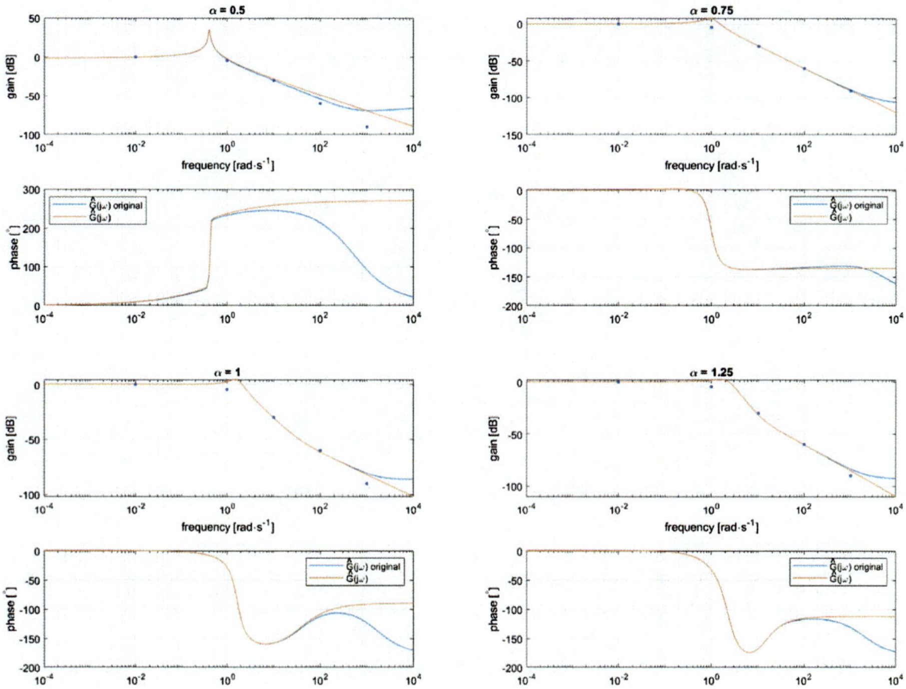

Fig. 5. Bode diagrams of the example of identification of a transfer function with two poles in Sect. 5.3.

constant values cannot be used, since numerical problems would follow; either a constant decrease or a random fluctuation can be used to avoid them.

Unstable transfer functions being quite useless as filters, it is a good idea to perform iterations, changing the random fluctuation around the desired values of the gain, until a stable filter is found.

Example. Data in Fig. 6 was used to obtain transfer function

$$\hat{G}(s) = \frac{0.003555\, s^{8.33} - 0.8298 s^{6.67} + 71.53\, s^5 - 2561 s^{3.33} - 26.46\, s^{1.67} - 0.009001}{s^{8.33} - 119.9\, s^{6.67} - 1060 s^5 - 2084 s^{3.33} - 32.74\, s^{1.67} - 0.003954} \quad (33)$$

that was then pruned with (26)–(28) and the elimination of close pairs of poles and zeros:

$$\hat{G}(s) = \frac{1}{0.06204\, s^{3.33} + 0.5194 s^{1.67} + 1} \quad (34)$$

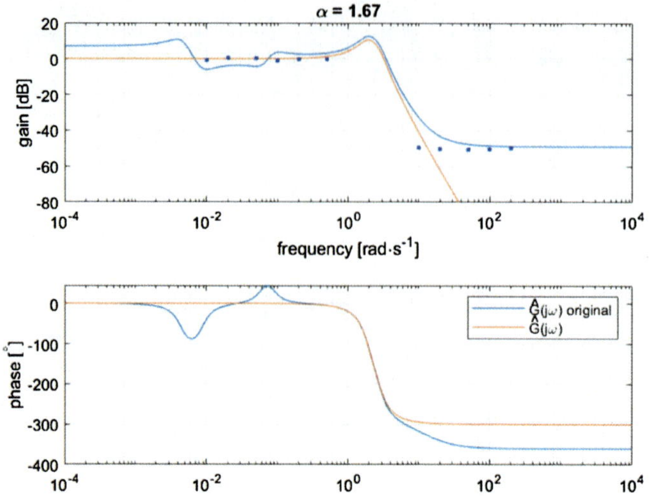

ω [rad/s]	0.01	0.02	0.05	0.1	0.2	0.5	10	20	50	100	200
gain [dB]	0.92	−0.51	−0.44	0.40	0.96	0.30	−49.74	−50.07	−50.41	−49.44	−49.13

Fig. 6. Bode diagrams of the example of identification of a filter in Sect. 5.4, and the data from which they were obtained.

6 Discussion and Conclusions

As always happens with identification methods, care is needed to select the data, apply the algorithm, and interpret the results. The examples given show that the Matsuda method works to identify models from data, and to obtain fractional order filters; its performance should be studied in the presence of noise, compared with the performance of the Levy method (even though the latter requires data on the phase, unnecessary with the Matsuda method), and assessed for applications with experimental data.

Acknowledgment. This work was supported by FCT, through IDMEC, under LAETA, project UIDB/50022/2020. Author M. D. Ortigueira was funded by Portuguese National Funds through the FCT—Foundation for Science and Technology within the scope of the CTS Research Unit—Center of Technology and Systems/UNINOVA/FCT/NOVA, under the reference UIDB/00066/2020.

References

1. Matsuda, K., Fujii, H.: $\mathcal{H}_\infty$ optimized wave-absorbing control – analytical and experimental results. J. Guid. Control. Dyn. **16**(6), 1146–1153 (1993)
2. Ortigueira, M.D., Valério, D.: Fractional Signals and Systems. De Gruyter, Berlin (2020)

3. Valério, D., Ortigueira, M.D., Sá da Costa, J.: Identifying a transfer function from a frequency response. ASME J. Comput. Nonlinear Dyn. **3**(2), 021207 (2008)
4. Valério, D., Sá da Costa, J.: Introduction to single-input, single-output fractional control. IET Control Theory Appl. **5**(8), 1033–1057 (2011)
5. Vinagre, B.M., Podlubny, I., Hernández, A., Feliu, V.: Some approximations of fractional order operators used in control theory and applications. Fract. Calc. Appl. Anal. **3**, 231–248 (2000)

An Existence Result for a Fractional Integral Boundary Value Problem of Mixed Type with Impulses

Sadek Habani[1] and Djamila Seba[2(✉)]

[1] Department of Mathematics, Saâd Dahlab University USDB, Blida 1, BP 270, 09000 Blida, Algeria
[2] Dynamic of Engines and Vibroacoustic Laboratory, University M'Hamed Bougara of Boumerdes, Boumerdes, Algeria
seba@univ-boumerdes.dz

Abstract. In this paper we investigate the existence of weak solutions for an impulsive fractional integral boundary value problem of mixed type in a Banach space. The results are obtained using Henstock-Kurzweil-Pettis integrals and De Blasi measure of weak noncompactness. An example is given to illustrate the theory.

Keywords: Fractional derivative · Weak solution · Fixed point

1 The Impulsive Problem

In this work, we are interested on an integral boundary value problem of nonlinear impulsive differential equations of mixed type of the form:

$$\begin{cases} {}^cD^r x(t) = \phi(t,\, x(t),\, \psi x(t),\, \varphi x(t)), \quad 1 < r \leq 2,\ t \in J', \\ \Delta x(t_k) = \chi_k(x(t_k),\, x'(t_k)), \quad \Delta x'(t_k) = \chi_k^*(x(t_k),\, x'(t_k)), \quad k = 0, 1, 2, ..., m, \\ \delta x(0) + \mu x'(0) = \int_0^b \sigma_1(x(s))ds, \quad \delta x(b) + \mu x'(b) = \int_0^b \sigma_2(x(s))ds, \end{cases} \tag{1}$$

where ${}^cD^r$ is the Caputo fractional derivative of order r, $J = [0, b]$, ϕ is a function from $J \times\ \mathcal{F} \times\ \mathcal{F} \times\ \mathcal{F}$ to $\mathcal{F}$, $\chi_k, \chi_k^* \in C(\mathcal{F} \times\ \mathcal{F}, \mathcal{F})$, $0 = t_0 < t_1 < ... < t_k < ... < t_{m-1} < t_m = b$, $J' = J/\{t_0, t_1, ..., t_m\}$, $\Delta x(t_k) = x(t_k^+) - x(t_k^-)$. The right and the left limits of $x(t)$ at $t = t_k$ $(k = 1, 2, ..., m)$ are respectively$x(t_k^+)$ and $x(t_k^-)$. $\Delta x'(t_k)$ have a similar meaning for $x'(t)$, $\sigma_1, \sigma_2 : \mathcal{F} \longrightarrow \mathcal{F}$, $\delta > 0$, $\mu \geq 0$ and $\mathcal{F}$ is a Banach space with norm $\|.\|$. Moreover $\psi x(t) = \int_0^t g(t,s)x(s)ds$, $\varphi x(t) = \int_0^t h(t,s)x(s)ds$ where g and h will be specified later.

Our main concern is to use Henstock-Kurzweil-Pettis integrals and De Blasi measure of weak nonompactness to prove the existence of weak solutions for the fractional boundary value problem (1).

In the next Section we present some basic definitions, notations, lemmas and theorems that are needed in the third Section. In this later, we first present the solution form of the studied problem with a detailed proof and then the main result is given where the demonstration is based on the properties of the Henstock-Kurzweil-Pettis integrals and the De Blasi measure of weak nonompactness.

A. Dzielinski et al. (Eds.): ICFDA 2021, LNNS 452, pp. 208–217, 2022.
https://doi.org/10.1007/978-3-031-04383-3_23

1.1 Preliminaries

$C(J,\mathcal{F}_w) = (C(J,\mathcal{F}), w)$ denotes the space of all continuous functions from J to $\mathcal{F}$ with the usual supremum norm $\|x\|_\infty = \sup_{t\in J} \|x(t)\|$ and the weak topology. $\mathcal{F}^*$ is the dual space of $\mathcal{F}$ and $\mathcal{F}_w = (\mathcal{F}, w) = (\mathcal{F}, \sigma(\mathcal{F}, \mathcal{F}^*))$ is the space $\mathcal{F}$ with its weak topology. Let the Banach space

$$PC(J,\mathcal{F}) = \{x : J \to \mathcal{F} :\ x \in C(J_k,\ \mathcal{F}),\ x(t_k^+) \text{ and } x(t_k^-) \text{ exist},\ k = 1,2,...,m\},$$

be equipped with $\|x\|_{PC} = \sup_{t\in J} \|x(t)\|$.

J_{k-1} here represents $(t_{k-1}, t_k]$, $k = 1,2,...,m+1$ with $t_0 = 0$ and $t_{m+1} = b$. We define the Banach space $\mathfrak{P} = PC^1(J,\mathcal{F}) = \{x \in PC(J,\mathcal{F}),\ x'(t_k^+),\ x'(t_k^-)$ exist and x' is left continuous at t_k, for , $k = 1,2,...,m\}$ with the norm $\|x\|_{\mathfrak{P}} = \sup_{t\in J}\{\|x(t)\|_{PC},\ \|x'(t)\|_{PC}\}$. In the sequel, we will make use of the definition of Henstock-Kurzweil-Pettis integrable functions which can be found in [4,7,11].

Theorem 1 [3]. *Let $\phi : J \to \mathcal{F}$ and assume that $\phi_n : J \to \mathcal{F}$, $n \in \mathbb{N}$, are Henstock-Kurzweil-Pettis integrable on J. For each $n \in \mathbb{N}$, let Φ_n be a primitive of ϕ_n. If we assume that:*

(i) $\forall x^* \in \mathcal{F}^*$, $x^*(\phi_n(t)) \to x^*(\phi(t))$ *a.e. on* J,
(ii) for each $x^* \in \mathcal{F}^*$, *the family* $G = \{x^*\Phi_n : n = 1,2,...\}$ *is uniformly generalized absolutely continuous on J (ie weakly uniformly generalized absolutely continuous on J),*
(iii) for each $x^* \in \mathcal{F}^*$, *the set G is equicontinuous on J,*

then ϕ is Henstock-Kurzweil-Pettis integrable on J and $\int_0^t \phi_n(s)ds$ tends weakly in $\mathcal{F}$ to $\int_0^t \phi(s)ds$ for each $t \in J$.

Definition 1 [7]. *Let $x : [d,\infty) \longrightarrow \mathcal{F}$ be a function. The fractional Henstock-Kurzweil-Pettis-integral of the function x of order $\alpha \in \mathbb{R}_+$ is defined by*

$$\mathfrak{I}^{\alpha}_{d+}x(t) = \int_d^t \frac{(t-s)^{\alpha-1}}{\Gamma(\alpha)} x(s)ds,$$

where the sign $\int$ denotes the Henstock-Kurzweil-Pettis-integral and Γ is the Gamma function.

For details on the properties of the De Blasi measure of noncompactness along with its definition, which are crutial in our proofs, the reader can see [5,8] and the references therein

Lemma 1 [8]. *Let $\mathcal{D}$ be a subset of $C(J,\mathcal{F})$ and $u_0 \in C(J,\mathcal{F})$.*

- *If $\mathcal{D}$ is equicontinuous, then $\overline{co}\{\mathcal{D}, u_0\}$ is equicontinuous in $C(J,\mathcal{F})$.*
- *If $\mathcal{D}$ is bounded and equicontinuous, then $\beta(\mathcal{D}) = \max_{t\in J} \beta(\mathcal{D}(t))$. Moreover $\beta(\mathcal{D}(t)) \in C(J,\mathbb{R}^+)$ and $\beta(\mathcal{D}(s)ds) \le \beta(\mathcal{D}(s))ds$, $\forall t \in J$.*

Now, let us recall that for a subset W of a Banach space $\mathcal{F}$, if W is nonempty, closed, convex and bounded, $G : W \to W$ is weakly sequentially continuous (Shortly (WSC)) and there exist a vector $x_0 \in W$ and an integer n_0 for which G is power-convex condensing ((PCC) for short). Then and according to [[1], Theorem 2.1], G has at least one fixed point in W.

2 Main Result

Here, we give the definition of a solution of problem (1) followed by a lemma that provides its form.

Definition 2. *A map $x \in PC^1(J, \mathcal{F})$ is called a solution of problem (1) if it satisfies the equation ${}^cD^r x(t) = \phi(t, x(t), \psi x(t), \varphi x(t))$ on J' and the conditions $\delta x(0) + \mu x'(0) = \int_0^b \sigma_1(x(s))ds$, $\delta x(b) + \mu x'(b) = \int_0^b \sigma_2(x(s))ds$, $\Delta x(t_k) = \chi_k(x(t_k), x'(t_k))$, and $\Delta x'(t_k) = \chi_k^*(x(t_k), x'(t_k))$, for $k = 0, 1, 2, ..., m$.*

Lemma 2. *For any $\sigma, \sigma_1, \sigma_2 \in C(J, \mathcal{F})$, a function x is a solution of the impulsive boundary value problem (2)–(5):*

$$ {}^cD^r x(t) = \sigma(t), \quad 1 < r \leq 2, \ k = 0, 1, ..., m, \ t \in J', \tag{2} $$

$$ \Delta x(t_k) = \chi_k(x(t_k), x'(t_k)), \quad \Delta x'(t_k) = \chi_k^*(x(t_k), x'(t_k)), \quad k = 1, 2, ..., m, \tag{3} $$

$$ \delta x(0) + \mu x'(0) = \int_0^b \sigma_1(s)ds, \tag{4} $$

$$ \delta x(b) + \mu x'(b) = \int_0^b \sigma_2(s)ds, \tag{5} $$

if and only if x is a solution of the following impulsive fractional integral equation

$$ x(t) = \begin{cases} \frac{1}{\Gamma(r)} \int_0^t (t-s)^{r-1}\sigma(s)ds + \mathcal{A}, & t \in J_0 \\ \frac{1}{\Gamma(r)} \int_{t_k}^t (t-s)^{r-1}\sigma(s)ds + \mathcal{A} \\ + \sum\limits_{0<t_k<t} \left(\frac{1}{\Gamma(r)} \left[\int_{t_{k-1}}^{t_k} (t_k - s)^{r-1}\sigma(s)ds + \chi_k(x(t_k), x'(t_k)) \right] \right) \\ + \sum\limits_{0<t_k<t} \left(\frac{(t-t_k)}{\Gamma(r-1)} \left[\int_{t_{k-1}}^{t_k} (t_k - s)^{r-2}\sigma(s)ds + \chi_k^*(x(t_k), x'(t_k)) \right] \right), & t \in J_k \end{cases} \tag{6} $$

where
$$ \mathcal{A} = +\frac{1}{b}\left(\frac{\mu}{\delta} - t\right)\left(\frac{1}{\Gamma(r)} \int_{t_k}^b (b-s)^{r-1}\sigma(s)ds + \frac{\mu}{\delta}\frac{1}{\Gamma(r-1)} \int_{t_k}^b (b-s)^{r-2}\sigma(s)ds \right. $$
$$ + \sum_{0<t_k<b} \left[\frac{1}{\Gamma(r)} \int_{t_{k-1}}^{t_k} (t_k - s)^{r-1}\sigma(s)ds + \chi_k(x(t_k), x'(t_k)) \right] $$
$$ \left. + \sum_{0<t_k<b} \left[\frac{(\frac{\mu}{\delta} + b - t_k)}{\Gamma(r-1)} \int_{t_{k-1}}^{t_k} (t_k - s)^{r-2}\sigma(s)ds + \chi_k^*(x(t_k), x'(t_k)) \right] \right) $$
$$ + \frac{1}{b\delta^2}\left[(\delta(b-t) + \mu) \int_0^b \sigma_1(s)ds + (\delta t - \mu) \int_0^b \sigma_2(s)ds \right] $$

Proof. By Lemma 1 in [11] and for $a_0, a_1 \in \mathfrak{P}$, the solution of (2) can be written

$$ \text{as } x(t) = \mathfrak{I}^r\sigma(t) - a_0 - a_1 t = \frac{1}{\Gamma(r)} \int_0^t (t-s)^{r-1}\sigma(s)ds - a_0 - a_1 t, \quad t \in [0, t_1] $$

in view of the relations ${}^cD^{r_1}\mathfrak{I}^{r_1}x(t) = x(t)$ and $\mathfrak{I}^{r_1}\mathfrak{I}^{r_2}x(t) = \mathfrak{I}^{r_1+r_2}x(t)$ for $r_1, r_2 > 0$, we obtain $x'(t) = \frac{1}{\Gamma(r-1)} \int_0^t (t-s)^{r-2}\sigma(s)ds - a_1$.
Applying the boundary condition (4), we find that

$$ x(t) = \frac{1}{\Gamma(r)} \int_0^t (t-s)^{r-1}\sigma(s)ds + a_1\left(\frac{\mu}{\delta} - t\right) + \frac{1}{\delta} \int_0^b \sigma_1(s)ds, \quad t \in [0, t_1]. \tag{7} $$

If $t \in (t_1, t_2]$, then we have $x(t) = \frac{1}{\Gamma(r)} \int_{t_1}^{t} (t-s)^{r-1}\sigma(s)ds - b_0 - b_1(t-t_1)$, where $b_0, b_1 \in \mathfrak{P}$.
Using the impulse conditions (3), we have

$$\begin{aligned}
-b_0 =& \frac{1}{\Gamma(r)} \int_0^{t_1} (t_1 - s)^{r-1}\sigma(s)ds + a_1(\frac{\mu}{\delta} - t_1) + \frac{1}{\delta}\int_0^b \sigma_1(s)ds + \chi_1(x(t_1), x'(t_1))\\
-b_1 =& \frac{1}{\Gamma(r)} \int_0^{t_1} (t_1 - s)^{r-2}\sigma(s)ds - a_1 + \chi_1^*(x(t_1), x'(t_1)).
\end{aligned}$$

Thus, for $t \in (t_1, t_2]$

$$\begin{aligned}
x(t) =& \frac{1}{\Gamma(r)} \int_{t_1}^{t} (t - s)^{r-1}\sigma(s)ds + \frac{1}{\Gamma(r)} \int_0^{t_1} (t_1 - s)^{r-1}\sigma(s)ds\\
&+ \frac{1}{\delta}\int_0^b \sigma_1(s)ds + \chi_1(x(t_1), x'(t_1))\\
&+ a_1(\frac{\mu}{\delta} - t) + (t - t_1)\left[\frac{1}{\Gamma(r-1)} \int_0^{t_1} (t_1 - s)^{r-1}\sigma(s)ds + \chi_1^*(x(t_1), x'(t_1))\right].
\end{aligned}$$

Repeating the process in this way, the solution $x(t)$ for $t \in (t_k, t_{k+1}]$ can be written as

$$\begin{aligned}
x(t) =& \frac{1}{\Gamma(r)} \int_{t_k}^{t} (t - s)^{r-1}\sigma(s)ds + a_1(\frac{\mu}{\delta} - t) + \frac{1}{\delta}\int_0^1 \sigma_1(s)ds\\
&+ \sum_{0<t_k<t} \left[\frac{1}{\Gamma(r)} \int_{t_{k-1}}^{t_k} (t_k - s)^{r-1}\sigma(s)ds + \chi_k(x(t_k), x'(t_k))\right]\\
&+ \sum_{0<t_k<t} (t - t_k)\left[\frac{1}{\Gamma(r-1)} \int_{t_{k-1}}^{t_k} (t_k - s)^{r-2}\sigma(s)ds + \chi_k^*(x(t_k), x'(t_k))\right] \qquad (8)
\end{aligned}$$

Applying the boundary condition (5), we find that

$$\begin{aligned}
a_1 =& \frac{1}{b}\Bigg[\frac{1}{\Gamma(r)} \int_{t_k}^{b} (b - s)^{r-1}\sigma(s)ds + \frac{\mu}{\delta}\frac{1}{\Gamma(r-1)} \int_{t_k}^{b} (b - s)^{r-2}\sigma(s)ds\\
&+ \sum_{0<t_k<b} \left(\frac{1}{\Gamma(r)} \int_{t_{k-1}}^{t_k} (t_k - s)^{r-1}\sigma(s)ds + \chi_k(x(t_k), x'(t_k))\right)\\
&+ \frac{1}{\delta}\left(\int_0^b \sigma_1(s)ds - \int_0^b \sigma_2(s)ds\right)\\
&+ \sum_{0<t_k<b} \frac{(\frac{\mu}{\delta} + b - t_k)}{\Gamma(r-1)} \int_{t_{k-1}}^{t_k} (t_k - s)^{r-2}\sigma(s)ds + \chi_k^*(x(t_k), x'(t_k))\Bigg]
\end{aligned}$$

Substituting the value of a_1 in (7) and (8), we obtain (6). This completes the proof.

Let us impose the following assumptions:generalized absolutely continuous

(A1) For each uniformly generalized absolutely continuous function $x : J \to \mathcal{F}$, ϕ is (WSC) and $g(t,.)x(.)$, $h(t,.)x(.)$, $\phi(.,x(.),\psi x(.),\varphi x(.))$ are Henstock-Kurzweil-Pettis integrable functions.

(A2) For any $\alpha > 0$, there exists an Henstock-Kurzweil-integrable function $M_\alpha : J \to \mathbb{R}^+$ and a non-decreasing continuous function $\Omega : [0, \infty) \to [0, \infty)$ such that $\|\phi(t, x, y, z)\| \le M_\alpha(t)\Omega(\alpha)$ for all $t \in J$, $(x, y, z) \in D_\alpha \times D_\alpha \times D_\alpha$, and there exist positive constants L_1 and L_2 such that $|\chi_k(x(.), x'(.))| \le L_1$, $|\chi_k^*(x(.), x'(.))| \le L_2$ for $t \in J$, $x \in D_\alpha$, $k = 1, ..., m$.

(A3) There exists constants $c_1, c_2 > 0$ such that $\|\sigma_1\| \le c_1$, $\|\sigma_2\| \le c_2$

(A4) For each bounded sets $V, U, W \subset D_\alpha$ and for each closed interval $I \subset J$, $t \in I$, there exist positive constants $\zeta_1, \zeta_2 \in J$:
$\beta(g(I, I)V(I)) \le \zeta_1\beta(V(I))$, $\beta(h(I, I)V(I)) \le \zeta_2\beta(V(I))$
and
$\beta(\phi(t, V, U, W)) \le M_\alpha(t)\max(\beta(V), \beta(U), \beta(W))$.

(A5) The family $\{x^*\phi(., x(.), \psi x(.), \varphi x(.)) : x^* \in \mathcal{F}^*, \|x^*\| \le 1\}$ is uniformly Henstock-Kurzweil-integrable over J for every $x \in D_r$.

(A6) The families $\left\{x^* \int\limits_{J_k} g(t, s)x_n(s)ds\right\}_{n=0}^{\infty}$, $\left\{x^* \int\limits_{J_k} h(t, s)x_n(s)ds\right\}_{n=0}^{\infty}$ and $\left\{x^* \int\limits_{J_k} G(t, s)\phi(., x_n(s), \psi x_n(s), \varphi x_n(s))ds\right\}_{n=0}^{\infty}$ are uniformly generalized absolutely continuous and equicontinuous on J for every $t \in J$.

Our main result reads as follow

Theorem 2. *Suppose that the assumptions (A1)-(A6) Hold. If*

$$\|M_\alpha\| \max(1, b\zeta_1, b\zeta_2)\frac{b^*}{\Gamma(r)}\left((m+1)(\frac{\mu}{\delta})^2 + (3m+2)\frac{\mu}{\delta} + 4m + 2\right) < 1, \quad (9)$$

Then the boundary value problem (1) has at least one solution.

Proof. Let $T : PC(J, \mathcal{F}) \longrightarrow PC(J, \mathcal{F})$

$$\begin{aligned}
Tx(t) &= \frac{1}{\Gamma(r)}\int_0^t (t-s)^{r-1}\phi(s, x(s), \psi x(s), \varphi x(s))ds \\
&+ \frac{1}{b}\left(\frac{\mu}{\delta} - t\right)\left(\frac{1}{\Gamma(r)}\int_{t_k}^b (b-s)^{r-1}\phi(s, x(s), \psi x(s), \varphi x(s))ds\right. \\
&+ \frac{\mu}{\delta}\frac{1}{\Gamma(r-1)}\int_{t_k}^b (b-s)^{r-2}\phi(s, x(s), \psi x(s), \varphi x(s))ds \\
&+ \sum_{0<t_k<b}\left[\frac{1}{\Gamma(r)}\int_{t_{k-1}}^{t_k} (t_k-s)^{r-1}\phi(s, x(s), \psi x(s), \varphi x(s))ds + \chi_k(x(t_k), x'(t_k))\right] \\
&+ \left.\sum_{0<t_k<b}\left[\frac{(\frac{\mu}{\delta} + b - t_k)}{\Gamma(r-1)}\int_{t_{k-1}}^{t_k} (t_k-s)^{r-2}\phi(s, x(s), \psi x(s), \varphi x(s))ds + \chi_k^*(x(t_k), x'(t_k))\right]\right) \\
&+ \sum_{0<t_k<t}\left(\frac{1}{\Gamma(r)}\left[\int_{t_{k-1}}^{t_k} (t_k-s)^{r-1}\phi(s, x(s), \psi x(s), \varphi x(s))ds + \chi_k(x(t_k), x'(t_k))\right]\right) \\
&+ \sum_{0<t_k<t}\left(\frac{(t-t_k)}{\Gamma(r-1)}\left[\int_{t_{k-1}}^{t_k} (t_k-s)^{r-2}\phi(s, x(s), \psi x(s), \varphi x(s))ds + \chi_k^*(x(t_k), x'(t_k))\right]\right) \\
&+ \frac{1}{b\delta^2}\left[(\delta(b-t)+\mu)\int_0^b \sigma_1(s)ds + (\delta t - \mu)\int_0^b \sigma_2(s)ds\right]
\end{aligned}$$

be an operator defined in view of Lemma 2. The solutions of the problem (1) are the fixed points of the operator T . For convenience, we will give some notations: Let $\aleph$ be a family of functions which are uniformly Henstock-Kurzweil-integrable over an interval J. Then following the proof given in [9], Theorem 4.28, $\aleph$ satisfies uniformly the Cauchy criterion over any closed sub-interval $I \subset J$. Analogously to [9], Theorem 4.27, the condition

$\forall\varepsilon > 0$; $\exists$gauge γ; on J; $\forall P_1, P_2$ γ fine-partitions; $\forall\phi \in \aleph$; $|S(\phi, P_1) - S(\phi, P_2)| < \varepsilon$, implies

$\forall\varepsilon > 0$; $\exists$gauge γ; on J; $\forall P_1, P_2$ γ fine-partitions; $\forall\phi \in \aleph$;
$\left|S(\phi, P) - (HK)\int_J f(t)dt\right| < \varepsilon$.

Therefore, if $\aleph$ is a family of uniformly Henstock-Kurzweil-integrable functions over the interval $[0, N]$, then the family $\aleph$ is uniformly Henstock-Kurzweil-integrable over $[0, \tau]$ for every $\tau < N$. Consequently, in view of assumption (A4) the family

$$\{x^*\phi(., x(.), \psi x(.), \varphi x(.)) : \; x^* \in \mathcal{F}^*, \; \|x^*\| \leq 1\}$$

is uniformly Henstock-Kurzweil-integrable over any sub-interval $[0, \tau] \subset J$, for every $x \in D_\alpha$. This entails the weak*-continuity of the linear functional

$$x^* \in \mathcal{F}^* \to (HK)\int_0^\tau G(t, s)x^*\phi(s, x(s), \psi x(s), \varphi x(s))ds, \; \forall\tau \in J.$$

This latter means that there is $x_{t,\tau} \in \mathcal{F}$ such that

$$x^*x_{t,\tau} = (HK)\int_0^\tau x^*G(t, s)\phi(s, x(s), \psi x(s), \varphi x(s))ds, \; \forall x^* \in \mathcal{F}^*.$$

That is the function $G(t, .)\phi(, x(.), \psi x(.), \varphi x(.))$ is Henstock-Kurzweil-Pettis-integrable on J and thus the operator T makes sense.

Consider the set $D = D_\alpha \bigcap \{x \in PC(J, \mathcal{F}) : \forall\tau_1, \tau_2 \in J, \tau_1 < \tau_2 : \|x(\tau_2) - x(\tau_1)\| \leq P + \frac{1}{\delta}(\tau_2 - \tau_1)(c_1 + c_2)\}$.

where $P = \frac{\|M_\alpha\|\Omega(\alpha)}{\Gamma(r)}\left[(\tau_2^r - \tau_1^r) + (\tau_2 - \tau_1)b^*\left(1 + 5m + (m+1)\frac{\mu}{\delta}\right)\right]$.

Clearly, the subset D is closed, convex and equicontinuous.

We shall show that T satisfies all the conditions of the fixed point theorem recalled in the preliminaries as [[1], Theorem 2.1]. The proof will be given in several steps.

- First we prove that T maps D into itself:

Take $x \in D$ and $t \in J$ and assume that $Tx(t) \neq 0$. By the Hahn-Banach theorem, there exists $x^* \in \mathcal{F}^*$ with $\|x^*\| = 1$ and $\|Tx(t)\| = |x^*(Tx(t))|$. Thus

$$\|Tx(t)\| \leq \Bigg(\frac{1}{\Gamma(r)} \int_0^t (t-s)^{r-1} |x^*(\phi(s,x(s),\psi x(s),\varphi x(s)))| ds$$
$$+ \frac{1}{b} \left| \frac{\mu}{\delta} - t \right| \left[\frac{1}{\Gamma(r)} \int_{t_k}^b (b-s)^{r-1} |x^*(\phi(s,\, x(s),\, \psi x(s),\, \varphi x(s)))| ds \right.$$
$$+ \frac{\mu}{\delta} \frac{1}{\Gamma(r-1)} \int_{t_k}^b (b-s)^{r-2} |x^*(\phi(s,x(s),\psi x(s),\varphi x(s)))| ds$$
$$+ \sum_{0<t_k<b} \left(\frac{1}{\Gamma(r)} \int_{t_{k-1}}^{t_k} (t_k-s)^{r-1} |x^*(\phi(s,x(s),\psi x(s),\varphi x(s)))| ds + |\chi_k(x(t_k),x'(t_k))| \right)$$
$$\left. + \sum_{0<t_k<b} (\frac{\mu}{\delta} + b) \left(\int_{t_{k-1}}^{t_k} \frac{(t_k-s)^{r-2}}{\Gamma(r-1)} |x^*(\phi(s,x(s),\psi x(s),\, \varphi x(s)))| ds + |\chi_k^*(x(t_k),\, x'(t_k))| \right) \right]$$
$$+ \sum_{0<t_k<t} \left(\frac{1}{\Gamma(r)} \int_{t_{k-1}}^{t_k} (t_k-s)^{r-1} |x^*(\phi(s,\, x(s),\, \psi x(s),\, \varphi x(s)))| ds + |\chi_k(x(t_k),\, x'(t_k))| \right)$$
$$+ \sum_{0<t_k<t} (t-t_k) \left(\int_{t_{k-1}}^{t_k} \frac{(t_k-s)^{r-2}}{\Gamma(r-1)} |x^*(\phi(s,\, x(s),\, \psi x(s),\, \varphi x(s)))| ds + |\chi_k^*(x(t_k),\, x'(t_k))| \right)$$
$$+ \frac{1}{b\delta^2} \left[(\delta(b-t)+\mu) \int_0^b |\sigma_1(s)| ds + |\delta t - \mu| \int_0^b |\sigma_2(s)| ds \right] \Bigg)$$
$$\leq ||M_\alpha|| \Omega(\alpha) \frac{b^*}{\Gamma(r)} \Bigg(((m+1)(\frac{\mu}{\delta})^2 + (3m+2)\frac{\mu}{\delta} + 4m+2) + m(\frac{\mu}{\delta} + 2T^*)L_1$$
$$+ m((\frac{\mu}{\delta} + b^*)^2 + b^*)L_2 \Bigg) + \frac{\delta b^* + \mu}{\delta^2} [c_1 + c_2] \leq r_0.$$

Let $\tau_1, \tau_2 \in J_k$, $k = 1, ..., m$, $\tau_1 < \tau_2$, $x \in D$, so $Tx(\tau_2) - Tx(\tau_1) \neq 0$. Then there exists $x^* \in \mathcal{F}^*$ such that $\|Tx(\tau_2) - Tx(\tau_1)\| = |x^*(Tx(\tau_2) - Tx(\tau_1))|$. Thus

$$\|Tx(\tau_2) - Tx(\tau_1)\| \leq ||M_\alpha|| \Omega(\alpha) \frac{1}{\Gamma(r)} \left[(\tau_2^r - \tau_1^r) + (\tau_2 - \tau_1) b^* \left(1 + 5m + (m+1)\frac{\mu}{\delta} \right) \right]$$
$$+ \frac{1}{\delta}(\tau_2 - \tau_1)(c_1 + c_2) \leq P + \frac{1}{\delta}(\tau_2 - \tau_1)(c_1 + c_2)$$

Hence $T(D) \subset D$.
- Then we show that T is (WSC).

Let (x_n) be a sequence in D and let $x_n(t) \to x(t)$ in $C(J, \mathcal{F}_w)$. By (A6), we have
$\int_0^b g(t,s)x_n(s)ds \to \int_0^b g(t,s)x(s)ds$ and $\int_0^b h(t,s)x_n(s)ds \to \int_0^b h(t,s)x(s)ds$ weakly in in $\mathcal{F}$, for each $t \in J$ and $\psi x_n(t) \to \psi x(t)$, $\varphi x_n(t) \to \varphi x(t)$. Therefore, the operators ψ, φ are (WSC) in D. Moreover, because ϕ is (WSC),

$$\phi(t, x_n(t), \psi x_n(t), \varphi x_n(t)) \to \phi(t, x(t), \psi x(t), \varphi x(t))$$

weakly in $\mathcal{F}$, for each $t \in J$. Using (A4), Theorem 5 in [6] and Lemma 25 in [10], the function $G(t,.)\phi(., x_n(.), \psi x_n(.), \varphi x_n(.))$ is Henstock-Kurzweil-Pettis-integrable on J for every $n \geq 1$. By Theorem 1 and assumption (A6), we have

$$\lim_{n\to\infty} Tx_n(t) = Tx(t)$$

i.e. $Tx_n(t) \to Tx(t)$ in $PC(J, \mathcal{F})$.

- After that we shall show that $T : D \to D$ is (PCC).

Let $B = \overline{\text{co}}T(D) \subset D$. Note that B is bounded, convex, closed, and $T(\overline{\text{co}}T(D)) \subset T(B) \subset \overline{\text{co}}T(D)$, i.e. $T : B \to B$. From Lemma 1, B is equicontinuous in $C(J, \mathcal{F}_w)$. Obviously, T is bounded and continuous. Set $x_0 \in B$, we will prove that for all $n \in \mathbb{N}$ and for any bounded $V \subset B$,

$$\beta(T^{(n_0,x_0)}(V)) \leq \beta(V).$$

By $V \subset B \subset D$, $T(V)$ is equicontinuous. Then $T^{(2,x_0)}(V)$ is equicontinuous from $T^{(2,x_0)}(V) = T(\overline{\text{co}}T(D), x_0) \subset T(D)$. Generally, $\forall n \in \mathbb{N}$, $T^{(n,x_0)}(V)$ is equicontinuous. Since $T^{(n,x_0)}(V)$ is bounded, By Lemma 1

$$\beta(T^{(n,x_0)}(V)) = \max_{t\in J}(T^{(n,x_0)}(V)(t)), \quad n = 2, 3, ... \tag{10}$$

We have

$$\begin{aligned}
\beta(T^{(1,x_0)}(V)(t)) &= \beta(TV(t)) \\
&\leq \int_0^t \frac{(t-s)^{r-1}}{\Gamma(r)} \beta(\phi(s, V(s), \psi V(s), \varphi V(s)))ds \\
&+ \frac{1}{b}(\frac{\mu}{\delta} - t)\Bigg[\int_{t_k}^b \frac{(b-s)^{r-1}}{\Gamma(r)} \beta(\phi(s,\ V(s),\ \psi V(s),\ \varphi V(s)))ds \\
&+ \frac{\mu}{\delta}\int_{t_k}^b \frac{(b-s)^{r-2}}{\Gamma(r-1)} \beta(\phi(s, V(s), \psi V(s), \varphi V(s)))ds \\
&+ \sum_{0<t_k<b} \int_{t_{k-1}}^{t_k} \frac{(t_k-s)^{r-1}}{\Gamma(r)} \beta(\phi(s, V(s), \psi V(s), \varphi V(s)))ds \\
&+ \sum_{0<t_k<b} (\frac{\mu}{\delta} + b - t_k) \int_{t_{k-1}}^{t_k} \frac{(t_k-s)^{r-2}}{\Gamma(r-1)} \beta(\phi(s, V(s), \psi V(s), \varphi V(s)))ds\Bigg] \\
&+ \sum_{0<t_k<t} \int_{t_{k-1}}^{t_k} \frac{(t_k-s)^{r-1}}{\Gamma(r)} \beta(\phi(s, V(s), \psi V(s), \varphi V(s)))ds \\
&+ \sum_{0<t_k<t} (t-t_k) \int_{t_{k-1}}^{t_k} \frac{(t_k-s)^{r-2}}{\Gamma(r-1)} \beta(\phi(s, V(s), \psi V(s), \varphi V(s)))ds \\
&\leq \|M_\alpha\| \max(1, b\zeta_1, b\zeta_2)\frac{b^*}{\Gamma(r)}\left((m+1)(\frac{\mu}{\delta})^2 + (3m+2)\frac{\mu}{\delta} + 4m + 2\right)\beta(V).
\end{aligned}$$

By the equicontinuity of $T^{(1,x_0)}(V)(t) = T(V)$, $G(t,s)\phi(s,\overline{\text{co}}\{T(V)(s),x_0\}$, $\psi\overline{\text{co}}\{T(V)(s),x_0\}$ and $\varphi\overline{\text{co}}\{T(V)(s),x_0\})$ are equicontinuous. Therefore,

$$\begin{aligned}
&\beta(T^{(2,x_0)}(V)(t)) = \beta(T\overline{\text{co}}\{T^{(1,x_0)}(V)(t),x_0\})\\
&\leq \|M_\alpha\|\max(1,b\zeta_1,b\zeta_2)B(T(V))\Bigg(\frac{b^r}{\Gamma(r+1)}+\frac{1}{b}(\frac{\mu}{\delta}+b)\Bigg[\frac{b^r}{\Gamma(r+1)}+\frac{\mu}{\delta}\frac{b^{r-1}}{\Gamma(r)}\\
&+mT\frac{b^r}{\Gamma(r+1)}+mT(\frac{\mu}{\delta}+b)\frac{b^{r-1}}{\Gamma(r)}\Bigg]+mT\frac{b^r}{\Gamma(r+1)}+mT^2\frac{b^{r-1}}{\Gamma(r)}\Bigg)\\
&\leq \left[\|M_\alpha\|\max(1,b\zeta_1,b\zeta_2)\frac{b^*}{\Gamma(r)}\left((m+1)(\frac{\mu}{\delta})^2+(3m+2)\frac{\mu}{\delta}+4m+2\right)\right]^2\beta(V).
\end{aligned}$$

Generally, $\beta(T^{(n,x_0)}(V)(t))$ is equal to
$\left[\|M_\alpha\|\max(1,b\zeta_1,b\zeta_2)\frac{b^*}{\Gamma(r)}\left((m+1)(\frac{\mu}{\delta})^2+(3m+2)\frac{\mu}{\delta}+4m+2\right)\right]^n\beta(V)\leq\beta(V).$
By (10), $\beta(T^{(n,x_0)}(V))\leq\beta(V)$. Therefore $T: V\to V$ is (PCC).

Applying the fixed point theorem [[1] Theorem 2.1], we conclude that T has a fixed point which is a solution of the problem (1).

3 Example

For $r=\frac{3}{2}$, $\mu=2,\delta=3,t_1=\frac{1}{2}$, $g(t,s)=2\frac{(t-s)^{\frac{3}{2}}}{\Gamma(\frac{3}{2})}$, $h(t,s)=\frac{(t-s)^{\frac{1}{2}}}{\Gamma(\frac{1}{2})}$, $\sigma_1(t)=\frac{x(t)}{12+||x(t)||}$ and $\sigma_2(t)=\frac{x(t)}{15+||x(t)||}$, we consider the following impulsive fractional differential equation of mixed type:

$$\begin{cases}
{}^cD^{\frac{3}{2}}x(t)=\frac{1}{20+e^t}(\frac{x(t)}{1+||x(t)||})+\int_0^t g(t,s)x(t)ds+\int_0^t h(t,s)x(t)ds,\ t\in[0,1],\ t\neq\frac{1}{2},\ k=0,1\\
\Delta x(\frac{1}{2})=\frac{2\left(x(\frac{1}{2})+x'(\frac{1}{2})\right)}{1+||x||+||x'||},\quad \Delta x'(\frac{1}{2})=\frac{\left(x(\frac{1}{2})+x'(\frac{1}{2})\right)}{1+||x||+||x'||}\\
3x(0)+2x'(0)=\int_0^1\sigma_1(x(s))ds,\quad 3x(1)+2x'(1)=\int_0^1\sigma_2(x(s))ds,
\end{cases}\tag{11}$$

This problems satisfies all the above stated hypotheses and therefore our main result.

References

1. Agarwal, R.P., O'Regan, D., Taoudi, M.A.: Fixed point theorems for convex-power condensing operators relative to the weak topology and applications to Volterra integral equations. J. Integral Equ. Appl. **24**(2), 167–181 (2012). https://doi.org/10.1216/JIE-2012-24-2-167
2. Agarwal, R.P., O'Regan, D., Sikorska Nowak, A.: The set of solutions of integro-differential equations and the Henstock-Kurzweil-Pettis integral in Banach spaces. Bull. Aust. Math. Soc. **78**, 507–522 (2008)
3. Cichoń, M.: Convergence theorem for the Henstock-Kurwzeil-Pettis integrals. Acta Math. Hung. **92**, 75–82 (2001)

4. Cichon, M., Kubiaczyk, I., Sikorska Nowak, A.: The Henstock-Kurzweil-Pettis integrals and existence theorems for the Cauchy problem. Czechoslov. Math. J. **54**, 279–289 (2004)
5. De Blasi, F.S.: On a property of the unit sphere in Banach space. Bull. Math. Soc. Sci. Math. Roum. **21**, 259–262 (1977)
6. Di Piazza, L.: Kurzweil-Henstock type integration on Banach spaces. Real Anal. Exch. **29**(2), 543–555 (2003)
7. Gou, H., Li, B.: Weak solutions for fractional differential equations via Henstock-Kurzweil-Pettis integrals. Int. J. Nonlinear Sci. Num. Simul. **21** (2019). https://doi.org/10.1515/ijnsns-2018-0174
8. Hussain, N., Taoudi, M.A.: Krasnoselskii-type fixed point theorems with applications to Volterra integral equations. Fixed Point Theory Appl. **196** (2013)
9. Kurtz, D., Swartz, C.: Theories of Integration, The Integrals of Riemann, Lebesgue. Henstock-Kurzweil and McShane. World Scientific, Singapore (2004)
10. Satco, B.: A Kolmós-type theorem for the set-valued Henstock-Kurzweil-Pettis integrals and applications. Czechoslov. Math. J **56**(131), 1029–1047 (2006)
11. Seba, D., Habani, S., Benaissa, A., Rebai, H.: The Henstock-Kurzweil-Pettis integral and multiorders fractional differential equations with impulses and multipoint fractional integral boundary conditions in Banach spaces. Math. Meth. Appl. Sci. (2019). https://doi.org/10.1002/mma.6020
12. Stamova, I., Stamov, I.: Functional and Impulsive Differential Equations of Fractional Order Qualitative Analysis and Applications, CRC Press, Boca Raton (2021). ISBN: 9780367782726

Pattern Formation in Activator-Inhibitor Fractional Reaction-Diffusion Systems

Bohdan Datsko[1,2(✉)], Cristina Naconechna[3], and Vasyl Gafiychuk[2]

[1] Rzeszow University of Technology, 35-959 Rzeszow, Poland
datskob@prz.edu.pl
[2] IAPMM, National Academy of Sciences of Ukraine, Lviv, Ukraine
[3] Danylo Halytsky Lviv National Medical University, Lviv, Ukraine

Abstract. Pattern formation scenarios in nonlinear two-component incommensurate fractional reaction-diffusion systems (FRDS) are investigated. It is shown that the stability and subsequent evolution of steady-state solutions in generalized time-space FRDS are mainly determined by the eigenvalue spectrum of the corresponding linearized system and the values of fractional derivative orders. The results of the linear stability analysis are confirmed by computer simulations. It is demonstrated that the conditions of instability and pattern dynamics in FRDS can be quite different from the standard ones with integer derivatives. As a result, new qualitative properties for pattern formation and richer and more complicated spatiotemporal dynamics take place in fractional reaction-diffusion systems.

Keywords: Fractional reaction-diffusion system · Pattern formation · Self-organization phenomena · Anomalous diffusion

1 Generalized Basic Two-Component Activator-Inhibitor System

Among the fractional differential models [1–3], the activator-inhibitor systems rightfully play an important role [3–6]. On the basis of mathematical modeling of the standard activator-inhibitor reaction-diffusion systems (RDS), a lot of amazing nonlinear self-organization phenomena in physical, biological, and chemical systems have been explained [7,8]. Therefore, investigation of fractional models for such media, as granular and porous materials, various materials with memory, complex chemical environments, and biological tissues, where self-organization phenomena are observed and diffusion has essentially an anomalous character, represents a great interest.

The starting point of our consideration is the one-dimensional two-component fractional RDS with different orders of fractional derivatives

$$\tau_1 \frac{\partial^{\alpha_1} u_1(x,t)}{\partial t^{\alpha_1}} = l_1^2 \frac{\partial^{\gamma_1} u_1(x,t)}{\partial x^{\gamma_1}} + P(u_1, u_2, \mathcal{A}), \qquad (1)$$

A. Dzielinski et al. (Eds.): ICFDA 2021, LNNS 452, pp. 218–224, 2022.
https://doi.org/10.1007/978-3-031-04383-3_24

$$\tau_2 \frac{\partial^{\alpha_2} u_2(x,t)}{\partial t^{\alpha_2}} = l_2^2 \frac{\partial^{\gamma_2} u_2(x,t)}{\partial x^{\gamma_2}} + Q(u_1, u_2, \mathcal{A}) \tag{2}$$

subject to Neumann $\partial u_i/\partial x|_{x=0} = \partial u_i/\partial x|_{x=\pounds_x} = 0$, $i = 1,2$ or periodic $u_i|_{x=0} = u_i|_{x=\pounds_x}$, $\partial u_i/\partial x|_{x=0} = \partial u_i/\partial x|_{x=\pounds_x}$, $i = 1,2$ boundary conditions and with certain initial conditions. Here $u_1(x,t)$, $u_2(x,t)$ are the activator and inhibitor variables, $0 \le x \le \pounds_x$, $\tau_1 = \tilde{\tau}_1^{\alpha_1}, \tau_2 = \tilde{\tau}_2^{\alpha_2}, l_1 = \tilde{l}_1^{\gamma_1}, l_2 = \tilde{l}_2^{\gamma_2}$, $\tilde{\tau}_1, \tilde{\tau}_2, \tilde{l}_1, \tilde{l}_2$ are the characteristic times and lengths of the system, correspondingly, $\mathcal{A}$ is an external parameter and $P(u_1, u_2, \mathcal{A}), Q(u_1, u_2, \mathcal{A})$ are nonlinear sources of the system modeling their production rates.

Time derivatives $\partial^{\alpha_i} u_i(x,t)/\partial t^{\alpha_i}$ on the left hand side of equations (1), (2) instead of standard ones are the Caputo fractional derivatives in time of the order $0 < \alpha < 2$ and are represented as [9,10]

$$\frac{\partial^\alpha}{\partial t^\alpha} f(x,t) := \frac{1}{\Gamma(m-\alpha)} \int_0^t \frac{f^{(m)}(x,\tau)}{(t-\tau)^{\alpha+1-m}} d\tau, \tag{3}$$

where $m - 1 < \alpha < m$, $m = 1, 2$.

Space derivatives $\partial^{\gamma_i} u_i(x,t)/\partial x^{\gamma_i}$ on the right hand side of equations (1), (2) instead of standard ones for $1 < \gamma < 2$ are determined as [9,10]

$$\frac{\partial^\gamma f(x,t)}{\partial |x|^\gamma} = -\frac{1}{2\cos(\pi\gamma/2)} \left[D_+^\gamma f(x,t) + D_-^\gamma f(x,t) \right], \tag{4}$$

where the fractional operators $D_+^\gamma f(x,t)$ and $D_-^\gamma f(x,t)$ are defined as

$$D_+^\gamma f(x,t) = \frac{1}{(2-\gamma)} \frac{d^2}{dx^2} \int_{-\infty}^{x} \frac{f(\xi,t)}{(x-\xi)^{\gamma-1}} d\xi, \tag{5}$$

$$D_-^\gamma f(x,t) = \frac{1}{(2-\gamma)} \frac{d^2}{dx^2} \int_{x}^{\infty} \frac{f(\xi,t)}{(\xi-x)^{\gamma-1}} d\xi. \tag{6}$$

In this case, we can consider the system (1), (2) as a generalized system, which describes all combinations of subdifussive and superdiffusie processes [1–3].

2 Linear Stability Analysis

2.1 Standard Reaction-Diffusion System

For standard RD system ($\alpha_1 = \alpha_2 = 1$, $\gamma_1 = \gamma_2 = 2$), it is convenient to analyze null-clines of the system (1), (2):

$$P = P(u_1, u_2, \mathcal{A}) = 0, \quad Q = Q(u_1, u_2, \mathcal{A}) = 0. \tag{7}$$

Simultaneous solution of the two equations (7) leads to homogeneous solution $\bar{\mathbf{u}}^s = (\bar{u}_1^s, \bar{u}_2^s)^T$. The stability of such steady-state solution corresponding to the homogeneous equilibrium state can be analyzed by linearization of the system nearby this solution $\bar{\mathbf{u}}^s = (\bar{u}_1^s, \bar{u}_2^s)^T$. As a result, we can write the expression for eigenvalues of the linearized system

$$\lambda_{1,2} = \frac{1}{2}(trF \pm \sqrt{tr^2 F - 4\det F}), \tag{8}$$

where

$$F(k) \equiv L|_{(\bar{u}_1^s, \bar{u}_2^s)} = \begin{pmatrix} (a_{11} - k^2 l_1^2)/\tau_1 & a_{12}/\tau_1 \\ a_{21}/\tau_2 & (a_{22} - k^2 l_2^2)/\tau_2 \end{pmatrix}, \tag{9}$$

$$a_{11} = \frac{\partial P}{\partial u_1}|_{(\bar{u}_1^s, \bar{u}_2^s)},\ a_{12} = \frac{\partial P}{\partial u_2}|_{(\bar{u}_1^s, \bar{u}_2^s)},\ a_{21} = \frac{\partial Q}{\partial u_1}|_{(\bar{u}_1^s, \bar{u}_2^s)},\ a_{22} = \frac{\partial Q}{\partial u_2}|_{(\bar{u}_1^s, \bar{u}_2^s)},$$

$$\mathrm{tr}F(k) = (a_{11} - k^2 l_1^2)/\tau_1 + (a_{22} - k^2 l_2^2)/\tau_2, \tag{10}$$

$$\det F(k) = (a_{11} - k^2 l_1^2)(a_{22} - k^2 l_2^2)/(\tau_1\tau_2) - (a_{12}a_{21})/(\tau_1\tau_2), \tag{11}$$

$k = (\pi j/\mathcal{L})$, $j = 1, 2, \ldots$.

The eigenvalues (8) mainly define the stability and nonlinear dynamics of the system (1), (2) [11]. In this case, for $k = 0$ at conditions

$$\mathrm{tr}F(0) > 0, \quad \det F(0) > 0 \tag{12}$$

we have a Hopf bifurcation, and for $k \neq 0$ at conditions

$$\mathrm{tr}F < 0, \quad \det F(0) > 0, \quad \det F(k_0) < 0 \tag{13}$$

we have a Turing one.

The instability conditions for these two types of bifurcation are realized due to positive feedback in the system ($a_{11} > 0$) and at the conditions $\tau_1/\tau_2 \to 0$ or $l_1/l_2 \to 0$ [7]. As a result, we obtain nonlinear dynamics leading to spatially-homogeneous oscillatory solutions or spatially-inhomogeneous stationary DS. When conditions for both instabilities arise, we can expect more complex dynamics. In the case of a fractional reaction-diffusion (FRD) system, the dynamics can be much more complex [6,12].

2.2 Space-Fractional Reaction-Diffusion System

In the case of space-fractional derivatives ($\alpha_1 = \alpha_2 = 1,\ 1 < \gamma_1 < 2, 1 < \gamma_2 < 2$), we can apply to the linearized problem (1), (2) Fourier transform with respect to the space coordinate. As a result, for the linearized system, we obtain the coefficients of characteristic equation, which include the space-fractional orders of derivatives γ_1, γ_2 as parameters

$$\mathrm{tr}F(k) = (a_{11} - k^{\gamma_1} l_1^2)/\tau_1 + (a_{22} - k^{\gamma_2} l_2^2)/\tau_2, \tag{14}$$

$$\det F(k) = (a_{11} - k^{\gamma_1} l_1^2)(a_{22} - k^{\gamma_2} l_2^2)/(\tau_1\tau_2) - (a_{12}a_{21})/(\tau_1\tau_2). \tag{15}$$

Even in the commensurate case $\gamma = \gamma_1 = \gamma_2$ we can see that space-fractional order does not change the conditions of stability for the spatially homogeneous state ($k = 0$). But due to the fact that the critical wave number $k_0 = [(l_1^2 a_{22} + l_2^2 a_{11})/(2 l_1^2 l_2^2)]^{1/\gamma}$ depends on space fractional derivative order, the Turing bifurcation occurs with another spatial period in comparison to standard systems. That means that the space-fractional derivatives can mainly influence spatial location, smooth and mobility of appearing dissipative structures (DS).

2.3 Time-Fractional Reaction-Diffusion System

Time fractional derivatives change qualitatively the stability conditions the spatially homogeneous state. In this case ($0 < \alpha_1 < 2, 0 < \alpha_2 < 2, \gamma_1 = \gamma_2 = 2$), for linearized system we must take into account relation between imaginary and real parts of the eigenvalues. In the case $\alpha_1 \neq \alpha_2$ the system (1), (2) can be transformed to the equivalent special type system with the same "effective order" [6,9]. If the fractional derivative index is the "greatest common factor" α of the values $\alpha_1 = p\alpha$ and $\alpha_2 = r\alpha$, $p, r \in N$, we obtain the commensurate system of $p + r$ differential equations of special structure

$$\tau_1 \frac{\partial^\alpha w_p(x,t)}{\partial t^\alpha} = l_1^2 \frac{\partial^2 u_1(x,t)}{\partial x^2} + P(u_1, u_2, \mathcal{A}), \tag{16}$$

$$\tau_2 \frac{\partial^\alpha v_r(x,t)}{\partial t^\alpha} = l_2^2 \frac{\partial^2 u_2(x,t)}{\partial x^2} + Q(u_1, u_2, \mathcal{A}), \tag{17}$$

$$\frac{\partial^\alpha w_{i-1}(x,t)}{\partial t^\alpha} = w_i(x,t), \quad p \geq i > 1, \quad w_1 \equiv u_1, \tag{18}$$

$$\frac{\partial^\alpha v_{j-1}(x,t)}{\partial t^\alpha} = v_j(x,t), \quad r \geq j > 1, \quad v_1 \equiv u_2, \tag{19}$$

where $i = p, p-1, ...1; \quad j = r, r-1, ...1.$

Such presentation of the system (1), (2) makes it possible to apply the Matignon stability theorem [9] and write down explicitly the characteristic equation for any relation between derivative orders [6]

$$(-\lambda)^{p+r} + (-1)^{r-1} \tfrac{a_{22}(k)}{\tau_2} (-\lambda)^p + (-1)^{p+1} \tfrac{a_{11}(k)}{\tau_1} (-\lambda)^r + (-1)^{p+r} \det F = 0, \tag{20}$$

where $a_{11}(k) = a_{11} - k^2 l_1^2$, $a_{22}(k) = a_{22} - k^2 l_2^2$, coefficients a_{ij} are defined as before. The eigenvalue spectrum, that is determined by the Eq. (20) allows us to get instability conditions for any stationary spatially homogeneous solution

$$\min_i \{|\mathrm{Arg}(\lambda_i)|\} < \pi\alpha/2, \quad i = 1, 2, \ldots, p + r \tag{21}$$

depending on the orders of fractional derivatives and any other system parameters. The view of the characteristic equation shows that its roots distribution and thus the stability properties of the linearized system strongly depend both on orders of the fractional derivatives and on the ratio between them.

3 Pattern Formation in Fractional RDS

3.1 Basic Activator-Inhibitor Mathematical Models

The existence of the positive feedback in the first variable u_1 and the negative one in the second variable u_2

$$\partial P/\partial u_1 > 0, \quad \partial Q/\partial u_2 < 0 \tag{22}$$

in a certain range of change in the bifurcation parameter $\mathcal{A}$, together with the relation between $\theta = l_1/l_2$ and $\epsilon = \tau_1/\tau_2$, cause the instability of equilibrium states of the system $(\overline{u}_1^s, \overline{u}_2^s)$ which are determined as intersection points of its null-clines. For auto-wave systems, which satisfy the conditions (22), the qualitative form of null-clines can be established from general reasoning and it can be shown that the null-cline $P(u_1, u_2) = 0$ must have N -like, И-like, Λ-like or V-like shape [7]. Because of it, to illustrate the influence of fractional derivative orders on stability properties and dynamics in two-component fractional systems, we consider two classical nonlinear models which represent these dependencies. The first one is the Bonhoeffer-van der Pol model [7,8] with the source terms

$$P(u_1, u_2, \mathcal{A}, B) = u_1 - u_1^3/3 - u_2, \qquad Q(u_1, u_2, \mathcal{A}, B) = -u_2 + Bu_1 + \mathcal{A}. \tag{23}$$

The second one is the Brusselator model [7,8] with the nonlinearities

$$P(u_1, u_2, \mathcal{A}, B) = \mathcal{A} + u_1 u_2^2 - (B+1)u_2, \qquad Q(u_1, u_2, \mathcal{A}, B) = Bu_2 - u_1 u_2^2, \tag{24}$$

where $\mathcal{A}, B$ are some external parameters.

3.2 Complex Spatio-Temporal Dynamics in Fractional RDS

The solution of characteristic equation allows getting the eigenvalues spectrum of the linearized system for each stationary point at any ratio of system parameters. As a consequence, we can construct the instability regions for any values of the fractional orders α_1, α_2 and establish the most significant dependencies between them and other system parameters [11,12].

Analysis of stability domains for the considered typical mathematical models allows to notice some general properties of basic two-component fractional activator-inhibitor systems:

a) The effect of time fractional derivatives is most significant when the ratio of characteristic times $\varepsilon = \tau_1/\tau_2$ is close to 1. In other words, for the systems with relaxation self-excited oscillations at $\varepsilon = \tau_1/\tau_2 \ll 1$, the stability regions for integer and fractional systems are practically the same.
b) The slope of the null-clines in the point of intersection also leads to an intense variation of instability domain. If the slope of null-clines in the intersection point is practically the same and the values of system variables in the intersection point of null-clines is close, then conditions of instability for fractional system coincide with integer one. In another case, the fractional system can by unstable not only for $\tau_1/\tau_2 \leq 1$ (as in integer case) but also for $\tau_1/\tau_2 \geq 1$.

Formation of the final attractor in auto-wave systems is mainly determined by the spectrum of eigenvalues of the linearized system for all wave numbers. In two different limiting cases RDS possesses spatially homogeneous oscillations ($\tau_1/\tau_2 \ll 1, l_1/l_2 \geq 1$) or stationary spatially inhomogeneous dissipative structures ($\tau_1/\tau_2 \geq 1, l_1/l_2 \ll 1$). The most complex spatio-temporal dynamics arises when $\tau_1/\tau_2 \ll 1$ and $l_1/l_2 \ll 1$. In other words, if the linearized system has pure real roots, then the order of fractional derivative has no additional effect on the development of instability. The order of fractional derivatives, in this case, changes the spatial (amplitude) and temporal (frequency) characteristics of system solutions. The presence of complex eigenvalues can change qualitatively the type of instability depending on the order of fractional derivative. In this case, increasing or decreasing the order of fractional derivative can change the type of bifurcation and thereby change the type of system solutions from one to another.

Variations in the ratio of the derivative orders can change qualitatively the spectrum of the linearized system as well as the type of system solutions. Domination of the activator derivative with respect to inhibitor one leads to the domination of the modes $k \neq 0$ against to $k = 0$. This, in turn, leads to the domination of spatially inhomogeneous solutions with respect to homogeneous oscillations. The occurrence of different types of bifurcation simultaneously with the memory effect inherent to the FRDS leads to very complex dynamics which in standard systems is practically not realized. Moreover, this type of dynamics takes place in a wide range of bifurcation parameter variation and ratio between the orders of fractional derivatives.

The effect of space-fractional derivatives is most significant when the ratio of characteristic lengths $\zeta = l_1/l_2$ of RDS is close to 1. Space-fractional derivatives primarily affect the spatial characteristics of the appearing space-inhomogeneous solutions: period, smoothness, mobility, size. These regularities are of a general nature and can be applied not only to auto-waves occurring in the area of system instability but also to auto-solitons that occur in the stability regions of a system [12].

References

1. Uchaikin, V.: Fractional Derivatives for Physicists and Engineers. Springer, Berlin (2013). https://doi.org/10.1007/978-3-642-33911-0
2. Povstenko, Y.: Linear Fractional Diffusion-Wave Equation for Scientists and Engineers. Birkhauser, New York (2015)
3. Mendez, V., Fedotov, S., Horsthemke, W.: Reaction-Transport Systems: Mesoscopic Foundations, Fronts, and Spatial Instabilities. Springer, New York (2009). https://doi.org/10.1007/978-3-642-11443-4
4. Hornung, G., Berkowitz, B., Barkai, N.: Morphogen gradient formation in a complex environment: an anomalous diffusion model. Phys. Rev. E **72**, 041916-1-10 (2005)
5. Langlands, T., Henry, B., Wearne, S.: Anomalous subdiffusion with multispecies linear reaction dynamics. Phys. Rev. E **77**, 021111-1-9 (2008)
6. Gafiychuk, V., Datsko, B.: Spatiotemporal pattern formation in fractional reaction-diffusion systems with indices of different order. Phys. Rev. E **77**, 066210-1-9 (2008)

7. Kerner, B., Osipov, V.: Autosolitons. Kluwer, Dordrecht (1994)
8. Cross, M., Hohenberg, P.: Pattern formation out of equilibrium. Rev. Mod. Phys. **65**, 851–1112 (1993)
9. Podlubny, I.: Fractional Differential Equations. Academic Press, San Diego (1999)
10. Samko, S., Kilbas, A., Marichev, O.: Fractional Integrals and Derivatives: Theory and Applications. Gordon and Breach, San Diego (1993)
11. Datsko, B., Kutniv, M., Wloch, A.: Mathematical modelling of pattern formation in activator-inhibitor reaction-diffusion systems with anomalous diffusion. J. Math. Chem. **58**, 612–631 (2020)
12. Datsko, B., Gafiychuk, V.: Complex spatio-temporal solutions in fractional reaction-diffusion systems near a bifurcation point. Fract. Calc. Appl. Anal. **21**, 237–253 (2018)

Stability Results for Two-Term Fractional-Order Difference Equations

Oana Brandibur(✉) and Eva Kaslik

Department of Mathematics and Computer Science, West University of Timişoara, Timişoara, Romania
oana.brandibur@e-uvt.ro

Abstract. Fractional-order difference equations of Caputo-type with two subunitary fractional orders are analyzed with respect to their asymptotic stability and instability properties. In particular, the asymptotic stability of two-term fractional-order difference equations is investigated, in terms of the fractional orders of the considered Caputo-type differences and in terms of the coefficients of the equation and the discretization step size.

Keywords: Caputo-type fractional difference · Fractional difference equation · Stability · Instability

1 Introduction

Fractional calculus has numerous applications in modelling several phenomena from both scientific and engineering fields [3–5]. Due to memory and hereditary properties of fractional-order derivatives, recent research results indicate that fractional-order systems confer more accurate results in a broad range of practical applications.

Fractional-order difference equations have not been explored as often as their continuous-time counterparts [6–10]. Still, stability results for systems of fractional-order difference equations have been recently investigated in [2]. Recurrence formulas for the solutions of initial value problems of linear systems with Caputo fractional-order difference equations with variable orders of convolution type and their stability results have been presented in [13]. Other stability and linearization results for discrete-time fractional systems are given in [11, 14], while solutions of initial value problems involving fractional-order Caputo-type and Riemann-Liouville-type difference equations with positive orders have been exposed in [12].

In this paper we explore stability results for two-term fractional-order difference equations. We opted for the Caputo fractional h-difference operator, as it has nonzero equilibria, compared to Grunwald-Letnikov or Riemann-Liouville difference operators, that have only the origin as equilibrium points.

A. Dzielinski et al. (Eds.): ICFDA 2021, LNNS 452, pp. 225–230, 2022.
https://doi.org/10.1007/978-3-031-04383-3_25

The paper is organized as follows: in Sect. 2 we introduce some preliminary notions regarding fractional-order difference operators and the $\mathcal{Z}$ transform associated to the Caputo h-difference operator of order $q \in (0,1]$; in Sect. 3 we define the two-term fractional-order difference equation with Caputo forward difference operators, we determine the characteristic equation associated to the considered equation and we expose a theorem of characterization of the stability and instability properties of the considered equations with respect to the roots of the characteristic equation. Section 4 provides fractional-order dependent results, exploring stability results with respect to the coefficients of the equation, the two fractional orders and the step size. Conclusions are drawn in Sect. 5.

2 Preliminaries

Let $h > 0$, $(h\mathbb{N})_0 = \{0, h, 2h, ...\}$ and $x : (h\mathbb{N})_0 \to \mathbb{R}$ be an arbitrary function. Consider the forward h-difference operator defined by:

$$(\Delta_h x)(t) := \frac{x(t+h) - x(t)}{h}, \quad t \in (h\mathbb{N})_0.$$

We introduce a family of binomial functions on $\mathbb{Z}$ parameterized by $q > 0$ and given by

$$a_j^{(-q)} = \binom{j+q-1}{j} = (-1)^j \binom{-q}{j}, \text{ for } q > 0, \quad a_j^{(-q)} = 0, \text{ if } q < 0,$$

with $a_0^{(-q)} = 1$.

Definition 1.

- *Let $x : (h\mathbb{N})_0 \to \mathbb{R}$ be a function. The fractional h-sum of order $q > 0$ is defined by*

$$(\Delta_h^{-q} x)(t) := h^q (a^{(-q)} * \bar{x})(j), \quad t = jh, \quad \bar{x}(s) := x(sh), \quad s \in \mathbb{N},$$

  where "$*$" denotes the convolution, i.e. $(a^{(-q)} * \bar{x})(j) := \sum_{s=0}^{j} \binom{j-s+q-1}{j-s} \bar{x}(s)$.
- *Let $q \in (0,1]$. The operator denoted by ${}^c\Delta^q$ is called the Caputo-type h-difference operator of order q for a function $x : (h\mathbb{N})_0 \to \mathbb{R}$ and is given by*

$$({}^c\Delta^q x)(t) = (\Delta_h^{-(1-q)}(\Delta_h x))(t), \quad t \in (h\mathbb{N})_0.$$

Proposition 1. *Let $q \in (0,1]$ and $y(j) = ({}^c\Delta^q x)(t)$, where $t \in (h\mathbb{N})_0$ and $t = jh$. The $\mathcal{Z}$-transform of $y(j)$ is:*

$$\mathcal{Z}[y](z) = h^{-q} z (1 - z^{-1})^q \bar{X}(z) - h^{-q} z (1 - z^{-1})^{q-1} x_0,$$

where $\bar{X}(z) = \mathcal{Z}[\bar{x}](z)$, $\bar{x}(j) = x(jh)$ and $x_0 = x(0)$.

3 Stability Results for Two-Term Fractional-Order Difference Equations

Consider the following two-term fractional-order difference equations with Caputo forward difference operators:

$$^{c}\Delta^{q_1}x(nh) + a^{c}\Delta^{q_2}x(nh) + bx(nh) = 0, \tag{1}$$

where h is the discretization step and $0 < q_1 \leq q_2 \leq 1$ are the fractional orders of Caputo forward difference operators.

Applying the $\mathcal{Z}$ transform to Eq. (1), we obtain

$$\left(h^{-q_1}z(1-z^{-1})^{q_1} + ah^{-q_2}z(1-z^{-1})^{q_2} + b\right)X(z) = F(z),$$

where the $\mathcal{Z}$-transform of x is denoted by $\mathcal{Z}[x] = X$ and

$$F(z) = h^{-q_1}z(1-z^{-1})^{q_1-1}x_0 + ah^{-q_2}z(1-z^{-1})^{q_2-1}x_0.$$

We obtain the following characteristic equation associated to (1):

$$h^{-q_1}z(1-z^{-1})^{q_1} + ah^{-q_2}z(1-z^{-1})^{q_2} + b = 0. \tag{2}$$

Denoting $\Delta(z) = h^{-q_1}z(1-z^{-1})^{q_1} + ah^{-q_2}z(1-z^{-1})^{q_2} + b$, we expose the following characterization of the stability and instability properties of equation (1) in terms of the distributions of the roots of the characteristic equation with respect to the unit circle, which extends to the fractional-order case, the well-known classical results from the stability theory of discrete-time equations and systems. The proof of this theorem can be obtained as in the continuous-time counterpart [1].

Theorem 1. *i. Equation* (1) *is* $\mathcal{O}(n^{-q})$*-globally asymptotically stable if and only if all the roots of* Δ *are inside the unit circle* $|z| < 1$*, where* $q = \min\{q_1, q_2\}$.

ii. If $b \neq 0$ *and* Δ *has at least one root outside the closed unit circle* $|z| \geq 1$*, equation* (1) *is unstable.*

4 Fractional-Order Dependent Stability Results

In what follows, we consider $b \neq 0$. Dividing the characteristic equation (2) by b, and denoting $\frac{1}{b} = \alpha_1$ and $\frac{a}{b} = \alpha_2$, the characteristic equation becomes:

$$\Delta(z) := \alpha_1 h^{-q_1}z(1-z^{-1})^{q_1} + \alpha_2 h^{-q_2}z(1-z^{-1})^{q_2} + 1 = 0. \tag{3}$$

Lemma 1. *Let* $0 < q_1 < q_2 \leq 1$*. The characteristic function* $\Delta(z)$ *has a root* $z = -1$ *if and only if* (α_1, α_2) *belongs to the line* $l(q_1, q_2, h)$ *defined by:*

$$l(q_1, q_2, h): \quad \alpha_1 \left(\frac{2}{h}\right)^{q_1} + \alpha_2 \left(\frac{2}{h}\right)^{q_2} - 1 = 0$$

Proof. Using basic mathematical tools, equation $\Delta(-1) = 0$ is equivalent to

$$-\alpha_1 h^{-q_1} \cdot 2^{q_1} - \alpha_2 h^{-q_2} \cdot 2^{q_2} + 1 = 0,$$

which leads to (α_1, α_2) belonging to the line $l(q_1, q_2, h)$. □

Lemma 2. *Let $0 < q_1 < q_2 \leq 1$. The characteristic function $\Delta(z)$ has a root z on the unit circle ($|z| = 1$), with $z \notin \{-1, 1\}$ if and only if (α_1, α_2) belongs to the parametric curve $\Gamma(q_1, q_2, h)$ defined by:*

$$\Gamma(q_1, q_2, h)\colon \quad \begin{cases} \alpha_1 = h^{q_1}(2\cos\theta)^{-q_1}\rho(q_2, q_1, \theta) \\ \alpha_2 = h^{q_2}(2\cos\theta)^{-q_2}\rho(q_1, q_2, \theta) \end{cases},$$

where $\theta \in \left(0, \frac{\pi}{2}\right)$, $\rho(q_1, q_2, \theta) = \frac{\sin(2-q_1)\theta}{\sin(q_2-q_1)\theta}$.

Proof. The function $\Delta(z)$ has a root on the unit circle if and only if there exists $\theta \in \left(0, \frac{\pi}{2}\right)$ such that $\Delta(e^{i(\pi-2\theta)}) = 0$. For $z = e^{i(\pi-2\theta)} = -e^{-2i\theta}$, we compute

$$1 - z^{-1} = 1 + e^{2i\theta} = 1 + \cos 2\theta + i \sin 2\theta = 2\cos\theta \cdot e^{i\theta}.$$

Therefore, the equation $\Delta(-e^{-2i\theta}) = 0$ is equivalent to

$$-\alpha_1 h^{-q_1}(2\cos\theta)^{q_1} e^{i(q_1-2)\theta} - \alpha_2 h^{-q_2}(2\cos\theta)^{q_2} e^{i(q_2-2)\theta} + 1 = 0.$$

Taking the real and imaginary parts of the previous relation, the following system is obtained:

$$\begin{cases} -\alpha_1 h^{-q_1}(2\cos\theta)^{q_1}\cos(q_1-2)\theta - \alpha_2 h^{-q_2}(2\cos\theta)^{q_2}\cos(q_2-2)\theta + 1 = 0 \\ -\alpha_1 h^{-q_1}(2\cos\theta)^{q_1}\sin(q_1-2)\theta - \alpha_2 h^{-q_2}(2\cos\theta)^{q_2}\sin(q_2-2)\theta = 0 \end{cases}$$

By solving the above system with respect to the unknown term α_1 and α_2, it results that $(\alpha_1, \alpha_2) \in \Gamma(q_1, q_2, h)$. □

It is interesting to note that the curve $\Gamma(q_1, q_2, h)$ is included in the second quadrant of the (α_1, α_2)-plane.

We recall the following result which was rigorously proved in [2].

Lemma 3. *Let $\alpha \in \mathbb{R}$ and $q \in (0, 1)$. All the roots of the equation*

$$z(1 - z^{-1})^q + \alpha = 0 \tag{4}$$

are in the interior of the unit disc if and only if $\alpha \in (0, 2^q)$.

Remark 1. In the limiting case $q_1 = q_2 = q$, the characteristic equation (3) reduces to:

$$z(1 - z^{-1})^q + \frac{h^q}{\alpha_1 + \alpha_2} = 0,$$

and hence, all the roots of the characteristic equation are in the interior of the unit disc if and only if $\alpha_1 + \alpha_2 > \left(\frac{h}{2}\right)^q$. Furthermore, in the integer-order case $q = 1$, we obtain the stability condition $\alpha_1 + \alpha_2 > \frac{h}{2}$.

We can now state the main result of this section.

Theorem 2. *(Fractional-order dependent stability results)*
Let $0 < q_1 < q_2 \leq 1$. Equation (1) is asymptotically stable if and only if (α_1, α_2) belongs to the connected component of the (α_1, α_2)-plane bounded by the line $l(q_1, q_2, h)$ defined in Lemma 1 and the curve $\Gamma(q_1.q_2, h)$ introduced in Lemma 2, containing the half-lines $l_1(q_1.q_2, h)$ and $l_2(q_1.q_2, h)$, where

$$l_1(q_1.q_2, h) = \left\{(\alpha_1, \alpha_2) \in \mathbb{R}^2 : \alpha_2 = 0,\ \alpha_1 > \left(\frac{h}{2}\right)^{q_1}\right\}$$
$$l_2(q_1.q_2, h) = \left\{(\alpha_1, \alpha_2) \in \mathbb{R}^2 : \alpha_1 = 0,\ \alpha_2 > \left(\frac{h}{2}\right)^{q_2}\right\}$$

Proof. The proof results from Lemma 1, Lemma 2 and Lemma 3. Indeed, if $\alpha_1 = 0$ in the characteristic equation (3), based on Lemma 3, all the roots of $\Delta(z)$ are in the interior of the unit disc if and only if $\frac{h^{q_2}}{\alpha_2} \in (0, 2^{q_2})$, i.e. $\alpha_2 > \left(\frac{h}{2}\right)^{q_2}$, meaning that $(\alpha_1, \alpha_2) \in l_2(q_1, q_2, h)$. A similar reasoning holds for the case when $\alpha_2 = 0$: all the roots of $\Delta(z)$ are in the interior of the unit disc if and only if $(\alpha_1, \alpha_2) \in l_1(q_1, q_2, h)$. When α_1 and α_2 are varied, the number of roots of $\Delta(z)$ from the exterior of the unit circle can change if and only if (α_1, α_2) cross the line $l(q_1, q_2, h)$ defined in Lemma 1 or the curve $\Gamma(q_1.q_2, h)$ given by Lemma 2. Based on Theorem 1, we obtain the required result. □

The stability region provided by Theorem 2 is exemplified in Fig. 1, considering the stepsize $h = 1$ and different values for the fractional orders q_1, q_2.

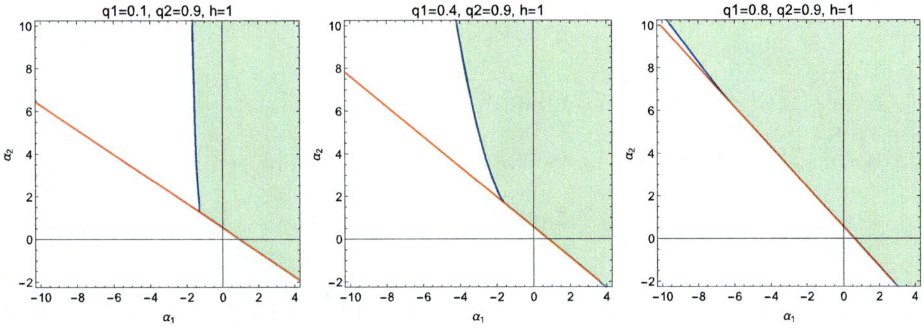

Fig. 1. Stability region of Eq. (1) in the (α_1, α_2)-plane, where $\alpha_1 = \frac{1}{b}$ and $\alpha_2 = \frac{a}{b}$.

5 Conclusions

Stability analysis properties for fractional-order difference equations of Caputo-type with two subunitary fractional orders have been investigated. The stability and instability conditions for the considered equations have been explored both in terms of the position of the roots of the corresponding characteristic equation towards the unit circle and in terms of the fractional orders of the considered Caputo-type differences and of the coefficients of the equation and the discretization step size. Figure 1 has been provided in order to exemplify the fractional-order dependent stability conditions drawn in Theorem 2. A possible application to the theoretical findings explored in the present work could be the analysis of the stability conditions in the discrete-time counterpart of the fractional-order Basset equation.

References

1. Brandibur, O., Kaslik, E.: Stability of two-component incommensurate fractional-order systems and applications to the investigation of a FitzHugh-Nagumo neuronal model. Math. Methods Appl. Sci. **41**(17), 7182–7194 (2018)
2. Brandibur, O., Kaslik, E., Mozyrska, D., Wyrwas, M.: Stability results for two-dimensional systems of fractional-order difference equations. Mathematics **8**(10), 1751 (2020)
3. Henry, B.I., Wearne, S.L.: Existence of Turing instabilities in a two-species fractional reaction-diffusion system. SIAM J. Appl. Math. **62**, 870–887 (2002)
4. Mainardi, F.: Fractional relaxation-oscillation and fractional phenomena. Chaos Solitons Fractals **7**(9), 1461–1477 (1996)
5. Ionescu, C., Lopes, A., Copot, D., Machado, J.A.T., Bates, J.H.T.: The role of fractional calculus in modeling biological phenomena: a review. Commun. Nonlinear Sci. Numer. Simul. **51**, 141–159 (2017)
6. Sabatier, J., Farges, C.: On stability of commensurate fractional order systems. Int. J. Bifurcat. Chaos **22**(04), 1250084 (2012)
7. Li, C., Ma, Y.: Fractional dynamical system and its linearization theorem. Nonlinear Dyn. **71**(4), 621–633 (2013)
8. Čermák, J., Kisela, T.: Stability properties of two-term fractional differential equations. Nonlinear Dyn. **80**(4), 1673–1684 (2015)
9. Wang, Z., Yang, D., Zhang, H.: Stability analysis on a class of nonlinear fractional-order systems. Nonlinear Dyn. **86**(2), 1023–1033 (2016). https://doi.org/10.1007/s11071-016-2943-6
10. Kukushkin, M.V.: On one method of studying spectral properties of non-selfadjoint operators. In: Abstract and Applied Analysis 2020 (2020)
11. Mozyrska, D., Wyrwas, M.: Explicit criteria for stability of fractional h-difference two-dimensional systems. Int. J. Dyn. Control **5**(1), 4–9 (2017). https://doi.org/10.1007/s40435-016-0239-9
12. Mozyrska, D., Wyrwas, M.: Fractional linear equations with discrete operators of positive order. In: Latawiec, K.J., Łukaniszyn, M., Stanisławski, R. (eds.) Advances in Modelling and Control of Non-integer-Order Systems. LNEE, vol. 320, pp. 47–58. Springer, Cham (2015). https://doi.org/10.1007/978-3-319-09900-2_5
13. Mozyrska, D., Wyrwas, M.: Stability of linear systems with caputo fractional-, variable-order difference operator of convolution type. In: 41st International Conference on Telecommunications and Signal Processing (TSP), pp. 1–4 (2018)
14. Mozyrska, D., Wyrwas, M.: Stability by linear approximation and the relation between the stability of difference and differential fractional systems. Math. Methods Appl. Sci. **40**(11), 4080–4091 (2017)

Global Thermal Modeling of Lung Heat Transfer with Blood Perfusion

Jean-François Duhé[1], Stéphane Victor[1(✉)], Pierre Melchior[1], Youssef Abdelmounen[2,3], and François Roubertie[2,3]

[1] Univ. Bordeaux, CNRS, IMS-UMR 5218, Bordeaux INP/enseirb-matmeca, 351 cours de la Libération, 33405 Talence Cedex, France
{jean-francois.duhe,stephane.victor,pierre.melchior}@ims-bordeaux.fr

[2] IHU Liryc, Electrophysiology and Heart Modeling Institute, Fondation Bordeaux Université, 33000 Bordeaux Cedex, France
{youssef.abdelmounen,francois.roubertie}@u-bordeaux.fr

[3] INSERM, Centre de recherche Cardio-Thoracique de Bordeaux, 33000 Bordeaux Cedex, France

Abstract. Thermal models often consider low-frequency approximations and are largely based upon RC circuits. If temperature fluctuations are non-negligible, a more accurate model is needed and is proposed through thermal two-port network which is direct extension of the heat equation into matrix formalism. As blood flow plays a major role for heat transfers in biological tissue, a second model is obtained from the bio-heat equation. These two type of models are combined into a global thermal modeling of human lung by considering blood perfusion.

Keywords: Heat equation · Bio-heat equation · Blood perfusion · Fractional calculus · Fractional order system · Human lung

1 Introduction

There is a wide range of applications in which temperature is critical, such as chip cooling, power systems, building comfort or biomedical systems. Very accurate models can be obtained via finite element methods, but they are computationally demanding and not well-suited for real-time applications.

RC circuits are simple and well-known models. They can be used for power electronics [1], human heat losses [2] or bio-impedance models [3]. Most of these applications assume low-frequency behavior. The thermal two-port network is introduced [4] in order to obtain a circuit model directly derived from the heat equation. Such a model greatly improves the accuracy of the time response and widens the frequency validity range.

On the other hand, this paper is focused on heat transfers in human lungs. During cardiac surgery, lungs will be under extra corporeal circulation (ECC). Organs are therefore exposed to harsh conditions and can get irreversible damages [6]. The therapeutic effect of a controlled hypothermia to protect the lungs

A. Dzielinski et al. (Eds.): ICFDA 2021, LNNS 452, pp. 231–238, 2022.
https://doi.org/10.1007/978-3-031-04383-3_26

has been studied [7] and usually real-life scenarios involve the used of thermal pads over the lung exposed surface. However, these techniques are empirical and a better understanding of thermal transient will allow a better thermal control to minimize injuries.

Lungs exhibit complex geometry and its fractal nature was originally proposed by Mandelbrot [8]. Moreover, mechanical impedance models were explored in [9], where fractional operators were used to characterize lung fractality. Blood flow prevents tissue injuries as well as temperature regulation. Perfusion is taken into account thanks to Pennes' bio-heat equation, which is described as the following:

$$\rho c \frac{\partial T}{\partial t} = k\nabla^2 T + \rho_b c_b \omega_b (T_a - T) + q_m + q_{ext}, \tag{1}$$

where T and T_a respectively stand for tissue and blood temperature. Equation (1) introduces blood perfusion coefficient ω_b as well as a metabolic term q_m to take into account cell breathing and q_{ext} to include external heat sources (expressed as power per volume unit).

The paper is organized as follows: Sect. 2 presents thermal two-port networks, the global lung model is presented in Sect. 3, simulations and results are presented in Sect. 4 and finally conclusions are given in Sect. 5.

2 Two-Port Network Models

Consider a thermal wall with 1D heat conduction as shown in Fig. 1, with section S_w and length L. Its properties are given by conductivity k, density ρ and specific heat capacity c. The blood thermal properties are given by ρ_b, c_b and the perfusion coefficient is ω_b.

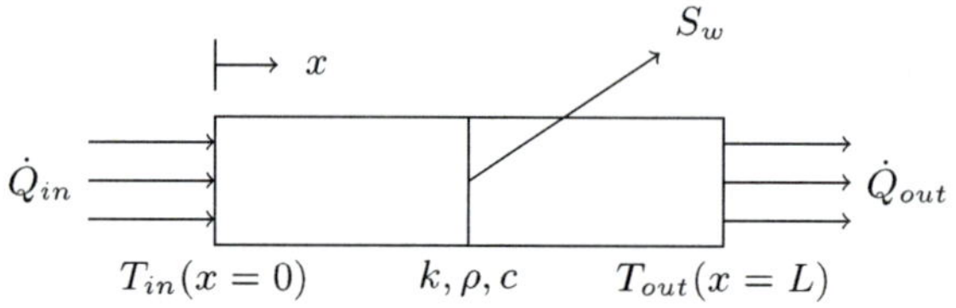

Fig. 1. 1D thermal system

2.1 Classic Two-Port Network

The thermal two-port network [4] allows to relate inputs and outputs of a thermal system, namely the heat equation, into a matrix form. The considered variables are the temperature T and the heat flow $\dot{Q}$. Put under matrix form, the two-port network yields (see [5]):

$$\begin{bmatrix} T_{in}(s) \\ \dot{Q}_{in}(s) \end{bmatrix} = \boldsymbol{M} \begin{bmatrix} T_{out}(s) \\ \dot{Q}_{out}(s) \end{bmatrix}, \quad \text{with} \quad \boldsymbol{M} = \begin{bmatrix} \cosh(\delta_c L) & \frac{1}{kS_w\delta_c}\sinh(\delta_c L) \\ kS_w\delta_c \sinh(\delta L) & \cosh(\delta_c L) \end{bmatrix} \tag{2}$$

where $\delta_c = \sqrt{\frac{s}{a}}$ and $a = \frac{k}{\rho c}$.

2.2 Bio-heat Two-Port Network

Bio-heat equation is used to take into account the blood influence. Metabolic heat rate and external sources are considered negligible. Variable change $\tilde{T} = T - T_b$ is introduced to analyze deviations around the body core temperature T_b. The blood temperature T_a is supposed constant and equal to the body core one. This assumption lead to the same two-port network as in Eq. (2), with the following difference:

$$\delta_{bio} = \sqrt{\frac{s}{a} + h_b} \tag{3}$$

where $h_b = \frac{\rho_b c_b \omega_b}{k}$.

Note that δ involves the natural presence of fractional order derivatives.

2.3 T Circuit

Matrix $\boldsymbol{M}$ obtained for both the classic and bio-heat two-port network leads to a T circuit representation of the system (see Fig. 2).

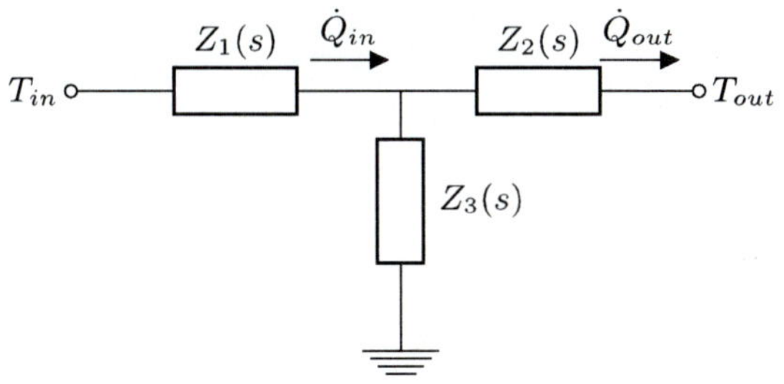

Fig. 2. Thermal two-port network

Impedances are given by:

$$Z_1(s) = Z_2(s) = \frac{1}{kS_w\delta}[\coth(\delta L) - \operatorname{csch}(\delta L)] \tag{4}$$

and

$$Z_3(s) = \frac{1}{kS_w\delta}\operatorname{csch}(\delta L) \tag{5}$$

where coth and csch respectively are the hyperbolic cotangent and cosecant functions. The variable $\delta = \delta_c = \sqrt{\frac{s}{a}}$ will be used for systems modelled from the classic heat equation and as $\delta = \delta_{bio} = \sqrt{\frac{s}{a} + h_b}$ if the model is based upon the bio-heat equation.

3 Lung Global Model

Figure 3 shows a lung scheme. Each branch n has a length L_n as well as its cross-section S_n. A combination of Murray's law with a mechanical impedance model [10] allows to establish a relationship between two consecutive cross-sections:

$$\frac{S_{n+1}}{S_n} = 2^{-\frac{2}{3}}, \tag{6}$$

as well as two consecutive lengths:

$$\frac{L_{n+1}}{L_n} = 2^{-\frac{1}{3}}. \tag{7}$$

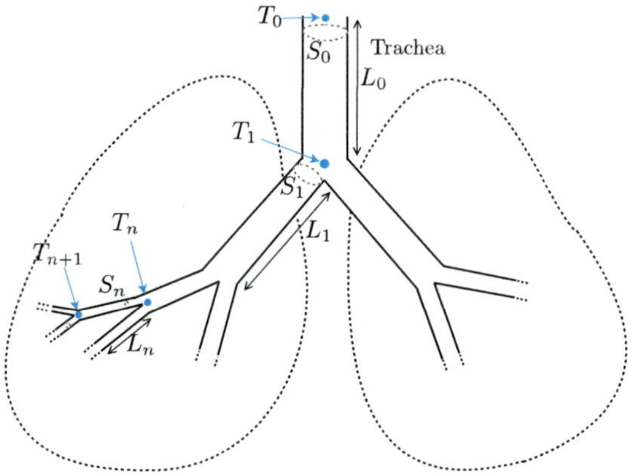

Fig. 3. Lung scheme

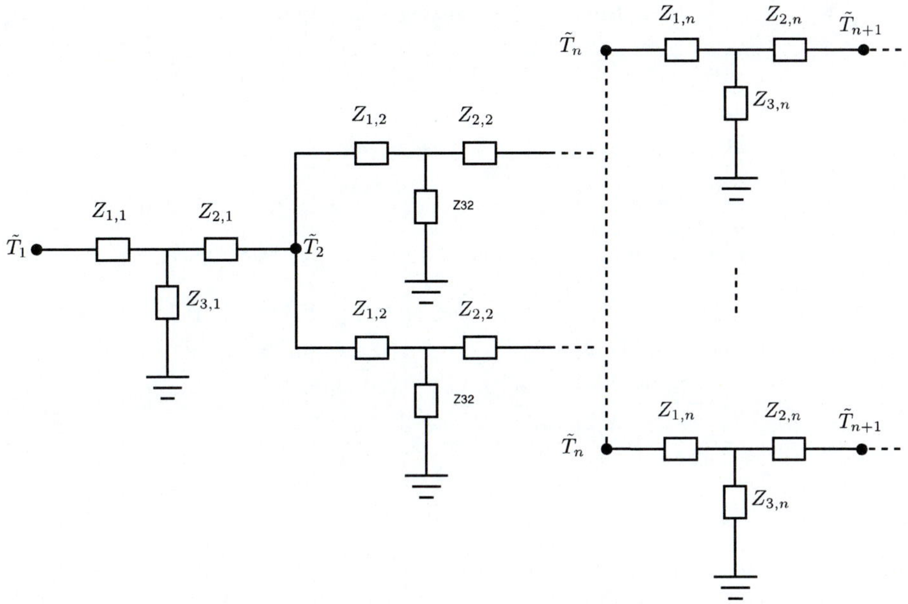

Fig. 4. Global circuit model for a lung

A symmetric lung model may be represented as the circuit shown in Fig. 4. On a same branching level n, the equivalent impedance is the parallel combination of Z_1 and Z_3, each lung level having 2^n branches which give:

$$Z_{i-eq,n}(s) = \frac{Z_{i,n}(s)}{2^n} \quad \text{for} \quad i = 1, 2, 3 \quad \text{and} \quad n = 1, 2, \ldots, 23. \tag{8}$$

4 Simulations and Results

Lung blood circulation is usually non-uniform. Normally the first branches are considered as a dead space used only for passive air transport. The gas exchange zone begins around the 15-th branch. Therefore, three different models are simulated for comparison (see Table 1). Simulation parameters are shown in Table 2, where the subscript t stands for tissue, b for blood and a for air. Figure 5 shows the equivalent impedance for the three models from the trachea and from $n = 13$ (interior of the lungs).

As observed, air model behaves as a capacitance for low-frequencies, which would allow slow enough thermal disturbances to fully propagate through the lungs. Mixed and biological models have a constant static gain and similar behavior, though the higher perfusion in the biological model led to a lower static gain. High-frequency behavior coincides for air and mixed model, which means dead

Table 1. Global model propositions

Model	Dead space ($n < 15$)	Exchange space ($n \geq 15$)
Air	Classic	Classic
Mix	Classic	Bio-heat
Bio	Bio-heat	Bio-heat

Table 2. Simulation parameters for the lung impedances

a_a	2.2e−5 [m^2][s$^-$1]	k_a	0.025 [W][m$^-$1][K$^-$1]
ρ_t	550 [kg][m$^-$3]	ρ_b	1060 [kg][m$^-$3]
c_t	3718 [J][kg$^-$1][K$^-$1]	c_b	3900 [J][kg$^-$1][K$^-$1]
k_t	0.28 [W][m$^-$1][K$^-$1]	ω_b	0.0043 s$^-$1
L_0	0.1 m	d_0	1.4 cm

space acts as a bypass at such frequencies. The equivalent impedance for the 13-th level is shown in Fig. 5 to illustrate the differences in the interior of the lungs. As one approaches the exchange zone, capacitance behavior is more present in the air model, which implies more heat accumulation. All proposed models show a slope of −10 dB/dec for the high frequency limit, which means the three equivalent impedances have a fractional-order component. By observing Fig. 5, one can derive the general structure of impedance for both models (mix and bio models exhibit the same mathematical structure):

$$Z_{eq-air}(s) = \frac{K(1+as)}{s(1+\tau\sqrt{s})} \tag{9}$$

and

$$Z_{eq-mix-bio}(s) = \frac{K}{1+\tau\sqrt{s}}. \tag{10}$$

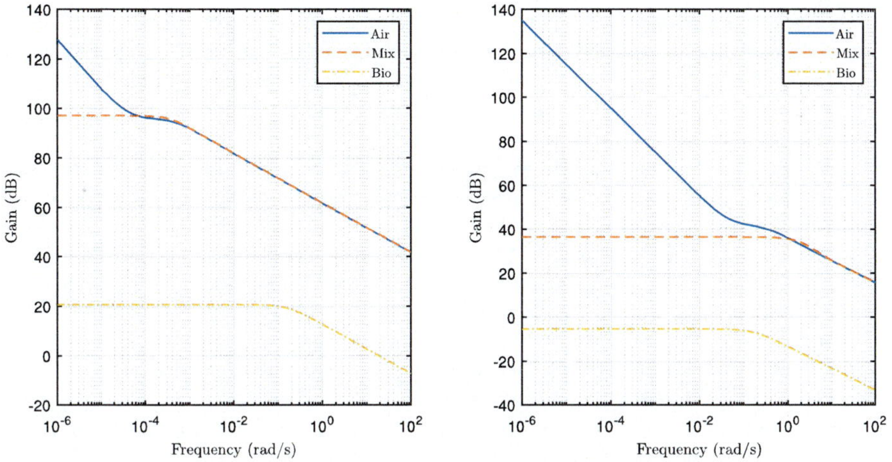

Fig. 5. Input equivalent impedance for lung system at $n = 1$ (left) and $n = 13$ (right)

5 Conclusions

A circuit approach for a global model of heat transfer in human lung was presented and three different models were obtained. The equivalent impedance of the models were compared. The air model implies that as one approaches gas exchange zone, capacitance behavior becomes dominant. This neglects dissipation and implies strong energy storage on alveolar level. On the other hand, slow temperature fluctuations may propagate through the whole lung which is not realistic. Mix and bio models have a static gain which is inversely proportional to blood perfusion. This means that higher perfusion would require a larger heat input in order to significantly increase temperature in the lung, which is coherent with blood thermal regulation role. Research perspectives may include analyzing more complex scenarios (as respiratory obstructions or illness) to relate impedance parameter variations to different medical conditions. This could lead to a medical application of equivalent thermal impedance identification.

References

1. Kunzi, R.: Thermal design of power electronic circuits. In: Proceedings of the CAS-CERN Accelerator School: Power Converters, Baden, Switzerland (2015)
2. Ismail, N., Ghaddar, N., Ghali, K.: Electric circuit analogy of heat losses of clothed walking human body in windy environment. Int. J. Thermal Sci. **127**, 105–116 (2018). https://doi.org/10.1016/j.ijthermalsci.2018.01.025
3. Chinen, K., Kinjo, I., Zamami, A., Irei, K., Nagayama, K.: New equivalent-electrical circuit model and a practical measurement method for human body impedance. Bio-Med. Mater. Eng. **26**, S779–S786 (2015)
4. Maillet, D., Andre, S., Batsale, J., Degiovanni, A., Moyne, C.: Thermal Quadrupoles: Solving the Heat Equation Through Integral Transforms. John Wiley & Sons, Hoboken (2000)

5. Duhé, J.-F., Victor, S., Melchior, P., Abdelmoumen, Y., Roubertie, F.: Thermal modeling using two-port network impedance fractional-order approximations, In: 17th ASME/IEEE MESA International Design Engineering Technical Conferences and Computers and Information in Engineering Conference, IDETC/CIE 2021 (2021)
6. Badenes, R., Lozano, A., Belda, F.: Postoperative pulmonary dysfunction and mechanical ventilation in cardiac surgery. Critical Care Res. Pract. **2015**, 420513 (2015). https://doi.org/10.1155/2015/420513
7. Giesinger, R., McNamara, P.: The impact of therapeutic hypothermia on pulmonary hemodynamics of meconium aspiration syndrome. Am. J. Resp. Crit. Care Med. **198**(2), 286–287 (2018). https://doi.org/10.1164/rccm.201802-0207LE
8. Weibel, E. R.: mandelbrot's fractals and the geometry of life: a tribute to Benoît Mandelbrot on his 80th birthday. In: Fourth International Symposium on Fractals in Biology and Medicine (2014)
9. Ionescu, C.M., Copot, D., De Keyser, R.: A fractional order impedance model to capture the structural changes in lungs. IFAC Proc. Vol. **47**(3), 53635368 (2014). https://doi.org/10.3182/20140824-6-ZA-1003.01124
10. Kuwahara, F., Sano, Y., Liu, J., Nakayama, A.: A porous media approach for bifurcating flow and mass transfer in a human lung. J. Heat Transf. **131**(10), 1–5 (2009). https://doi.org/10.1115/1.3180699

Finite Difference Schemes with Non-uniform Time Meshes for Distributed-Order Diffusion Equations

M. L. Morgado[1], M. Rebelo[2(✉)], and L. L. Ferrás[3]

[1] Department of Mathematics, University of Trás-os-Montes and Alto Douro, Vila Real, and Center for Computational and Stochastic Mathematics, Instituto Superior Técnico, Lisboa, Portugal
luisam@utad.pt

[2] Center for Mathematics and Applications (CMA) and Department of Mathematics, FCT NOVA, Lisboa, Portugal
msjr@fct.unl.pt

[3] Department of Mechanical Engineering - FEUP, University of Porto & CMAT - Centre of Mathematics, University of Minho, Braga, Portugal

Abstract. In this work, a stable and convergent numerical scheme on non-uniform time meshes is proposed, for the solution of distributed-order diffusion equations. A set of numerical results illustrates that the use of particular non-uniform time meshes provides more accurate results than the use of a uniform mesh, in the case of non-smooth solutions.

Keywords: Distributed-order derivatives · Finite differences · Diffusion equations · Non-uniform meshes

1 Introduction

Distributed-order differential equations have been investigated intensively in the last decades [1–7]. The concept of distributed-order derivative was developed by Caputo [1] and further developed by Caputo, Bagley, Torvik, Mainardi and Gorenflo (just to name a few) [2–8].

Diffusion equations with distributed-order may be regarded as a generalization of integer and fractional-order diffusion equations and they can involve distributed-order derivatives in time, or in space or both. In particular, Caputo time distributed-order diffusion equations, which are useful to describe ultra-slow diffusion processes ([8]) are given by:

$$\int_0^1 c(\alpha)\frac{\partial^\alpha u(x,t)}{\partial t^\alpha}\,d\alpha = \frac{\partial^2 u(x,t)}{\partial x^2} + f(x,t), \quad 0 < t \le T,\ 0 \le x \le L, \qquad (1)$$

where the weighting function of the orders of the time derivatives is such that $c(\alpha) \ge 0$ and $\displaystyle\int_0^1 c(\alpha)\,d\alpha = C > 0$, ([6,7]), and the fractional derivatives correspond to the Caputo derivatives, which for orders between 0 and 1 are given by:

A. Dzielinski et al. (Eds.): ICFDA 2021, LNNS 452, pp. 239–244, 2022.
https://doi.org/10.1007/978-3-031-04383-3_27

$$\frac{\partial^\alpha u(x,t)}{\partial t^\alpha} = \frac{1}{\Gamma(1-\alpha)} \int_0^t (t-s)^{-\alpha} \frac{\partial u(x,s)}{\partial s}\, ds. \tag{2}$$

In fact, we easily see that if $c(\alpha) = \delta(\alpha - \beta)$ then (1) becomes the singular-order differential equation:

$$\frac{\partial^\beta u(x,t)}{\partial t^\beta} = \frac{\partial^2 u(x,t)}{\partial x^2} + f(x,t), \quad 0 < t \leq T, \; 0 \leq x \leq L. \tag{3}$$

When $\beta = 1$ Eq. (3) reduces to the classical (integer-order) diffusion equation. In the case where $0 < \beta < 1$, (3) is called the subdiffusion equation and it has been the subject of investigation of many researchers.

With the growing interest on this type of equation, numerical methods started being developed. Diethelm and Ford [9] developed a basic framework for the numerical solution of distributed-order differential equations. Ford and Morgado [10] discussed the existence and uniqueness of solutions for this type of equation, and also proposed a numerical method for their approximation in the case where the initial conditions are not known. Morgado and Rebelo [12] developed an implicit finite difference scheme for the numerical approximation of the distributed order time-fractional reaction-diffusion equation with a nonlinear source term. Ye et al. [13] developed a numerical method based on a compact difference scheme for a distributed order time-fractional diffusion-wave equation. Wang et al. [14] derived and analysed a second-order accurate implicit numerical method for the Riesz space distributed-order advection-dispersion equation and Jin et al. [15] presented a numerical solution of an initial boundary value problem for the distributed order time fractional diffusion equation. They developed a space semidiscrete scheme based on the standard Galerkin finite element method, and established error estimates for both smooth and nonsmooth initial data.

These works struggled with the low convergence order of the numerical schemes, since the solution of fractional differential equations may be nonsmooth at $t = 0$, even if the data is infinitely smooth.

In [16] the authors have shown that typical solutions of the single-order differential equation (3) possess a singularity at the origin in time, in the sense that they satisfy:

$$\left| \frac{\partial^\ell u(x,t)}{\partial t^\ell} \right| \leq C(1 + t^{\beta-\ell}), \; \ell = 0, 1, 2, \; t \in (0, T], \tag{4}$$

for some positive constant C, a fact that brings serious difficulties in the derivation of accurate numerical methods. In order to overcome this problem, the authors propose the use of a time graded mesh in which the grading exponent is properly chosen to recover the optimal convergence order attained for sufficiently smooth solutions. Naturally, the grading exponent depends on the order of the time derivative β and we refer the interested reader to [16] for more details.

Estimates as (4) are not easy to achieve for distributed-order diffusion equations as (1) because the order of the time derivative is distributed over the interval $[0, 1]$, and, to the best of our knowledge, they have not been obtained so far. The use of time graded meshes without knowing how to properly choose the grading exponent is risky since, as explained also in [16], larger grading exponents will densely place more mesh points near the singularity which originates the appearance of round-off errors.

Hence, in this paper we consider initial-boundary value problems for (1), with the initial condition and boundary conditions of the form:

$$u(x, 0) = g(x), \tag{5}$$

$$u(0, t) = u_0, \quad u(L, t) = u_L, \tag{6}$$

where we assume that all the involved functions are continuous in their respective domains.

We propose a finite difference scheme on a non-uniform time mesh $0 = t_0 < t_1 < \cdots < t_n = T$, with $t_{i+1} = t_i + \tau_i$, $i = 0, 1, ..., n-1$. Since it is expected a singularity at the origin in time, we consider non-uniform meshes such that $\tau_i \leq \tau_{i+1}$. Here we will be interested in non-uniform meshes of the kind (see [17]):

$$\tau_i = t_{i+1} - t_i = (i+1)^\gamma \mu, \tag{7}$$

where γ is a positive integer and $\mu > 0$. Taking into account the summation formulas $\sum_{k=1}^{n} k = \frac{n(n+1)}{2}$, $\sum_{k=1}^{n} k^2 = \frac{n(n+1)(2n+1)}{6}$ and $\sum_{k=1}^{n} k^3 = \frac{n^2(n+1)^2}{4}$, we obtain, respectively:

$$\underline{\text{Mesh 1:}}\ (\gamma = 1)\ \mu = \frac{2T}{n(n+1)}, \tag{8}$$

$$\underline{\text{Mesh 2:}}\ (\gamma = 2)\ \mu = \frac{6T}{n(n+1)(2n+1)}, \tag{9}$$

$$\underline{\text{Mesh 3:}}\ (\gamma = 3)\ \mu = \frac{4T}{n^2(n+1)^2}. \tag{10}$$

The paper is organised as follows. In Sect. 2 we describe the numerical method and present some numerical results. We end with some conclusions and plans for further investigation.

2 Finite Difference Scheme

In order to numerically solve problem (1), (5), (6) we start with the discretization of the integral in (1). Let us consider a partition of the interval $[0, 1]$ into N subintervals, with equal length $h = 1/N$. Defining $\alpha_j = jh - \frac{h}{2}$, $j = 1, \ldots, N$, we can use the midpoint rule to obtain

$$\int_0^1 c(\alpha) \frac{\partial^\alpha u(x,t)}{\partial t^\alpha}\, d\alpha \approx h \sum_{j=1}^{N} c(\alpha_j) \frac{\partial^{\alpha_j} u(x,t)}{\partial t^{\alpha_j}}. \tag{11}$$

In order to approximate the spatial derivative, we consider a uniform mesh, on the interval $[0, L]$, defined by the grid points $x_i = i\Delta x$, $i = 0, 1, \dots, K$, where $\Delta x = \frac{L}{K}$, and use the second order formula:

$$\frac{\partial^2 u(x_i, t)}{\partial x^2} \approx \frac{u(x_{i+1}, t) - 2u(x_i, t) + u(x_{i-1}, t)}{(\Delta x)^2}. \tag{12}$$

Substituting (11) and (12) in (1), and denoting $U_i(t) \approx u(x_i, t)$, we obtain the semi-discretized scheme:

$$h \sum_{j=1}^{N} c(\alpha_j) \frac{\partial^{\alpha_j} U_i(t)}{\partial t^{\alpha_j}} = \frac{U_{i+1}(t) - 2U_i(t) + U_{i-1}(t)}{(\Delta x)^2} + f(x_i, t), \quad i = 1, \dots, K-1. \tag{13}$$

From (6), we have:

$$U_0(t) = u_0, \quad \text{and} \quad U_K(t) = u_L, \tag{14}$$

and from (5):

$$U_i(0) = g(x_i), \quad i = 1, \dots, K-1. \tag{15}$$

For the discretization of the fractional derivatives, we consider the non-uniform meshes (8), (9) and (10) and we use the L1 approximation formula (see [18]):

$$\frac{\partial^{\alpha_j} u(\cdot, t_k)}{\partial t^{\alpha_j}} \approx \frac{1}{\Gamma(2-\alpha_j)} \sum_{m=0}^{k-1} \tau_m^{-1} a_{m,k}^{\alpha_j} \left(u(\cdot, t_{m+1}) - u(\cdot, t_m) \right), \tag{16}$$

$$a_{m,k}^{\alpha_j} = (t_k - t_m)^{1-\alpha_j} - (t_k - t_{m+1})^{1-\alpha_j}, \tag{17}$$

with $m = 0, 1, \dots, k-1$, $k = 1, \dots, n$. Substituting in (13), and denoting by $U_i^k \approx u(x_i, t_k)$, we obtain the finite difference scheme:

$$h \sum_{j=1}^{N} c(\alpha_j) \frac{1}{\Gamma(2-\alpha_j)} \sum_{m=0}^{k-1} \tau_m^{-1} a_{m,k}^{\alpha_j} \left(U_i^{m+1} - U_i^m \right)$$
$$= \frac{U_{i+1}^k - 2U_i^k + U_{i-1}^k}{(\Delta x)^2} + f(x_i, t_k),\ i = 1, \dots, K-1, \quad k = 1, 2, \dots, n. \tag{18}$$

Hence, in order to obtain a numerical solution for (1), (5), (6), we need to solve the linear systems of equations (18) noting that, taking (14) and (15) into account, we have:

$$U_0^l = u_0, \quad U_L^l = u_L, \quad k = 1, 2, \dots, n \tag{19}$$
$$U_i^0 = g(x_i), \quad i = 1, \dots, K-1. \tag{20}$$

3 Numerical Results

In this section we use the numerical method (18)–(20) to approximate the solution of the following example.

Example 1
In (1) we have $c(\alpha) = \Gamma(-\alpha + 3/2)$ and the right-hand side function f is such that the analytical solution is $u(x,t) = \sqrt{t}(x-1)x, \quad t > 0,\ x \in [0,1]$.

In Table 1 we list the maximum of the absolute errors ($\|e\|_\infty$) and experimental orders of convergence (EOC) obtained with the numerical method (18) on uniform and on the non-uniform meshes defined by (8), (9) and (10).

Table 1. Example 1: maximum of errors and experimental convergence orders

		Uniform mesh			Mesh 1			Mesh 2			Mesh 3		
$N=K$	n	$\|e\|_\infty$	EOC_t	EOC_x	$\|e\|_\infty$	EOC_t	EOC_x	$\|e\|_\infty$	EOC_t	EOC_x	$\|e\|_\infty$	EOC_t	EOC_x
2	4	9.81e−3	–	–	7.57e−3	–	–	5.80e−3	–	–	3.53e−3	–	–
4	16	9.86e−3	(*)	(*)	5.84e−3	0.19	0.38	2.39e−3	0.64	1.28	1.36e−3	0.69	1.38
8	64	8.43e−3	0.11	0.23	2.39e−3	0.64	1.29	7.74e−4	0.81	1.63	3.77e−4	0.93	1.85
16	256	5.61e−3	0.29	0.59	8.46e−4	0.75	1.50	2.10e−4	0.94	1.88	8.72e−5	1.06	2.11
32	1024	3.16e−3	0.41	0.83	2.70e−4	0.82	1.64	5.21e−5	1.01	2.01	1.86e−5	1.11	2.23

The first derivative of the analytical solution of Example 1 has a singularity at $(x, 0)$ which implies a drop on the convergence orders of the numerical method (18)–(20) with uniform meshes, as we can be observed in the second column of Table 1. On the other hand, we observe that if we use the time mesh defined by Mesh 3 the temporal convergence order $EOC_t \sim 1 + h$ is recovered and also the experimental spatial convergence order $EOC_x \sim 2$.

4 Conclusions

A similar analysis as the one carried out in [11], to appear in a forthcoming paper, allows us to conclude that the numerical scheme is unconditionally stable and convergent with accuracy $O(N^{-2} + K^{-2} + n^{-1})$ for sufficiently smooth solutions. For nonsmooth solutions, finite difference schemes on uniform grids provide results with loss of convergence order, but the use of non-uniform meshes improves the results as illustrated in the numerical results above.

Aknowledgments. The first author acknowledges Fundação para a Ciência e Tecnologia within projects UIDB/04621/2020 and UIDP/04621/2020.

This work was also funded by national funds through the FCT - Fundação para a Ciência e a Tecnologia, I.P., under the scope of the project UIDB/00297/2020 (Center for Mathematics and Applications), and, through projects UIDB/00013/2020 and UIDP/00013/2020 (CMAT - Centre of Mathematics - University of Minho).

References

1. Caputo, M.: Elasticità e Dissipazione. Zanichelli, Bologna, Italy (1969)
2. Caputo, M.: Mean fractional-order-derivatives differential equations and filters, Annali dellUniversità di Ferrara. Nuova Serie. Sezione VII. Sci. Matematiche **41**, 73–84 (1995)
3. Caputo, M.: Distributed order differential equations modelling dielectric induction and diffusion. Fract. Calc. Appl. Anal. **4**, 421–442 (2001)
4. Bagley, R.L., Torvik, P.J.: On the existence of the order domain and the solution of distributed order equations. Int. J. Appl. Math. **I**, 865–882 (2000)
5. Bagley, R.L., Torvik, P.J.: On the existence of the order domain and the solution of distributed order equations. Int. J. Appl. Math. **II**, 965–987 (2000)
6. Gorenflo, R., Luchko, Y., Stojanovic, M.: Fundamental solution of a distributed order time-fractional diffusion-wave equation as probability density. Fract. Calc. Appl. Anal. **16**, 297–316 (2013)
7. Mainardi, F., Pagnini, G., Mura, A., Gorenflo, R.: Time-fractional diffusion of distributed order. J. Vib. Control **14**, 1267–1290 (2008)
8. Chechkin, A.V., Gorenflo, R., Sokolov, I.M.: Retarding subdiffusion and accelerating superdiffusion governed by distributed-order fractional diffusion equations. Phys. Rev. E **66**, 046129 (2002)
9. Diethelm, K., Ford, N.J.: Numerical solution methods for distributed order differential equations. Fract. Calc. Appl. Anal. **4**, 531–542 (2001)
10. Ford, N.J., Morgado, M.L.: Distributed order equations as boundary value problems. Comput. Math. Appl. **64**, 2973–2981 (2012)
11. Ford, N.J., Morgado, M.L., Rebelo, M.: An implicit finite difference approximation for the solution of the diffusion equation with distributed order in time. Electron. Trans. Numer. Anal. **44**, 289–305 (2015)
12. Morgado, M.L., Rebelo, M.: Numerical approximation of distributed order reaction-diffusion equations. J. Comput. Appl. Math. **275**, 216–227 (2015)
13. Ye, H., Liu, F., Anh, V.: Compact difference scheme for distributed-order time-fractional diffusion-wave equation on bounded domains. J. Comput. Phys. **298**, 652–660 (2015)
14. Wang, X., Liu, F., Chen, X.: Novel second-order accurate implicit numerical methods for the Riesz space distributed-order advection-dispersion equations. Adv. Math. Phys. (2015)
15. Jin, B., Lazarov, R., Sheen, D., Zhou, Z.: Error estimates for approximations of distributed order time fractional diffusion with nonsmooth data. Frac. Calc. Appl. Anal. **19**, 69–93 (2016)
16. Stynes, M., O'Riordan, E., Gracia, J.L.: Error analysis of a finite difference method on graded meshes for a time-fractional diffusion equation. SIAM J. Numer. Anal. **55**(2), 1057–1079 (2017)
17. Changpin, L., Qian, Y., Chen, A.: Finite difference methods with non-uniform meshes for nonlinear fractional differential equations. J. Comput. Phys. **316**, 614–631 (2016)
18. Morgado, M.L., Morgado, L.F.: Modeling transient currents in time-of-flight experiments with tempered time-fractional diffusion equations. Progr. Fract. Differ. Appl. **6**(1), 43–53 (2020). https://doi.org/10.18576/pfda/060105

CRONE Controller Synthesis Improvement by an Additional Feedback

Piotr Ostalczyk(✉)

Warsaw University of Technology, Warsaw, Poland
piotr.ostalczyk@ee.pw.edu.pl

Abstract. In the paper a method improving the closed-loop SISO system with the CRONE controller robustness due to the plant parameter changes is proposed. Among the closed-loop system stability the constant shape of transients is preserved (the maximum overshoot and the damping). The CRONE idea is about reshaping a frequency characteristic of an open-loop system to approximate the ideal fractional- real or complex order integrator. Zeros and poles should lie on the semi-straight lines with a vertex at the origin and symmetrical to the real axis. The plant with uncertain parameters is characterized by a set of Nyquist uncertainty. Every uncertainty area is characterized by an angle defined by two tangent semi-strait lines. The smaller the uncertainty angles, the better the CRONE robustness results are achieved. In the paper it is shown that the constant negative unity pre-feedback of the plant diminishes the uncertainty area angles. Then one evaluates the lower envelope of a set of uncertainty areas. Then, the CRONE controller synthesis comes down to the design such a transmittance that in a closed frequency interval it approximates the Nyquist plot of an ideal fractional-order integrator with assumed fractional order. The investigations are supported by numerical examples.

Keywords: Fractional calculus · CRONE control

1 Introduction

The robust control became an important task in a dynamical systems control with a plant uncertain parameters a few decades ago. Here, one can mention the most popular ones:

- H_2 method, [15,16]
- H_∞ method, [15,16]
- μ-synthesis method.

A significant method among the robust system synthesis [1,16] is the so called CRONE control (fr. *Commande Robuste d'Ordre non Entiere*) [10–12,14] basing on the Fractional Order Calculus [4,13]. In the case of linear systems described

A. Dzielinski et al. (Eds.): ICFDA 2021, LNNS 452, pp. 245–257, 2022.
https://doi.org/10.1007/978-3-031-04383-3_28

by a SISO transfer functions with uncertain parameters from known ranges for one chosen frequency instead of a point there is an uncertainty area (UA) of different shapes. The idea of the CRONE control lies on a specific shape of UAs with the CRONE controller related to frequencies from a finite range [5,6,9].

The application of the CRONE controller of the first, second and third generation preserves the closed-loop system transients properties. In addition to the guaranteed stability there is constant maximal overshoot and damping properties.

The mentioned above specific shape approximates the real or complex fractional order integrator [4,13]. In the fractional calculus there are several definitions of the fractional-order derivative and integrals.

The CRONE control has important practical applications.

In this paper a method transforming the UAs to improve the closed-loop system robustness is presented. It is proved that the preliminary unit negative feedback reshapes the UAs leading to a modified plant. Then a synthesis of the CRONE controller to this plant improves the robustness measures.

The paper is organized as follows. In Sect. 2 there is given selected information related to the fractional-order derivatives and integrals. In Sect. 3 a relation between the conformal map and the constant feedback is discussed. Section 4 serves a numerical support of Sect. 3.

2 Mathematical Preliminaries

In this Section notions used in the paper will be presented. The following notation will be used.

2.1 Fractional Integral and Fractional Derivative

Definition 1. *Let a real valued function f be defined on a finite interval $[t_0, t]$ on a real axis $\mathbb{R}$. The Riemann-Liouville left-hand integral ${}_{t_0}^{RL}I_t^{(\nu)f}$ of order where $\Re\{\nu\} > 0$ is defined as an integral*

$$\left({}_{t_0}^{RL}I_t^{(\nu)}f\right)(t) := \frac{1}{\Gamma(\nu)}\int_{t_0}^{t}\frac{f(t)}{(x-t)^{1-\nu}}dt \tag{1}$$

where $\Gamma(\nu)$ is the Euler Gamma function evaluated for ν and $t > t_0$.

For $\nu = n \in \mathbb{N}$ the Riemann-Liouville integral is a classical one

$$\left({}_{t_0}^{RL}I_t^{(n)}f\right)(t) = \int_{t_0}^{t}\cdots\int_{t_0}^{t_{n-1}} f\left(x_{n_1-1}\right)dx_{n-1}\cdots dx_1 = \frac{1}{(n-1)!}\int_{t_0}^{t}(t-x)^{n-1}f(x)dx \tag{2}$$

Definition 2. *The Riemann-Liouville left-hand derivative ${}_{t_0}^{RL}D_t^{(\nu)}f$ of order $\nu \in \mathbb{C}$ where $\Re\{\nu\} > 0$ is defined as an integral*

$$\left({}_{t_0}^{RL}D_t^{(\nu)}f\right)(t) := \left(\frac{d}{dt}\right)^n \left({}_{t_0}^{RL}I_t^{(\nu)}f\right)(t) = \frac{1}{\Gamma(\nu)}\left(\frac{d}{dt}\right)^n \int_{t_0}^{t} \frac{f(t)}{(x-t)^{1-\nu}}dt \quad (3)$$

where $n = [\Re\{\nu\}] \in \mathbb{N}$ and $t > t_0$. $[\Re\{\nu\}]$ denotes the integral part of $\Re\{\nu\}$.

As a special case one has

$$\left({}_{t_0}^{RL}D_t^{(0)}f\right)(t) := f(t) \quad (4)$$

$$\left({}_{t_0}^{RL}D_t^{(n)}f\right)(t) := \left(\frac{d^n}{dt^n}f\right)(t) \quad (5)$$

Now, one considers the exponential type function $f(t)$. Then, the one-sided Laplace transform of a fractional-order integral of a function f with the one-sided Laplace transform $(\mathcal{L}f)(s)$ is of the form $\nu > 0$ [2,13]

$$\left(\mathcal{L}_0^{RL}I_t^{(\nu)}f\right)(s) = \frac{(\mathcal{L}f))(t)}{s^\nu} \quad (6)$$

for $s \in \mathbb{C}$ with $\Re\{s\} > B$.

The one-sided Laplace transform of a fractional-order derivative of the function considered above is as follows [13]

$$\left(\mathcal{L}_0^{RL}D_t^{(\nu)}f\right)(s) = s^\nu(\mathcal{L}f)(t) - \sum_{k=0}^{n-1} s^{n-k-1}\left[{}_{t_0}^{RL}D_t^{(k)}\left({}_{t_0}^{RL}I_t^{(n-\nu)}\right)\right](0^+) \quad (7)$$

for $\Re\{s\} > B$.

3 Conformal Map and the Closed-Loop Dynamical System Nyquist Characteristic

In this Section there will be proven a relation between the conformal map [7,14] and the closed-loop Nyquist characteristics of the linear time invariant system with uncertainty parameters.

3.1 Conformal Map

The conformal map (CM) is defined as a following transformation of the areas $A \subset \mathbb{C}$ on the complex plane represented by its elements $s \in A$

$$f(s) = \frac{\alpha s + \beta}{\gamma s + \delta} \in \mathbb{C} \quad (8)$$

for $\alpha\delta - \beta\gamma \neq 0$.

3.2 Closed-Loop System with Unity Negative Feedback

For an open-loop transfer function $G_o(s)$ a closed-loop system transfer function with unity negative feedback $G_c(s)$ has a form

$$G_c(s) = \frac{G_o(s)}{1 + G_o(s)} \tag{9}$$

Remark 1. The closed-loop system configuration is a special case of the CM where $\alpha = 1$, $\beta = 0$, $\gamma = 1$ and $\delta = 1$. Evidently the transformation satisfies inequality $\alpha\delta - \beta\gamma = 1 \cdot 1 - 0 \cdot 1 = 1$.

The closed-loop configuration is presented in Fig. 1 where a vector $\mathbf{p}$ in a transfer function $G_o(s, \mathbf{p})$ denotes a vector of uncertainty parameters ranges.

Now, there will be recalled without a proof the fundamental property of CM.

Proposition 1. *A CM maps lines and circles to lines and circles.*

3.3 Conformal Map of a Circle Property

The main fundamental result of the paper will be stated. It evaluates parameter stated in Proposition 3.

Proposition 2. *Consider an uncertainty region described by a disc on a complex plane C_o enclosed by a circle of radius r_o centered at $s_o = x_o + +jy_o)$ where $x_o > r_o$. Hence, distanced from the origin by $R_o = \sqrt{x_o^2 + y_o^2}$. The cirlce C_c obtained by a special case of conformal map with $\alpha = 1, \beta = 0, \gamma = 1, \delta = 1$ where $ad - bc = 1 \neq 0$ of C_o may be expressed as*

$$C_c = \frac{C_o}{1 + C_o} \tag{10}$$

Then,

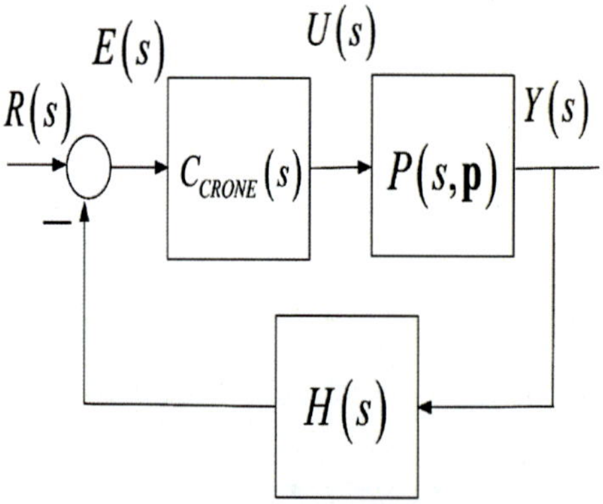

Fig. 1. Closed-loop system with uncertain plant and CRONE controller.

$$r_c < r_o \tag{11}$$

Proof. The conformal map transforms a circle to a circle as shown on Fig. 2. Disc C_o center $O_o = x_{co} + jy_{co}$ is transformed to center O_c of disc C_c

$$O_c = \frac{x_o + jy_o}{1 + x_o + jy_o} \tag{12}$$

Now one defines two points A_o and $A_c = G_c(A_o)$

$$A_o = x_o + r_o + jy_o \tag{13}$$

$$A_c = \frac{x_o + r_o + jy_o}{1 + x_o + r_o + jy_o} \tag{14}$$

The distance between A_o and O_o is a radius of circle

$$|A_o - O_o| = |(x_o + r_o + jy_o) - (x_o + jy_o)| = r_o \tag{15}$$

The distance between A_c and O_c

$$\begin{aligned} r_c = |A_c - O_c| &= |\frac{x_o + jy_o}{1 + x_o + jy_o} - \frac{x_o + r_o + jy_o}{1 + x_o + r_o + jy_o}| \\ &= |\frac{-r_o}{(1 + x_o + jy_o)(1 + x_o + r_o + jy_o)}| \\ &= \frac{r_o}{\sqrt{(1 + x_o)^2 + y_o^2}\sqrt{(1 + x_o + r_o)^2 + y_o^2}} < r_o \end{aligned} \tag{16}$$

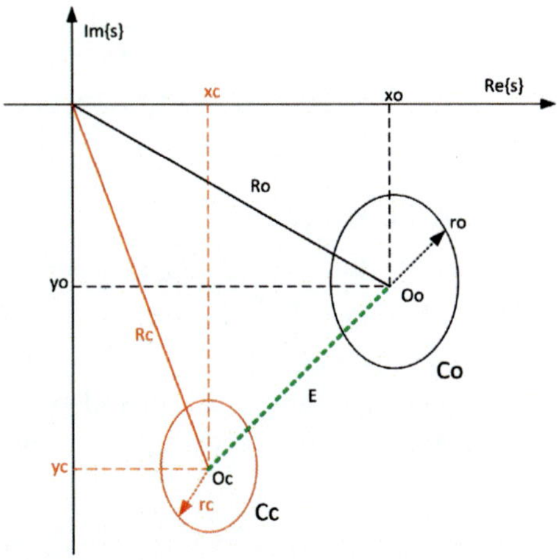

Fig. 2. Open and closed-loop uncertainty discs.

Now one defines second point B_o on the circle C_o and $B_c = G_s(B_o)$

$$B_o = x_o + j\,(y_o + r_o) \tag{17}$$

$$B_c = \frac{x_o + j\,(y_o + r_{co})}{1 + x_o + j\,(y_o + r_o)} \tag{18}$$

and the distance between the transformed point B_c and C_c equals

$$\begin{aligned} r_c = |B_o - B_c| &= |\frac{x_o + jy_o}{1 + x_o + jy_o} - \frac{x_o + jy_o + jr_c}{1 + x_o + jr_o + jy_o}| \\ &= |\frac{-r_o}{(1 + x_o + jy_o)\,(1 + x_o + r_o + jy_o)}| \\ &= \frac{r_o}{\sqrt{(1 + x_o)^2 + y_o^2}\sqrt{(1 + x_o)^2 + (r_o + y_o)^2}} < r_o \end{aligned} \tag{19}$$

Finally, the third point D_o on a circle D_o is defined

$$D_o = x_o - r_o + jy_o \tag{20}$$

$$D_c = G_s(D_o) = \frac{x_o - r_o + jy_o}{1 + x_{co} - r_o + jy_o} \tag{21}$$

The distance between the transformed point D_c and O_c equals

$$\begin{aligned} r_c = |D_c - O_c| &= |\frac{x_o + jy_o}{1 + x_o + jy_o} - \frac{x_o - r_o + jy_o}{1 + x_o - r_o + jy_o}| \\ &= |\frac{r_o}{(1 + x_o + jy_o)\,(1 + x_o - r_o + jy_o)}| \\ &= \frac{r_{co}}{\sqrt{(1 + x_o)^2 + y_o^2}\sqrt{(1 + x_o - r_o)^2 + y_o^2}} < r_o \end{aligned} \tag{22}$$

The above investigations are illustrated in Fig. 2. Three points lying on a circle are transformed by a conformal map to three points lying on a circle of a smaller radius. □

In practice, the measured data can be covered by a closed polygon forming a uncertainty area (UA) which further can be included in circles or ellipsoids. Exemplary circle and ellipsoids are presented in Figs. 3 and 4.

The angle α characterising the UA is defined as follows.

Definition 3. *The UA being the closed set on the complex plane s one defines an angle*

$$\alpha = \left| \underset{max}{arg}(UA) - \underset{min}{arg}(UA) \right| \tag{23}$$

Two opposite cases are presented in Fig. 4.

4 Open and Closed-Loop UAs Relations

According to the results stated in Proposition 3 one has $r_c < r_o$. This implies that $\alpha_c < \alpha_o$. Two forms of UAs are presented in Fig. 5. The left one presents "thin" uncertainty angle contrary to that one placed on the right. The first mentioned one in the CRONE robust control have much better robust properties.

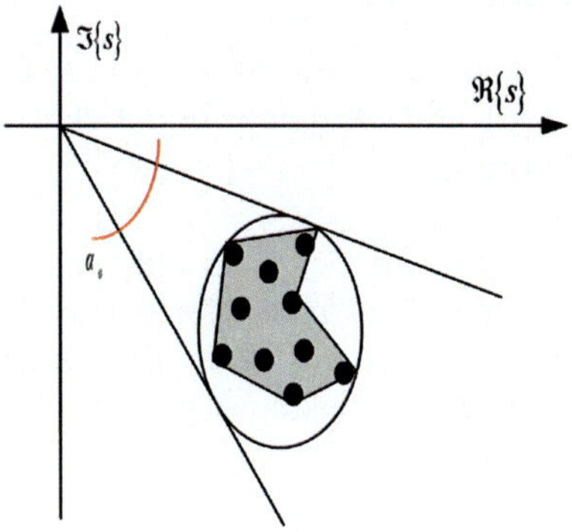

Fig. 3. Uncertainty disc parameters including measured data.

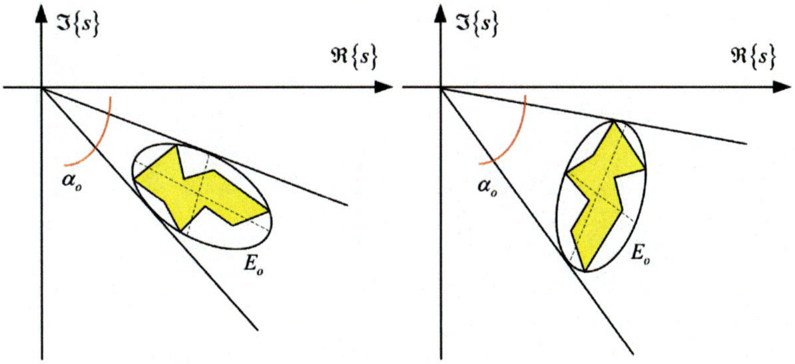

Fig. 4. Examples of an ellipsoidal UA.

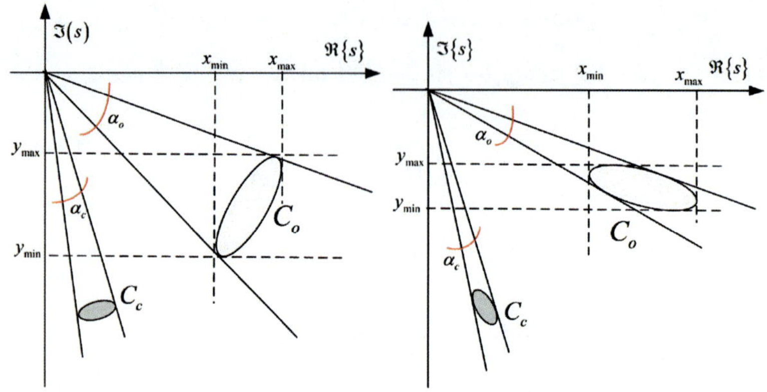

Fig. 5. Open- and closed-loop UAs cases.

4.1 Relation Between a Hypothetical Plant UA and Its UA Closed-Loop Counterpart

One considers a hypothetical plant whose Nyquist UAs are circles of constant radia $r_o < 1$ lying in the III and IV quadrants of the complex plane s are presented in Fig. 6a (BK). In the same figure related by the closed-loop configuration UAs are plotted (RD).
Angles: $\alpha_o(\omega)$ (in black) and $\alpha_c(\ omega)$ (in red) are shown in Fig. 6b.

The hypothetical plant frequency characteristic angle is constant whereas its closed-loop with unit negative feedback is tending sharply to zero. This proves the idea of the additional constant feedback proposed in this paper.

5 Plant Transfer Function with Uncertainty Parameters

Now, one considers a plant described by a transfer function

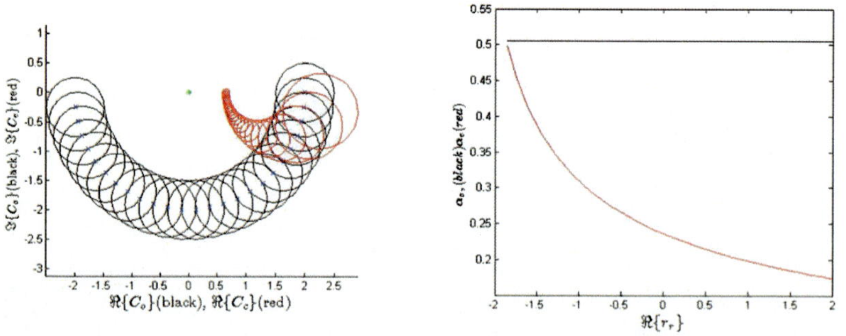

Fig. 6. Open (BK)and closed-loop (RD) uncertainty discs and appropriate angles (Color figure online)

$$G_o(s, \mathbf{p}) = \frac{\prod_{i=1}^{m}(s - z_i)}{\prod_{i=1}^{n}(s - p_i)} \tag{24}$$

where $m <= n$ and

$$p_i \in [p_{min,i}, p_{max,i}] \text{ for } i = 1, 2, \cdots, n \tag{25}$$

$$z_i \in [z_{min,i}, z_{max,i}] \text{ for } i = 1, 2, \cdots, m \tag{26}$$

The exemplary Nyquist UAs are considered in a following example.

5.1 Numerical Example

Consider a plant described by a transfer function

$$G(s, \mathbf{p}) = \frac{s - z_1}{s^3 - (p_1 + p_2 + p_3)\, s^2 + (p_1p_2 + p_1p_3 + p_2p_3)\, s - p_1p_2p_3} \tag{27}$$

where

$$\begin{aligned} p_1 &\in [-1.5, -0.5] \\ p_2 &\in [-2.5, -2.0] \\ p_3 &\in [-3.5, -3.0] \\ z_1 &\in [-5.0, -4.5] \end{aligned} \tag{28}$$

In Fig. 7 open- and closed-loop system UAs for frequency value $\omega = 1.0$ are given.

In Fig. 8a open- and closed-loop system UAs related to a discrete set of frequencies $\omega \in \{0.1, 0.2, \cdots, 2.3\}$ are presented. Related angles: $\alpha_o(\omega)$ (in black) and $\alpha_c(\omega)$ (in red) are plotted in Fig. 8b, respectively

Remark 2. The convexity of the UA set provides a segment joining the edges of the concavity.

Now, for the same plant as in Example 5.1 one considers a pair of complex conjugate poles with uncertainty values. The poles uncertainty range set is

$$\mathbf{p} = \begin{Bmatrix} p_1 : \text{for } p_1 \in [-1.0 + j1.0, -1 + j2.0] \\ p_2 : \text{for } p_1 \in [-1.0 - j2.0, -1 - j1.0] \\ p_3 : \text{for } \quad p_3 = \in [-2.5, -2.0] \\ z_1 : \text{for } \quad z_1 = \in [-5.0, -4.5] \end{Bmatrix}. \tag{29}$$

Related uncertainty areas are plotted in Fig. 9 for frequency values $\omega \in \{0.1, 0.2, \cdots, 2.3\}$

6 The CRONE Control

The CRONE control preserves the shape of the closed-loop system transients due to the plant parameters changes. The plant is described by the transfer function with known parameters changes intervals. The main idea is an approximation of the fractional real or complex ideal integrator [9,10] (Fig. 10).

Proposition 3. *For a transfer function with uncertainty parameters* (26) *and a controller described by a transfer function* $C(s)$ *the UAs satisfy condition*

$$\alpha\left(G(j\omega,\mathbf{p})\right)=\alpha\left(C(j\omega)G(j\omega,\mathbf{p})\right) \tag{30}$$

Proof. The controller frequency characteristic may be expressed in a form

$$C(j\omega)=\left|C(\omega)\right|e^{j\phi(\omega)} \tag{31}$$

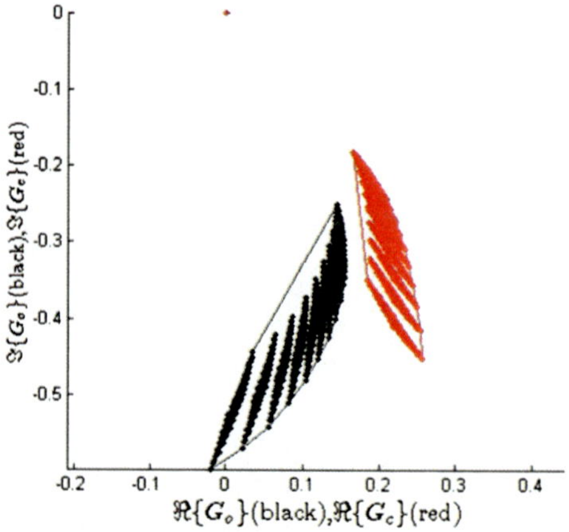

Fig. 7. Open (in black) and closed-loop (in red) uncertainty areas related to $\omega=1.0$ (Color figure online)

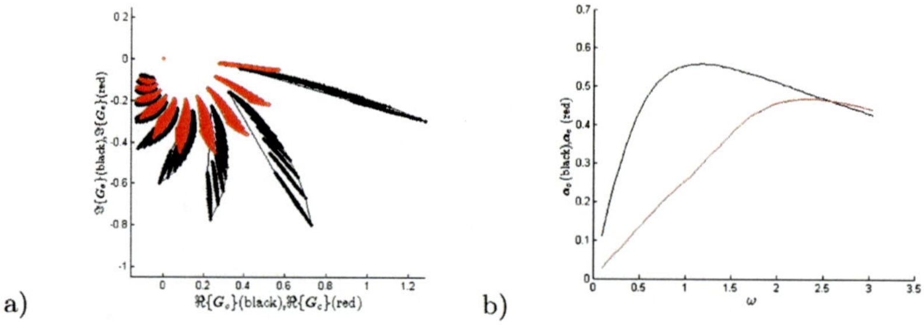

Fig. 8. a) Open (in black) and closed-loop (in red) UAs, b) Related angles α_o (in black) and α_c (in red). (Color figure online)

Hence, for every frequency ω the controller introduces two factors on the UA: multiplication by the controller gain $|C(\omega)|$ and a rotation about an angle $\phi(\omega)$. Both actions do not change the UA shape.

6.1 CRONE Improving Results Method

Summarizing the considerations given in the previous Sections the following closed-loop block diagram presented in Fig. 11 is proposed.
In practice, one assumes that $H_P(s) = H_P = const$

The CRONE controller synthesis may be performed due to the following procedure.

Algorithm

1. Evaluate the UA of the first feedback
2. Synthesise the CRONE controller

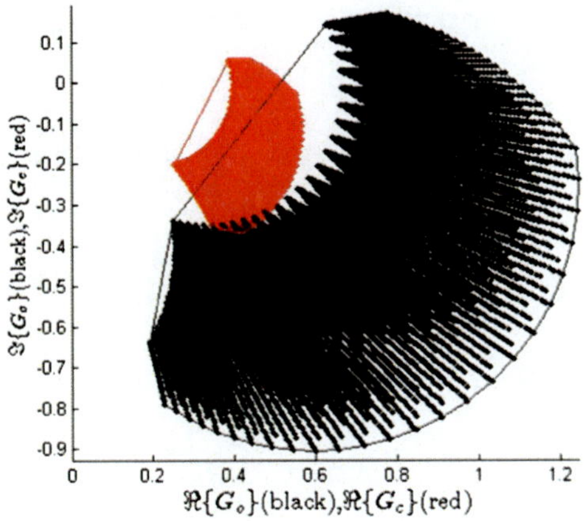

Fig. 9. Open (BK) and closed-loop (RD) UAs α_o (BK) and α_c (RD), for (29), respectively.

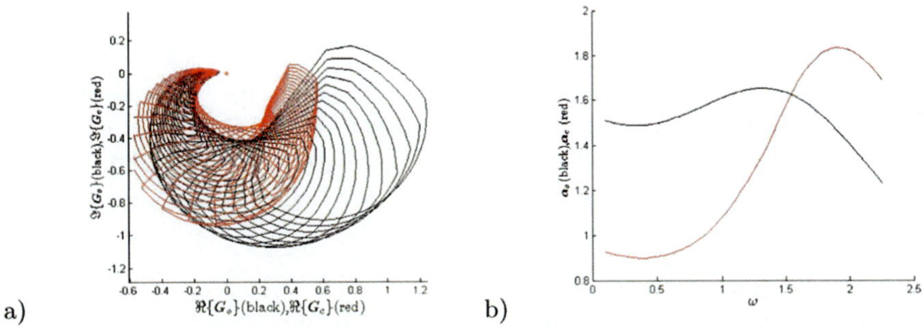

Fig. 10. Open (in black) and closed-loop (in red) UAs α_o (in black) and α_c (in red), respectively. (Color figure online)

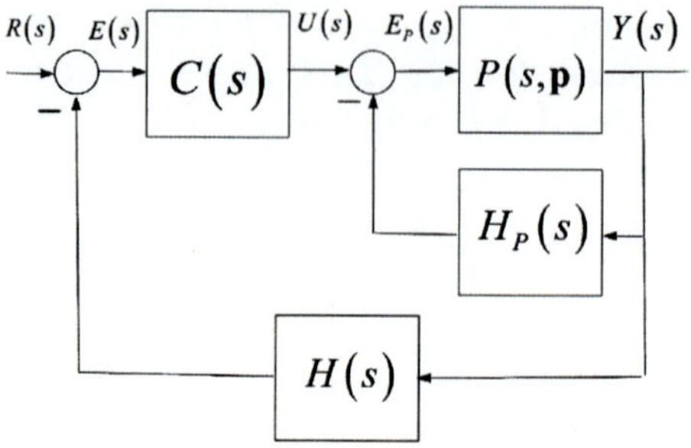

Fig. 11. Closed-loop system with CRONE controller

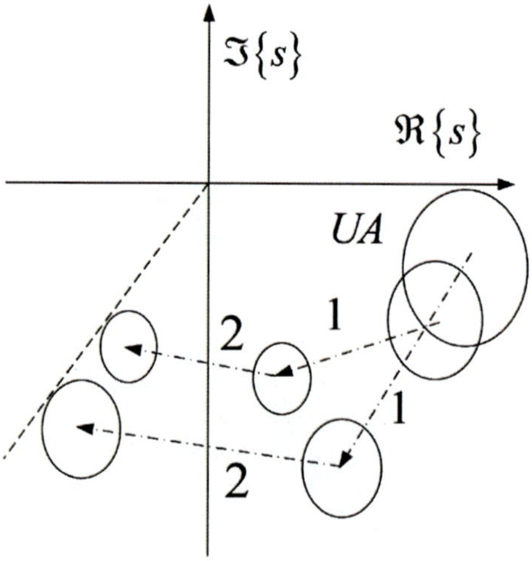

Fig. 12. Visualization of the CRONE synthesis algorithm

Figure 12 presents first step of the Algorithm proposed above.

7 Conclusions

In the paper a method reshaping the UAs of the plant with uncertainty parameters is proposed. It is proved that the constant negative feedback diminish the UAA leading to better robustness of the CRONE control. There are still open problems which will be mentioned below.

1. The CRONE control idea is still in progress especially for the complex, fractional-orders
2. Discrete-time controller needs to be evaluated
3. Time-variant uncertainty parameters plants, especially with parameters depending on frequencies ω.

The following abbreviations are used in this paper:

CRONE - Commande Robuste d'Ordre non Entiere
UA - uncertainty area
UAA - uncertainty areal angle
CM - Conformal map
UC - Unit circle

Acknowledgements. This work was supported by the National Science Centre Poland Grant Number 2016/23/B/St7/03686.

References

1. Chandrasekharan, P.C.: Robust Control of Linear Dynamical Systems. Academic Press Limited, London (1996)
2. Doetch, G.: Introduction to the Theory and Applicatio of the Laplace Transformation. Springer-Verlag, Heidelberg (1974). https://doi.org/10.1007/978-3-642-65690-3
3. Gatto, A.E.: Product rule and chain rule estimation for fractional derivatives on spaces that satisfy the doubling condition. J. Fract. Anal. **188**, 27–37 (2003). https://doi.org/10.1006/jfan.2001.3836
4. Kilbas, A.A., Srivastava, H.M. and Trujillo, J.J.: Theory and Applications of Fractional Differential Equations. Elsevier, North Holland Mathematics Studies 204, Amsterdam, The Netherland (2006)
5. Lanusse, P., Malti, R., Melchior, P.: CRONE control system design toolbox for the control engineering community: tutorial and case study. Philosophical transactions of the Royal Society A. **371**, 1–14 (2013). https://doi.org/10.1098/rsta.2012.0149
6. Lanusse, P., Sabatier, J.: PLC implementation of a CRONE controller. Fract. Calc. Appl. Anal. **14**(4), 505–522 (2011). https://doi.org/10.2478/s13540-011-0031-7
7. Maurin, K.: Analiza zespolona. In: Polish Wydawnictwo Naukowe PWN, Warszawa (2019)
8. Lanusse, P., Sabatieer, J., Oustaloup, A.: Fractional Order Differentiation and Robust Control Design. Intelligent Systems Control and Automation: Science and Engineering, vol. 77. Springer Science+Business Media, Dodrecht (2015). https://doi.org/10.1007/978-94-017-9807-5-2
9. Moreau, X., Daou, R.A.Z., Christophy, F.: Comparison between the second and third generations of the CRONE controller: application to the thermal diffusive interface medium. Fractal Fract. **2**(5), 1–19 (2018). https://doi.org/10.2290/fractalfract2010005
10. Oustaloup, A.: La robustesse. Analyse et synthèse de commandes robustes. Hermes, Traite des Nouvelles Technologies. Sèrie Automatique. Paris, France (1994)
11. Oustaloup, A.: CRONE Control. Hermes, Paris (1995)
12. Oustaloup, A.: Diversity and Non-integer Differentiation for System dynamics. Wiley, London (2014)
13. Podlubny, I.: Fractional Differential Equations. Academic Press, San Diego (1999)
14. Rudin, W.: Real and Complex Analysis, 3rd edn. McGraw-Hill, New York (1987)
15. Tøffner-Clausen, S.: System Identification and Robust Control. A Case Study. Approach Advances in Industrial Control, Springer-Verlag, Cham (1996). https://doi.org/10.1007/978-1-4471-1513-7
16. Zhou, K., Doyle, J.C., Glover, K.: Robust and Optimal Control. Prentice Hall, Upper Saddle River (1996)

Author Index

A. Dzielinski et al. (Eds.): ICFDA 2021, LNNS 452, pp. 259–260, 2022.
https://doi.org/10.1007/978-3-031-04383-3